Netty 权威指南（第2版）

李林锋 / 著

电子工业出版社
Publishing House of Electronics Industry
北京·BEIJING

内 容 简 介

《Netty 权威指南（第 2 版）》是异步非阻塞通信领域的经典之作，基于最新版本的 Netty 5.0 编写，是国内首本深入介绍 Netty 原理和架构的书籍，也是作者多年实战经验的总结和浓缩。内容不仅包含 Java NIO 入门知识、Netty 的基础功能开发指导、编解码框架定制等，还包括私有协议栈定制和开发、Netty 核心类库源码分析，以及 Netty 的架构剖析。

本书适合架构师、设计师、软件开发工程师、测试人员以及其他对 Java NIO 框架、Netty 感兴趣的相关人士阅读，通过本书的学习，读者不仅能够掌握 Netty 基础功能的使用和开发，更能够掌握 Netty 核心类库的原理和使用约束，从而在实际工作中更好地使用 Netty。

未经许可，不得以任何方式复制或抄袭本书之部分或全部内容。
版权所有，侵权必究。

图书在版编目（CIP）数据

Netty 权威指南 / 李林锋著. —2 版. —北京：电子工业出版社，2015.4
ISBN 978-7-121-25801-5

Ⅰ. ①N… Ⅱ. ①李… Ⅲ. ①JAVA 语言—程序设计—指南 Ⅳ. ①TP312-62

中国版本图书馆 CIP 数据核字（2015）第 067682 号

责任编辑：董 英
印　　刷：北京天宇星印刷厂
装　　订：北京天宇星印刷厂
出版发行：电子工业出版社
　　　　　北京市海淀区万寿路 173 信箱　邮编 100036
开　　本：787×980　1/16　印张：35.75　字数：758 千字
版　　次：2014 年 6 月第 1 版
　　　　　2015 年 4 月第 2 版
印　　次：2023 年 3 月第 20 次印刷
定　　价：89.00 元

凡所购买电子工业出版社图书有缺损问题，请向购买书店调换。若书店售缺，请与本社发行部联系，联系及邮购电话：（010）88254888。
质量投诉请发邮件至 zlts@phei.com.cn，盗版侵权举报请发邮件至 dbqq@phei.com.cn。
服务热线：（010）88258888。

前　言

　　2014年6月《Netty权威指南》第1版面世之后，很多读者通过邮件等方式向我表达了对本书的喜爱和赞誉。同时，对本书的一些瑕疵和不足也进行了指正，并给出了合理的建议。我对读者反馈的合理建议进行了记录和总结，以期在未来修订版或者第2版中能够修正这些问题。

　　大约在2014年11月份的时候，编辑与我协商出版《Netty权威指南（第2版）》的事宜，考虑到如下几个因素，最终我决定推出第2版：

- 第一版需要修正少部分印刷不太清晰的图片，这会改变后续章节的页码；
- 源码分析章节的代码希望重新编排一下，与前面的开发示例保持一致；
- 部分章节和内容需要优化调整；
- 部分读者对推出第2版的要求。

　　第2版的主要变更如下，删除第1版中的如下章节：

- 第7章：Java序列化；
- 第12章：UDP协议开发；
- 第13章：文件传输；
- 第22章：Netty行业应用。

　　新增本书中的如下章节：

- 第7章：MessagePack编解码；
- 第13章：服务端创建；
- 第14章：客户端创建；
- 第22章：高性能之道；
- 第23章：可靠性；

◎ 第 24 章：安全性。

第 1 版最初的想法是尽量照顾 NIO 编程和 Netty 初学者，因此入门和基础功能使用示例占了很大比例，涵盖的范围也很广。但事实上，由于 Netty 的功能过于庞杂，一本书很难涵盖 Netty 的所有功能点，因此，删除了不太常用的 Java 序列化、UDP 协议开发和文件传输。

Netty 行业应用的内容很多读者都很期望，希望能够展开详细讲解一下。我思索再三，忍痛割爱，不仅没有加强本章节，反而删除了它。为什么呢？对于真正想了解行业应用的读者，需要展开详细讲解才能够讲透，剖开 Netty 在 Spark、Hadoop 等大数据领域的应用不谈，即便是作为分布式服务框架的内部高性能通信组件，例如 Dubbo，没有大篇幅也很难讲透，与其一笔概括，泛泛而谈，还不如留给其他作者或者未来抽空单独梳理。

掌握 Netty 的基础功能使用比较容易，但是理解 Netty 底层的架构以及主要架构特性的设计理念却是件困难的事情，它需要长期的行业积累以及对 Netty 底层源码的透彻理解。应广大读者的要求，在第 2 版中新增了 Netty 的高性能、安全性和可靠性的架构剖析，通过这些章节的学习，读者可以更加清晰地理解 Netty 架构设计理念。

尽管我本人已经有 7 年的 NIO 编程和实战经验，在产品中也广泛应用了 Netty 和 Mina 等 NIO 框架。但是，受限于个人能力和水平，本书一定还有纰漏和不妥之处，希望广大读者能够批评指正。读者在阅读本书或者实际工作中如果有 Netty 相关的疑问，也可以直接联系我，我会尽量回复。我的联系方式如下：

◎ 邮箱：neu_lilinfeng@sina.com

◎ 新浪微博：Nettying

◎ 微信：Nettying。

《Netty 权威指南》第 1 版出版之后，很多读者来信咨询自己实际工作和学习中遇到的 Netty 问题和案例，有些案例和问题颇具典型性。我将这些案例进行了总结，在微信公众号"Netty 之家"中定期推送，希望广大读者可以关注。

感谢博文视点的小编丁一琼 MM 和幕后的美工，正是你们的辛苦工作才保证了本书能够顺利出版；感谢华为 IT PaaS 望岳、莫小君和 Digital SDP 集成开发部徐皓等领导对我的信任和支持；感谢我的老婆在我编辑第 2 版期间赦免了我做饭和刷碗的义务，我得以抽出时间安心写作。

最后感谢《Netty 权威指南》第 1 版的读者，你们的理解、鼓励和支持，使我有足够的勇气和动力继续前行。希望大家携起手来共同推动 NIO 编程和 Netty 在国内的应用和发展。

李林锋

2015 年 1 月 27 日于南京

第 1 版前言

大约在 2008 年的时候，我参与设计和开发的一个电信系统在月初出账期，总是发生大量的连接超时和读写超时异常，业务的失败率相比于平时高了很多，报表中的很多指标都差强人意。后来经过排查，发现问题主要出现在下游网元的处理性能上，月初的时候 BSS 出账，在出账期间 BSS 系统运行缓慢，由于双方采用了同步阻塞式的 HTTP+XML 进行通信，导致任何一方处理缓慢都会影响对方的处理性能。按照故障隔离的设计原则，对方处理速度慢或者不回应，不应该影响系统的其他功能模块或者协议栈，但是在同步阻塞 I/O 通信模型下，这种故障传播和相互影响是不可避免的，很难通过业务层面解决。

受限于当时 Tomcat 和 Servlet 的同步阻塞 I/O 模型，以及在 Java 领域异步 HTTP 协议栈的技术积累不足，当时我们并没有办法完全解决这个问题，只能通过调整线程池策略和 HTTP 超时时间来从业务层面做规避。

2009 年，由于对技术的热爱，我作为业务骨干被领导派去参加一个重点业务平台的研发工作，与两位资深的架构师（其中一位工作 20 年，做华为交换机出身）共同参与。这是我第一次全面接触异步 I/O 编程和高性能电信级协议栈的开发，眼界大开——异步高性能内部协议栈、异步 HTTP、异步 SOAP、异步 SMPP……所有的协议栈都是异步非阻塞的。后来的性能测试表明：基于 Reactor 模型统一调度的长连接和短连接协议栈，无论是性能、可靠性还是可维护性，都可以"秒杀"传统基于 BIO 开发的应用服务器和各种协议栈，这种差异本质上是一种代差。

在我从事异步 NIO 编程的 2009 年，业界还没有成熟的 NIO 框架，那个时候 Mina 刚刚开始起步，功能和性能都达不到商用标准。最困难的是，国内 Java 领域的异步通信还没有流行，整个业界的积累都非常少。那时资料匮乏，能够交流和探讨的圈内人很少，一旦踩住"地雷"，就需要夜以继日地维护。在随后 2 年多的时间里，经历了十多次的在通宵、凌晨被一线的运维人员电话吵醒等种种磨难之后，我们自研的 NIO 框架才逐渐稳定和成熟。期间，解决的 BUG 总计 20~30 个。

从 2004 年 JDK 1.4 首次提供 NIO 1.0 类库到现在，已经过去了整整 10 年。JSR 51 的设计初衷就是让 Java 能够提供非阻塞、具有弹性伸缩能力的异步 I/O 类库，从而结束了 Java 在高性能服务器领域的不利局面。然而，在相当长的一段时间里，Java 的 NIO 编程并没有流行起来，究其原因如下。

1．大多数高性能服务器，被 C 和 C++ 语言盘踞，由于它们可以直接使用操作系统的异步 I/O 能力，所以对 JDK 的 NIO 并不关心；

2．移动互联网尚未兴起，基于 Java 的大规模分布式系统极少，很多中小型应用服务对于异步 I/O 的诉求不是很强烈；

3．高性能、高可靠性领域，例如银行、证券、电信等，依然以 C++ 为主导，Java 充当打杂的角色，NIO 暂时没有用武之地；

4．当时主流的 J2EE 服务器，几乎全部基于同步阻塞 I/O 构建，例如 Servlet、Tomcat 等，由于它们应用广泛，如果这些容器不支持 NIO，用户很难具备独立构建异步协议栈的能力；

5．异步 NIO 编程门槛比较高，开发和维护一款基于 NIO 的协议栈对很多中小型公司来说像是一场噩梦；

6．业界 NIO 框架不成熟，很难商用；

7．国内研发界对 NIO 的陌生和认识不足，没有充分重视。

基于上述几种原因，NIO 编程的推广和发展长期滞后。值得欣慰的是，随着大规模分布式系统、大数据和流式计算框架的兴起，基于 Java 来构建这些系统已经成为主流，NIO 编程和 NIO 框架在此期间得到了大规模的商用。在互联网领域，阿里的分布式服务框架 Dubbo、RocketMQ，大数据的基础序列化和通信框架 Avro，以及很多开源的软件都已经开始使用 Netty 来构建高性能、分布式通信能力，Netty 社区的活跃度也名列前茅。根据目前的信息，Netty 已经在如下几个领域得到了大规模的商业应用。

1．互联网领域；

2．电信领域；

3．大数据领域；

4．银行、证券等金融领域；

5．游戏行业；

6．电力等企业市场。

2014 年春节前，我分享了一篇博文《Netty 5.0 架构剖析和源码解读》，短短 1 个月下载量达到了 4000 多。很多网友向我咨询 NIO 编程技术、NIO 框架如何选择等问题，也有一些圈内朋友和出版社邀请我写一本关于 Netty 的技术书籍。作为最流行、表现最优异的 NIO 框架，Netty 深受大家喜爱，但是长期以来除了 User Guide 之外，国内鲜有 Netty 相关的技术书籍供广大 NIO 编程爱好者学习和参考。由于 Netty 源码的复杂性和 NIO 编程本身的技术门槛限制，对于大多数读者而言，通过自己阅读和分析源码来深入掌握 Netty 的设计原理和实现细节是件困难的事情。从 2011 年开始我系统性地分析和应用了 Netty 和 Mina，转瞬间已经过去了 3 年多。在这 3 年的时间里，我们的系统经受了无数严苛的考验，在这个过程中，我对 Netty 和 Mina 有了更深刻的体验，也积累了丰富的运维和实战经验。我们都是开源框架 Netty 的受益者，为了让更多的朋友和同行能够了解 NIO 编程，深入学习和掌握 Netty 这个 NIO 利器，我打算将我的经验和大家分享，同时也结束国内尚无 Netty 学习教材的尴尬境地。

联系方式

尽管我也有技术洁癖，希望诸事完美，但是由于 Netty 代码的庞杂和涉及的知识点太多，一本书籍很难涵盖所有的功能点。如有遗漏或者错误，恳请大家能够及时批评和指正，如果你有好的建议或者想法，也可以联系我。我的联系方式如下。

邮箱：neu_lilinfeng@sina.com。

新浪微博：Nettying。

微信：Nettying。

致谢

如果说个人能够改变自己命运的话，对于程序员来说，唯有通过不断地学习和实践，努力提升自己的技能，才有可能找到更好的机会，充分发挥和体现自己的价值。我希望本书能够为你的成功助一臂之力。

感谢博文视点的策划编辑丁一琼和幕后的美编，正是你们的辛苦工作才保证了本书能够顺利出版；感谢华为 Netty 爱好者和关注本书的领导同事们的支持，你们的理解和鼓励为我提供了足够的勇气；感谢我的家人和老婆的支持，写书占用了我几乎所有的业余时间，没有你们的理解和支持，我很难安心写作。

最后感谢 Netty 中国社区的朋友，我的微博粉丝和所有喜欢 Netty 的朋友们，你们对技术的热情是鼓励我写书的最重要动力，没有你们，就没有本书。希望大家一如既往地喜欢 NIO 编程，喜欢 Netty，以及相互交流和分享，共同推动整个国内异步高性能通信领域的技术发展。

李林锋

2014 年 5 月 11 日于南京紫轩阁

目 录

基础篇 走进 Java NIO

第 1 章 Java 的 I/O 演进之路 2
- 1.1 I/O 基础入门 3
  - 1.1.1 Linux 网络 I/O 模型简介 3
  - 1.1.2 I/O 多路复用技术 6
- 1.2 Java 的 I/O 演进 8
- 1.3 总结 10

第 2 章 NIO 入门 11
- 2.1 传统的 BIO 编程 11
  - 2.1.1 BIO 通信模型图 12
  - 2.1.2 同步阻塞式 I/O 创建的 TimeServer 源码分析 13
  - 2.1.3 同步阻塞式 I/O 创建的 TimeClient 源码分析 16
- 2.2 伪异步 I/O 编程 18
  - 2.2.1 伪异步 I/O 模型图 19
  - 2.2.2 伪异步 I/O 创建的 TimeServer 源码分析 19
  - 2.2.3 伪异步 I/O 弊端分析 21
- 2.3 NIO 编程 24
  - 2.3.1 NIO 类库简介 24
  - 2.3.2 NIO 服务端序列图 28
  - 2.3.3 NIO 创建的 TimeServer 源码分析 30
  - 2.3.4 NIO 客户端序列图 36
  - 2.3.5 NIO 创建的 TimeClient 源码分析 39
- 2.4 AIO 编程 45
  - 2.4.1 AIO 创建的 TimeServer 源码分析 46
  - 2.4.2 AIO 创建的 TimeClient 源码分析 51

目录

- 2.4.3 AIO 版本时间服务器运行结果 ········· 56
- 2.5 4 种 I/O 的对比 ········· 58
  - 2.5.1 概念澄清 ········· 58
  - 2.5.2 不同 I/O 模型对比 ········· 59
- 2.6 选择 Netty 的理由 ········· 60
  - 2.6.1 不选择 Java 原生 NIO 编程的原因 ········· 61
  - 2.6.2 为什么选择 Netty ········· 62
- 2.7 总结 ········· 63

入门篇　Netty NIO 开发指南

第 3 章 Netty 入门应用 ········· 66
- 3.1 Netty 开发环境的搭建 ········· 66
  - 3.1.1 下载 Netty 的软件包 ········· 67
  - 3.1.2 搭建 Netty 应用工程 ········· 67
- 3.2 Netty 服务端开发 ········· 68
- 3.3 Netty 客户端开发 ········· 73
- 3.4 运行和调试 ········· 76
  - 3.4.1 服务端和客户端的运行 ········· 76
  - 3.4.2 打包和部署 ········· 77
- 3.5 总结 ········· 77

第 4 章 TCP 粘包/拆包问题的解决之道 ········· 79
- 4.1 TCP 粘包/拆包 ········· 79
  - 4.1.1 TCP 粘包/拆包问题说明 ········· 80
  - 4.1.2 TCP 粘包/拆包发生的原因 ········· 80
  - 4.1.3 粘包问题的解决策略 ········· 81
- 4.2 未考虑 TCP 粘包导致功能异常案例 ········· 82
  - 4.2.1 TimeServer 的改造 ········· 82
  - 4.2.2 TimeClient 的改造 ········· 83
  - 4.2.3 运行结果 ········· 84
- 4.3 利用 LineBasedFrameDecoder 解决 TCP 粘包问题 ········· 85
  - 4.3.1 支持 TCP 粘包的 TimeServer ········· 86
  - 4.3.2 支持 TCP 粘包的 TimeClient ········· 88
  - 4.3.3 运行支持 TCP 粘包的时间服务器程序 ········· 90

4.3.4　LineBasedFrameDecoder 和 StringDecoder 的原理分析	91

4.4　总结 …… 92

第 5 章　分隔符和定长解码器的应用 …… 93

5.1　DelimiterBasedFrameDecoder 应用开发 …… 94
　　5.1.1　DelimiterBasedFrameDecoder 服务端开发 …… 94
　　5.1.2　DelimiterBasedFrameDecoder 客户端开发 …… 97
　　5.1.3　运行 DelimiterBasedFrameDecoder 服务端和客户端 …… 99
5.2　FixedLengthFrameDecoder 应用开发 …… 101
　　5.2.1　FixedLengthFrameDecoder 服务端开发 …… 101
　　5.2.2　利用 telnet 命令行测试 EchoServer 服务端 …… 103
5.3　总结 …… 104

中级篇　Netty 编解码开发指南

第 6 章　编解码技术 …… 106

6.1　Java 序列化的缺点 …… 107
　　6.1.1　无法跨语言 …… 107
　　6.1.2　序列化后的码流太大 …… 107
　　6.1.3　序列化性能太低 …… 110
6.2　业界主流的编解码框架 …… 113
　　6.2.1　Google 的 Protobuf 介绍 …… 113
　　6.2.2　Facebook 的 Thrift 介绍 …… 115
　　6.2.3　JBoss Marshalling 介绍 …… 116
6.3　总结 …… 117

第 7 章　MessagePack 编解码 …… 118

7.1　MessagePack 介绍 …… 118
　　7.1.1　MessagePack 多语言支持 …… 119
　　7.1.2　MessagePack Java API 介绍 …… 119
　　7.1.3　MessagePack 开发包下载 …… 120
7.2　MessagePack 编码器和解码器开发 …… 120
　　7.2.1　MessagePack 编码器开发 …… 120
　　7.2.2　MessagePack 解码器开发 …… 121
　　7.2.3　功能测试 …… 121
7.3　粘包/半包支持 …… 124

| | 7.4 | 总结 | 127 |

第 8 章　Google Protobuf 编解码 … 128

- 8.1　Protobuf 的入门 … 129
  - 8.1.1　Protobuf 开发环境搭建 … 129
  - 8.1.2　Protobuf 编解码开发 … 131
  - 8.1.3　运行 Protobuf 例程 … 133
- 8.2　Netty 的 Protobuf 服务端开发 … 133
  - 8.2.1　Protobuf 版本的图书订购服务端开发 … 134
  - 8.2.2　Protobuf 版本的图书订购客户端开发 … 136
  - 8.2.3　Protobuf 版本的图书订购程序功能测试 … 139
- 8.3　Protobuf 的使用注意事项 … 140
- 8.4　总结 … 142

第 9 章　JBoss Marshalling 编解码 … 143

- 9.1　Marshalling 开发环境准备 … 143
- 9.2　Netty 的 Marshalling 服务端开发 … 144
- 9.3　Netty 的 Marshalling 客户端开发 … 147
- 9.4　运行 Marshalling 客户端和服务端例程 … 149
- 9.5　总结 … 150

高级篇　Netty 多协议开发和应用

第 10 章　HTTP 协议开发应用 … 154

- 10.1　HTTP 协议介绍 … 155
  - 10.1.1　HTTP 协议的 URL … 155
  - 10.1.2　HTTP 请求消息（HttpRequest） … 155
  - 10.1.3　HTTP 响应消息（HttpResponse） … 158
- 10.2　Netty HTTP 服务端入门开发 … 159
  - 10.2.1　HTTP 服务端例程场景描述 … 160
  - 10.2.2　HTTP 服务端开发 … 160
  - 10.2.3　Netty HTTP 文件服务器例程运行结果 … 166
- 10.3　Netty HTTP+XML 协议栈开发 … 170
  - 10.3.1　开发场景介绍 … 171
  - 10.3.2　HTTP+XML 协议栈设计 … 174
  - 10.3.3　高效的 XML 绑定框架 JiBx … 175

	10.3.4	HTTP+XML 编解码框架开发 ·································	183

- 10.3.4 HTTP+XML 编解码框架开发 ·· 183
- 10.3.5 HTTP+XML 协议栈测试 ·· 199
- 10.3.6 小结 ··· 201
- 10.4 总结 ·· 202

第 11 章 WebSocket 协议开发 ·· 203
- 11.1 HTTP 协议的弊端 ··· 204
- 11.2 WebSocket 入门 ·· 204
  - 11.2.1 WebSocket 背景 ··· 205
  - 11.2.2 WebSocket 连接建立 ·· 206
  - 11.2.3 WebSocket 生命周期 ·· 207
  - 11.2.4 WebSocket 连接关闭 ·· 208
- 11.3 Netty WebSocket 协议开发 ·· 209
  - 11.3.1 WebSocket 服务端功能介绍 ··· 209
  - 11.3.2 WebSocket 服务端开发 ··· 210
  - 11.3.3 运行 WebSocket 服务端 ·· 218
- 11.4 总结 ·· 219

第 12 章 私有协议栈开发 ·· 221
- 12.1 私有协议介绍 ·· 221
- 12.2 Netty 协议栈功能设计 ·· 223
  - 12.2.1 网络拓扑图 ··· 223
  - 12.2.2 协议栈功能描述 ··· 224
  - 12.2.3 通信模型 ··· 224
  - 12.2.4 消息定义 ··· 225
  - 12.2.5 Netty 协议支持的字段类型 ·· 226
  - 12.2.6 Netty 协议的编解码规范 ··· 227
  - 12.2.7 链路的建立 ··· 229
  - 12.2.8 链路的关闭 ··· 230
  - 12.2.9 可靠性设计 ··· 230
  - 12.2.10 安全性设计 ·· 232
  - 12.2.11 可扩展性设计 ··· 232
- 12.3 Netty 协议栈开发 ··· 233
  - 12.3.1 数据结构定义 ·· 233
  - 12.3.2 消息编解码 ··· 237

	12.3.3	握手和安全认证	241
	12.3.4	心跳检测机制	245
	12.3.5	断连重连	248
	12.3.6	客户端代码	249
	12.3.7	服务端代码	251
12.4	运行协议栈		252
	12.4.1	正常场景	252
	12.4.2	异常场景：服务端宕机重启	253
	12.4.3	异常场景：客户端宕机重启	256
12.5	总结		256

第 13 章 服务端创建 258

13.1	原生 NIO 类库的复杂性	259
13.2	Netty 服务端创建源码分析	259
	13.2.1 Netty 服务端创建时序图	260
	13.2.2 Netty 服务端创建源码分析	263
13.3	客户端接入源码分析	272
13.4	总结	275

第 14 章 客户端创建 276

14.1	Netty 客户端创建流程分析	276
	14.2.1 Netty 客户端创建时序图	276
	14.2.2 Netty 客户端创建流程分析	277
14.2	Netty 客户端创建源码分析	278
	14.2.1 客户端连接辅助类 Bootstrap	278
	14.2.2 客户端连接操作	281
	14.2.3 异步连接结果通知	283
	14.2.4 客户端连接超时机制	284
14.3	总结	286

源码分析篇 Netty 功能介绍和源码分析

第 15 章 ByteBuf 和相关辅助类 288

15.1	ByteBuf 功能说明	288
	15.1.1 ByteBuf 的工作原理	289
	15.1.2 ByteBuf 的功能介绍	294

- 15.2 ByteBuf 源码分析 ·················· 308
  - 15.2.1 ByteBuf 的主要类继承关系 ·················· 309
  - 15.2.2 AbstractByteBuf 源码分析 ·················· 310
  - 15.2.3 AbstractReferenceCountedByteBuf 源码分析 ·················· 319
  - 15.2.4 UnpooledHeapByteBuf 源码分析 ·················· 321
  - 15.2.5 PooledByteBuf 内存池原理分析 ·················· 326
  - 15.2.6 PooledDirectByteBuf 源码分析 ·················· 329
- 15.3 ByteBuf 相关的辅助类功能介绍 ·················· 332
  - 15.3.1 ByteBufHolder ·················· 332
  - 15.3.2 ByteBufAllocator ·················· 333
  - 15.3.3 CompositeByteBuf ·················· 334
  - 15.3.4 ByteBufUtil ·················· 336
- 15.4 总结 ·················· 337

第 16 章 Channel 和 Unsafe ·················· 338

- 16.1 Channel 功能说明 ·················· 338
  - 16.1.1 Channel 的工作原理 ·················· 339
  - 16.1.2 Channel 的功能介绍 ·················· 340
- 16.2 Channel 源码分析 ·················· 343
  - 16.2.1 Channel 的主要继承关系类图 ·················· 343
  - 16.2.2 AbstractChannel 源码分析 ·················· 344
  - 16.2.3 AbstractNioChannel 源码分析 ·················· 347
  - 16.2.4 AbstractNioByteChannel 源码分析 ·················· 350
  - 16.2.5 AbstractNioMessageChannel 源码分析 ·················· 353
  - 16.2.6 AbstractNioMessageServerChannel 源码分析 ·················· 354
  - 16.2.7 NioServerSocketChannel 源码分析 ·················· 355
  - 16.2.8 NioSocketChannel 源码分析 ·················· 358
- 16.3 Unsafe 功能说明 ·················· 364
- 16.4 Unsafe 源码分析 ·················· 365
  - 16.4.1 Unsafe 继承关系类图 ·················· 365
  - 16.4.2 AbstractUnsafe 源码分析 ·················· 366
  - 16.4.3 AbstractNioUnsafe 源码分析 ·················· 375
  - 16.4.4 NioByteUnsafe 源码分析 ·················· 379
- 16.5 总结 ·················· 387

第 17 章　ChannelPipeline 和 ChannelHandler············388

17.1　ChannelPipeline 功能说明············389
17.1.1　ChannelPipeline 的事件处理············389
17.1.2　自定义拦截器············391
17.1.3　构建 pipeline············392
17.1.4　ChannelPipeline 的主要特性············393

17.2　ChannelPipeline 源码分析············393
17.2.1　ChannelPipeline 的类继承关系图············393
17.2.2　ChannelPipeline 对 ChannelHandler 的管理············393
17.2.3　ChannelPipeline 的 inbound 事件············396
17.2.4　ChannelPipeline 的 outbound 事件············397

17.3　ChannelHandler 功能说明············398
17.3.1　ChannelHandlerAdapter 功能说明············399
17.3.2　ByteToMessageDecoder 功能说明············399
17.3.3　MessageToMessageDecoder 功能说明············400
17.3.4　LengthFieldBasedFrameDecoder 功能说明············400
17.3.5　MessageToByteEncoder 功能说明············404
17.3.6　MessageToMessageEncoder 功能说明············404
17.3.7　LengthFieldPrepender 功能说明············405

17.4　ChannelHandler 源码分析············406
17.4.1　ChannelHandler 的类继承关系图············406
17.4.2　ByteToMessageDecoder 源码分析············407
17.4.3　MessageToMessageDecoder 源码分析············410
17.4.4　LengthFieldBasedFrameDecoder 源码分析············411
17.4.5　MessageToByteEncoder 源码分析············415
17.4.6　MessageToMessageEncoder 源码分析············416
17.4.7　LengthFieldPrepender 源码分析············417

17.5　总结············418

第 18 章　EventLoop 和 EventLoopGroup············419

18.1　Netty 的线程模型············419
18.1.1　Reactor 单线程模型············420
18.1.2　Reactor 多线程模型············421
18.1.3　主从 Reactor 多线程模型············422

18.1.4 Netty 的线程模型 ·············· 423
18.1.5 最佳实践 ·············· 424
18.2 NioEventLoop 源码分析 ·············· 425
18.2.1 NioEventLoop 设计原理 ·············· 425
18.2.2 NioEventLoop 继承关系类图 ·············· 426
18.2.3 NioEventLoop ·············· 427
18.3 总结 ·············· 436

第 19 章 Future 和 Promise ·············· 438
19.1 Future 功能 ·············· 438
19.2 ChannelFuture 源码分析 ·············· 443
19.3 Promise 功能介绍 ·············· 445
19.4 Promise 源码分析 ·············· 447
19.4.1 Promise 继承关系图 ·············· 447
19.4.2 DefaultPromise ·············· 447
19.5 总结 ·············· 449

架构和行业应用篇　Netty 高级特性

第 20 章 Netty 架构剖析 ·············· 452
20.1 Netty 逻辑架构 ·············· 452
20.1.1 Reactor 通信调度层 ·············· 453
20.1.2 职责链 ChannelPipeline ·············· 453
20.1.3 业务逻辑编排层（Service ChannelHandler） ·············· 454
20.2 关键架构质量属性 ·············· 454
20.2.1 高性能 ·············· 454
20.2.2 可靠性 ·············· 457
20.2.3 可定制性 ·············· 460
20.2.4 可扩展性 ·············· 460
20.3 总结 ·············· 460

第 21 章 Java 多线程编程在 Netty 中的应用 ·············· 461
21.1 Java 内存模型与多线程编程 ·············· 461
21.1.1 硬件的发展和多任务处理 ·············· 461
21.1.2 Java 内存模型 ·············· 462
21.2 Netty 的并发编程实践 ·············· 464

- 21.2.1 对共享的可变数据进行正确的同步 ·········· 464
- 21.2.2 正确使用锁 ·········· 465
- 21.2.3 volatile 的正确使用 ·········· 467
- 21.2.4 CAS 指令和原子类 ·········· 470
- 21.2.5 线程安全类的应用 ·········· 472
- 21.2.6 读写锁的应用 ·········· 476
- 21.2.7 线程安全性文档说明 ·········· 477
- 21.2.8 不要依赖线程优先级 ·········· 478
- 21.3 总结 ·········· 479

第 22 章 高性能之道 ·········· 480

- 22.1 RPC 调用性能模型分析 ·········· 480
  - 22.1.1 传统 RPC 调用性能差的三宗罪 ·········· 480
  - 22.1.2 I/O 通信性能三原则 ·········· 481
- 22.2 Netty 高性能之道 ·········· 482
  - 22.2.1 异步非阻塞通信 ·········· 482
  - 22.2.2 高效的 Reactor 线程模型 ·········· 482
  - 22.2.3 无锁化的串行设计 ·········· 485
  - 22.2.4 高效的并发编程 ·········· 486
  - 22.2.5 高性能的序列化框架 ·········· 486
  - 22.2.6 零拷贝 ·········· 487
  - 22.2.7 内存池 ·········· 491
  - 22.2.8 灵活的 TCP 参数配置能力 ·········· 494
- 22.3 主流 NIO 框架性能对比 ·········· 495
- 22.4 总结 ·········· 497

第 23 章 可靠性 ·········· 498

- 23.1 可靠性需求 ·········· 498
  - 23.1.1 宕机的代价 ·········· 498
  - 23.1.2 Netty 可靠性需求 ·········· 499
- 23.2 Netty 高可靠性设计 ·········· 500
  - 23.2.1 网络通信类故障 ·········· 500
  - 23.2.2 链路的有效性检测 ·········· 507
  - 23.2.3 Reactor 线程的保护 ·········· 510
  - 23.2.4 内存保护 ·········· 513

  23.2.5 流量整形 ···················· 516
  23.2.6 优雅停机接口 ················ 519
 23.3 优化建议 ······················ 520
  23.3.1 发送队列容量上限控制 ············ 520
  23.3.2 回推发送失败的消息 ············· 521
 23.4 总结 ························ 521

第 24 章 安全性 ······················ 522
 24.1 严峻的安全形势 ·················· 522
  24.1.1 OpenSSL Heart bleed 漏洞 ·········· 522
  24.1.2 安全漏洞的代价 ··············· 523
  24.1.3 Netty 面临的安全风险 ············ 523
 24.2 Netty SSL 安全特性 ················ 525
  24.2.1 SSL 单向认证 ················ 525
  24.2.2 SSL 双向认证 ················ 532
  24.2.3 第三方 CA 认证 ··············· 536
 24.3 Netty SSL 源码分析 ················ 538
  24.3.1 客户端 ··················· 538
  24.3.2 服务端 ··················· 541
  24.3.3 消息读取 ·················· 544
  24.3.4 消息发送 ·················· 545
 24.4 Netty 扩展的安全特性 ··············· 546
  24.4.1 IP 地址黑名单机制 ·············· 547
  24.4.2 接入认证 ·················· 548
 24.4 总结 ························ 550

第 25 章 Netty 未来展望 ················· 551
 25.1 应用范围 ······················ 551
 25.2 技术演进 ······················ 552
 25.3 社区活跃度 ····················· 552
 25.4 Road Map ····················· 552
 25.5 总结 ························ 553

附录 A Netty 参数配置表 ················· 554

基础篇

走进 Java NIO

第 1 章　Java 的 I/O 演进之路
第 2 章　NIO 入门

第 1 章

Java 的 I/O 演进之路

Java 是由 Sun Microsystems 公司在 1995 年首先发布的编程语言和计算平台。这项基础技术支持最新的程序，包括实用程序、游戏和业务应用程序。Java 在世界各地的 8.5 亿多台个人计算机和数十亿套设备上运行着，其中包括移动设备和电视设备。

Java 之所以能够得到如此广泛的应用，除了摆脱硬件平台的依赖具有"一次编写、到处运行"的平台无关性特性之外，另一个重要原因是：其丰富而强大的类库以及众多第三方开源类库使得基于 Java 语言的开发更加简单和便捷。

但是，对于一些经验丰富的程序员来说，Java 的一些类库在早期设计中功能并不完善或者存在一些缺陷，其中最令人恼火的就是基于同步 I/O 的 Socket 通信类库。直到 2002 年 2 月 13 日 JDK1.4 Merlin 的发布，Java 才第一次支持非阻塞 I/O，这个类库的提供为 JDK 的通信模型带来了翻天覆地的变化。

在开始学习 Netty 之前，我们首先对 UNIX 系统常用的 I/O 模型进行介绍，然后对 Java 的 I/O 历史演进进行简单说明。通过本章节的学习，希望读者对同步和异步 I/O 以及 Java 的 I/O 类库发展有个直观的了解，方便后续章节的学习。如果你已经熟练 NIO 编程或者从事过 UNIX 网络编程，希望直接学习 Java 的 NIO 和 Netty，那就可以直接跳到第 2 章进行学习。

本章主要内容包括：

◎ I/O 基础入门

◎ Java 的 I/O 演进

1.1 I/O 基础入门

Java1.4 之前的早期版本，Java 对 I/O 的支持并不完善，开发人员在开发高性能 I/O 程序的时候，会面临一些巨大的挑战和困难，主要问题如下。

- 没有数据缓冲区，I/O 性能存在问题；
- 没有 C 或者 C++中的 Channel 概念，只有输入和输出流；
- 同步阻塞式 I/O 通信（BIO），通常会导致通信线程被长时间阻塞；
- 支持的字符集有限，硬件可移植性不好。

在 Java 支持异步 I/O 之前的很长一段时间里，高性能服务端开发领域一直被 C++和 C 长期占据，Java 的同步阻塞 I/O 被大家所诟病。

1.1.1 Linux 网络 I/O 模型简介

Linux 的内核将所有外部设备都看做一个文件来操作，对一个文件的读写操作会调用内核提供的系统命令，返回一个 file descriptor（fd，文件描述符）。而对一个 socket 的读写也会有相应的描述符，称为 socketfd（socket 描述符），描述符就是一个数字，它指向内核中的一个结构体（文件路径，数据区等一些属性）。

根据 UNIX 网络编程对 I/O 模型的分类，UNIX 提供了 5 种 I/O 模型，分别如下。

（1）阻塞 I/O 模型：最常用的 I/O 模型就是阻塞 I/O 模型，缺省情形下，所有文件操作都是阻塞的。我们以套接字接口为例来讲解此模型：在进程空间中调用 recvfrom，其系统调用直到数据包到达且被复制到应用进程的缓冲区中或者发生错误时才返回，在此期间一直会等待，进程在从调用 recvfrom 开始到它返回的整段时间内都是被阻塞的，因此被称为阻塞 I/O 模型，如图 1-1 所示。

（2）非阻塞 I/O 模型：recvfrom 从应用层到内核的时候，如果该缓冲区没有数据的话，就直接返回一个 EWOULDBLOCK 错误，一般都对非阻塞 I/O 模型进行轮询检查这个状态，看内核是不是有数据到来，如图 1-2 所示。

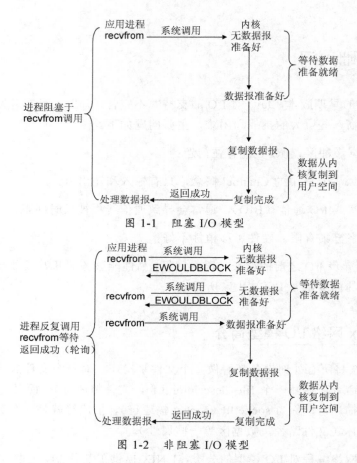

图 1-1 阻塞 I/O 模型

图 1-2 非阻塞 I/O 模型

（3）I/O 复用模型：Linux 提供 select/poll，进程通过将一个或多个 fd 传递给 select 或 poll 系统调用，阻塞在 select 操作上，这样 select/poll 可以帮我们侦测多个 fd 是否处于就绪状态。select/poll 是顺序扫描 fd 是否就绪，而且支持的 fd 数量有限，因此它的使用受到了一些制约。Linux 还提供了一个 epoll 系统调用，epoll 使用基于事件驱动方式代替顺序扫描，因此性能更高。当有 fd 就绪时，立即回调函数 rollback，如图 1-3 所示。

（4）信号驱动 I/O 模型：首先开启套接口信号驱动 I/O 功能，并通过系统调用 sigaction 执行一个信号处理函数（此系统调用立即返回，进程继续工作，它是非阻塞的）。当数据准备就绪时，就为该进程生成一个 SIGIO 信号，通过信号回调通知应用程序调用 recvfrom 来读取数据，并通知主循环函数处理数据，如图 1-4 所示。

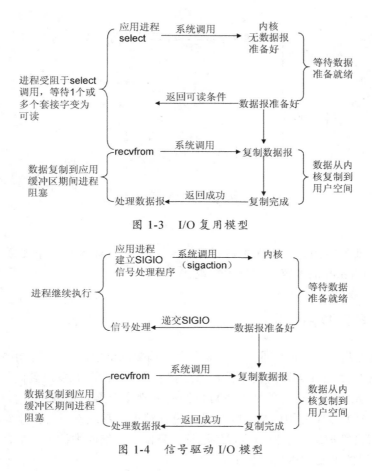

图 1-3　I/O 复用模型

图 1-4　信号驱动 I/O 模型

（5）异步 I/O：告知内核启动某个操作，并让内核在整个操作完成后（包括将数据从内核复制到用户自己的缓冲区）通知我们。这种模型与信号驱动模型的主要区别是：信号驱动 I/O 由内核通知我们何时可以开始一个 I/O 操作；异步 I/O 模型由内核通知我们 I/O 操作何时已经完成，如图 1-5 所示。

如果想要了解更多的 UNIX 系统网络编程知识，可以阅读《UNIX 网络编程》，里面有非常详细的原理和 API 介绍。对于大多数 Java 程序员来说，不需要了解网络编程的底层细节，大家只需要有个概念，知道对于操作系统而言，底层是支持异步 I/O 通信的。只不过在很长一段时间 Java 并没有提供异步 I/O 通信的类库，导致很多原生的 Java 程序员对这块儿比较陌生。当你了解了网络编程的基础知识后，理解 Java 的 NIO 类库就会更加容易一些。

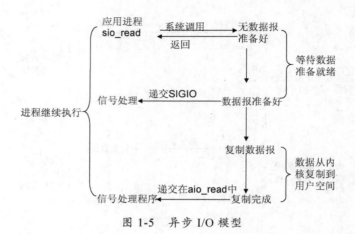

图 1-5 异步 I/O 模型

下一个小结我们重点讲下 I/O 多路复用技术，因为 Java NIO 的核心类库多路复用器 Selector 就是基于 epoll 的多路复用技术实现。

1.1.2 I/O 多路复用技术

在 I/O 编程过程中，当需要同时处理多个客户端接入请求时，可以利用多线程或者 I/O 多路复用技术进行处理。I/O 多路复用技术通过把多个 I/O 的阻塞复用到同一个 select 的阻塞上，从而使得系统在单线程的情况下可以同时处理多个客户端请求。与传统的多线程/多进程模型比，I/O 多路复用的最大优势是系统开销小，系统不需要创建新的额外进程或者线程，也不需要维护这些进程和线程的运行，降低了系统的维护工作量，节省了系统资源，I/O 多路复用的主要应用场景如下。

◎ 服务器需要同时处理多个处于监听状态或者多个连接状态的套接字；

◎ 服务器需要同时处理多种网络协议的套接字。

目前支持 I/O 多路复用的系统调用有 select、pselect、poll、epoll，在 Linux 网络编程过程中，很长一段时间都使用 select 做轮询和网络事件通知，然而 select 的一些固有缺陷导致了它的应用受到了很大的限制，最终 Linux 不得不在新的内核版本中寻找 select 的替代方案，最终选择了 epoll。epoll 与 select 的原理比较类似，为了克服 select 的缺点，epoll 作了很多重大改进，现总结如下。

1. 支持一个进程打开的 socket 描述符（FD）不受限制（仅受限于操作系统的最大文件句柄数）。

select 最大的缺陷就是单个进程所打开的 FD 是有一定限制的，它由 FD_SETSIZE 设置，默认值是 1024。对于那些需要支持上万个 TCP 连接的大型服务器来说显然太少了。可以选择修改这个宏然后重新编译内核，不过这会带来网络效率的下降。我们也可以通过选择多进程的方案（传统的 Apache 方案）解决这个问题，不过虽然在 Linux 上创建进程的代价比较小，但仍旧是不可忽视的。另外，进程间的数据交换非常麻烦，对于 Java 来说，由于没有共享内存，需要通过 Socket 通信或者其他方式进行数据同步，这带来了额外的性能损耗，增加了程序复杂度，所以也不是一种完美的解决方案。值得庆幸的是，epoll 并没有这个限制，它所支持的 FD 上限是操作系统的最大文件句柄数，这个数字远远大于 1024。例如，在 1GB 内存的机器上大约是 10 万个句柄左右，具体的值可以通过 cat /proc/sys/fs/file-max 察看，通常情况下这个值跟系统的内存关系比较大。

2. I/O 效率不会随着 FD 数目的增加而线性下降。

传统 select/poll 的另一个致命弱点，就是当你拥有一个很大的 socket 集合时，由于网络延时或者链路空闲，任一时刻只有少部分的 socket 是"活跃"的，但是 select/poll 每次调用都会线性扫描全部的集合，导致效率呈现线性下降。epoll 不存在这个问题，它只会对"活跃"的 socket 进行操作——这是因为在内核实现中，epoll 是根据每个 fd 上面的 callback 函数实现的。那么，只有"活跃"的 socket 才会去主动调用 callback 函数，其他 idle 状态的 socket 则不会。在这点上，epoll 实现了一个伪 AIO。针对 epoll 和 select 性能对比的 benchmark 测试表明：如果所有的 socket 都处于活跃态——例如一个高速 LAN 环境，epoll 并不比 select/poll 效率高太多；相反，如果过多使用 epoll_ctl，效率相比还有稍微地降低。但是一旦使用 idle connections 模拟 WAN 环境，epoll 的效率就远在 select/poll 之上了。

3. 使用 mmap 加速内核与用户空间的消息传递。

无论是 select、poll 还是 epoll 都需要内核把 FD 消息通知给用户空间，如何避免不必要的内存复制就显得非常重要，epoll 是通过内核和用户空间 mmap 同一块内存来实现的。

4. epoll 的 API 更加简单。

包括创建一个 epoll 描述符、添加监听事件、阻塞等待所监听的事件发生、关闭 epoll 描述符等。

值得说明的是，用来克服 select/poll 缺点的方法不只有 epoll，epoll 只是一种 Linux 的实现方案。在 freeBSD 下有 kqueue，而 dev/poll 是最古老的 Solaris 的方案，使用难度依次递增。kqueue 是 freebsd 的宠儿，它实际上是一个功能相当丰富的 kernel 事件队列，它不仅仅是 select/poll 的升级，而且可以处理 signal、目录结构变化、进程等多种事件。kqueue 是边缘触发的。/dev/poll 是 Solaris 的产物，是这一系列高性能 API 中最早出现的。Kernel 提供了一个特殊的设备文件/dev/poll，应用程序打开这个文件得到操作 fd_set 的句柄，通过写入 pollfd 来修改它，一个特殊的 ioctl 调用用来替换 select。不过由于出现的年代比较早，所以/dev/poll 的接口实现比较原始。

到这里，I/O 的基础知识已经介绍完毕。从 1.2 节开始介绍 Java 的 I/O 演进历史，从 BIO 到 NIO 是 Java 通信类库迈出的一小步，但却对 Java 在高性能通信领域的发展起到了关键性的推动作用。随着基于 NIO 的各类 NIO 框架的发展，以及基于 NIO 的 Web 服务器的发展，Java 在很多领域取代了 C 和 C++，成为企业服务端应用开发的首选语言。

1.2　Java 的 I/O 演进

在 JDK 1.4 推出 Java NIO 之前，基于 Java 的所有 Socket 通信都采用了同步阻塞模式（BIO），这种一请求一应答的通信模型简化了上层的应用开发，但是在性能和可靠性方面却存在着巨大的瓶颈。因此，在很长一段时间里，大型的应用服务器都采用 C 或者 C++ 语言开发，因为它们可以直接使用操作系统提供的异步 I/O 或者 AIO 能力。当并发访问量增大、响应时间延迟增大之后，采用 Java BIO 开发的服务端软件只有通过硬件的不断扩容来满足高并发和低时延，它极大地增加了企业的成本，并且随着集群规模的不断膨胀，系统的可维护性也面临巨大的挑战，只能通过采购性能更高的硬件服务器来解决问题，这会导致恶性循环。

正是由于 Java 传统 BIO 的拙劣表现，才使得 Java 支持非阻塞 I/O 的呼声日渐高涨，最终，JDK1.4 版本提供了新的 NIO 类库，Java 终于也可以支持非阻塞 I/O 了。

Java 的 I/O 发展简史

从 JDK1.0 到 JDK1.3，Java 的 I/O 类库都非常原始，很多 UNIX 网络编程中的概念或者接口在 I/O 类库中都没有体现，例如 Pipe、Channel、Buffer 和 Selector 等。2002 年发布

JDK1.4 时，NIO 以 JSR-51 的身份正式随 JDK 发布。它新增了个 java.nio 包，提供了很多进行异步 I/O 开发的 API 和类库，主要的类和接口如下。

- 进行异步 I/O 操作的缓冲区 ByteBuffer 等；
- 进行异步 I/O 操作的管道 Pipe；
- 进行各种 I/O 操作（异步或者同步）的 Channel，包括 ServerSocketChannel 和 SocketChannel；
- 多种字符集的编码能力和解码能力；
- 实现非阻塞 I/O 操作的多路复用器 selector；
- 基于流行的 Perl 实现的正则表达式类库；
- 文件通道 FileChannel。

新的 NIO 类库的提供，极大地促进了基于 Java 的异步非阻塞编程的发展和应用，但是，它依然有不完善的地方，特别是对文件系统的处理能力仍显不足，主要问题如下。

- 没有统一的文件属性（例如读写权限）；
- API 能力比较弱，例如目录的级联创建和递归遍历，往往需要自己实现；
- 底层存储系统的一些高级 API 无法使用；
- 所有的文件操作都是同步阻塞调用，不支持异步文件读写操作。

2011 年 7 月 28 日，JDK1.7 正式发布。它的一个比较大的亮点就是将原来的 NIO 类库进行了升级，被称为 NIO2.0。NIO2.0 由 JSR-203 演进而来，它主要提供了如下三个方面的改进。

- 提供能够批量获取文件属性的 API，这些 API 具有平台无关性，不与特性的文件系统相耦合。另外它还提供了标准文件系统的 SPI，供各个服务提供商扩展实现；
- 提供 AIO 功能，支持基于文件的异步 I/O 操作和针对网络套接字的异步操作；
- 完成 JSR-51 定义的通道功能，包括对配置和多播数据报的支持等。

1.3 总结

通过本章的学习，我们了解了 UNIX 网络编程的 5 种 I/O 模型，学习了 I/O 多路复用技术的基础知识。通过对 Java I/O 演进历史的总结和介绍，相信大家对 Java 的 I/O 演进有了一个更加直观的认识。后面的第 2 章节会对阻塞 I/O 和非阻塞 I/O 进行详细讲解，同时给出代码示例。相信学完第 2 章之后，大家就能够对传统的阻塞 I/O 的弊端和非阻塞 I/O 的优点有更加深刻的体会。好，稍微休息片刻，我们继续畅游在 NIO 编程的快乐海洋中！

第 2 章
NIO 入门

在本章中,我们会分别对 JDK 的 BIO、NIO 和 JDK 1.7 最新提供的 NIO 2.0 的使用进行详细说明,通过流程图和代码讲解,让大家体会到:随着 Java I/O 类库的不断发展和改进,基于 Java 的网络编程会变得越来越简单;随着异步 I/O 功能的增强,基于 Java NIO 开发的网络服务器甚至不逊色于采用 C++开发的网络程序。

本章主要内容包括:

◎ 传统的同步阻塞式 I/O 编程

◎ 基于 NIO 的非阻塞编程

◎ 基于 NIO2.0 的异步非阻塞(AIO)编程

◎ 为什么要使用 NIO 编程

◎ 为什么选择 Netty

2.1 传统的 BIO 编程

网络编程的基本模型是 Client/Server 模型,也就是两个进程之间进行相互通信,其中服务端提供位置信息(绑定的 IP 地址和监听端口),客户端通过连接操作向服务端监听的地址发起连接请求，通过三次握手建立连接，如果连接建立成功，双方就可以通过网络套接字（Socket）进行通信。

在基于传统同步阻塞模型开发中，ServerSocket 负责绑定 IP 地址，启动监听端口；Socket 负责发起连接操作。连接成功之后，双方通过输入和输出流进行同步阻塞式通信。

下面，我们就以经典的时间服务器（TimeServer）为例，通过代码分析来回顾和熟悉 BIO 编程。

2.1.1 BIO 通信模型图

首先，我们通过图 2-1 所示的通信模型图来熟悉 BIO 的服务端通信模型：采用 BIO 通信模型的服务端，通常由一个独立的 Acceptor 线程负责监听客户端的连接，它接收到客户端连接请求之后为每个客户端创建一个新的线程进行链路处理，处理完成之后，通过输出流返回应答给客户端，线程销毁。这就是典型的一请求一应答通信模型。

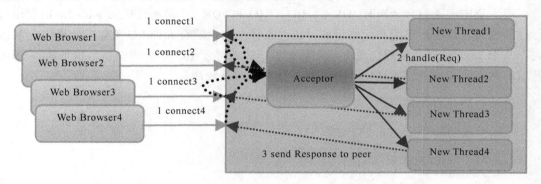

图 2-1　同步阻塞 I/O 服务端通信模型（一客户端一线程）

该模型最大的问题就是缺乏弹性伸缩能力，当客户端并发访问量增加后，服务端的线程个数和客户端并发访问数呈 1∶1 的正比关系，由于线程是 Java 虚拟机非常宝贵的系统资源，当线程数膨胀之后，系统的性能将急剧下降，随着并发访问量的继续增大，系统会发生线程堆栈溢出、创建新线程失败等问题，并最终导致进程宕机或者僵死，不能对外提供服务。

下面的两个小节，我们会分别对服务端和客户端进行源码分析，寻找同步阻塞 I/O 的弊端。

2.1.2 同步阻塞式 I/O 创建的 TimeServer 源码分析

代码清单 2-1 同步阻塞 I/O 的 TimeServer

（备注：以下代码行号均对应源代码中实际行号。）

```java
1.  package com.phei.netty.bio;
2.  import java.io.IOException;
3.  import java.net.ServerSocket;
4.  import java.net.Socket;
5.  /**
6.   * @author lilinfeng
7.   * @date 2014年2月14日
8.   * @version 1.0
9.   */
10. public class TimeServer {
11.
12.     /**
13.      * @param args
14.      * @throws IOException
15.      */
16.     public static void main(String[] args) throws IOException {
17.      int port = 8080;
18.      if (args != null && args.length > 0) {
19.
20.          try {
21.           port = Integer.valueOf(args[0]);
22.          } catch (NumberFormatException e) {
23.           // 采用默认值
24.          }
25.
26.      }
27.      ServerSocket server = null;
28.      try {
29.          server = new ServerSocket(port);
30.          System.out.println("The time server is start in port : " + port);
31.          Socket socket = null;
32.          while (true) {
33.           socket = server.accept();
34.           new Thread(new TimeServerHandler(socket)).start();
35.          }
36.      } finally {
```

```
37.         if (server != null) {
38.             System.out.println("The time server close");
39.             server.close();
40.             server = null;
41.         }
42.     }
43. }
44. }
```

TimeServer 根据传入的参数设置监听端口，如果没有入参，使用默认值 8080。第 29 行通过构造函数创建 ServerSocket，如果端口合法且没有被占用，服务端监听成功。第 32～35 行通过一个无限循环来监听客户端的连接，如果没有客户端接入，则主线程阻塞在 ServerSocket 的 accept 操作上。启动 TimeServer，通过 JvisualVM 打印线程堆栈，我们可以发现主程序确实阻塞在 accept 操作上，如图 2-2 所示。

```
"main" prio=6 tid=0x00879800 nid=0xb1c runnable [0x009af000]
   java.lang.Thread.State: RUNNABLE
       at java.net.TwoStacksPlainSocketImpl.socketAccept(Native Method)
       at java.net.AbstractPlainSocketImpl.accept(AbstractPlainSocketImpl.java:398)
       at java.net.PlainSocketImpl.accept(PlainSocketImpl.java:198)
       - locked <0x23021358> (a java.net.SocksSocketImpl)
       at java.net.ServerSocket.implAccept(ServerSocket.java:530)
       at java.net.ServerSocket.accept(ServerSocket.java:498)
       at com.phei.netty.bio.TimeServer.main(TimeServer.java:50)

   Locked ownable synchronizers:
       - None
```

图 2-2 主程序线程堆栈

当有新的客户端接入的时候，执行代码第 34 行，以 Socket 为参数构造 TimeServerHandler 对象，TimeServerHandler 是一个 Runnable，使用它为构造函数的参数创建一个新的客户端线程处理这条 Socket 链路。下面我们继续分析 TimeServerHandler 的代码。

代码清单 2-2　同步阻塞 I/O 的 TimeServerHandler

```
13. public class TimeServerHandler implements Runnable {
14.
15.     private Socket socket;
16.
17.     public TimeServerHandler(Socket socket) {
18.         this.socket = socket;
```

```
19.     }
20.
21.     /*
22.      * (non-Javadoc)
23.      *
24.      * @see java.lang.Runnable#run()
25.      */
26.     @Override
27.     public void run() {
28.      BufferedReader in = null;
29.      PrintWriter out = null;
30.      try {
31.          in = new BufferedReader(new InputStreamReader(
32.              this.socket.getInputStream()));
33.          out = new PrintWriter(this.socket.getOutputStream(), true);
34.          String currentTime = null;
35.          String body = null;
36.          while (true) {
37.           body = in.readLine();
38.           if (body == null)
39.              break;
40.           System.out.println("The time server receive order : " + body);
41.           currentTime = "QUERY TIME ORDER".equalsIgnoreCase(body) ? new java.util.Date(
42.              System.currentTimeMillis()).toString() : "BAD ORDER";
43.           out.println(currentTime);
44.          }
45.
46.      } catch (Exception e) {
47.          if (in != null) {
48.           try {
49.              in.close();
50.           } catch (IOException e1) {
51.              e1.printStackTrace();
52.           }
53.          }
54.          if (out != null) {
55.           out.close();
56.           out = null;
57.          }
58.          if (this.socket != null) {
```

```
59.        try {
60.            this.socket.close();
61.        } catch (IOException e1) {
62.            e1.printStackTrace();
63.        }
64.        this.socket = null;
65.    }
66.  }
67. }
68. }
```

第 37 行通过 BufferedReader 读取一行,如果已经读到了输入流的尾部,则返回值为 null,退出循环。如果读到了非空值,则对内容进行判断,如果请求消息为查询时间的指令"QUERY TIME ORDER",则获取当前最新的系统时间,通过 PrintWriter 的 println 函数发送给客户端,最后退出循环。代码第 47~64 行释放输入流、输出流和 Socket 套接字句柄资源,最后线程自动销毁并被虚拟机回收。

在下一个小结中,我们将介绍同步阻塞 I/O 的客户端代码,然后分别运行服务端和客户端,查看下程序的运行结果。

2.1.3　同步阻塞式 I/O 创建的 TimeClient 源码分析

客户端通过 Socket 创建,发送查询时间服务器的"QUERY TIME ORDER"指令,然后读取服务端的响应并将结果打印出来,随后关闭连接,释放资源,程序退出执行。

代码清单 2-3　同步阻塞 I/O 的 TimeClient

```
13. public class TimeClient {
14.
15.    /**
16.     * @param args
17.     */
18.    public static void main(String[] args) {
19.        int port = 8080;
20.        if (args != null && args.length > 0) {
21.            try {
22.                port = Integer.valueOf(args[0]);
23.            } catch (NumberFormatException e) {
24.                // 采用默认值
25.            }
```

```
26.        }
27.        Socket socket = null;
28.        BufferedReader in = null;
29.        PrintWriter out = null;
30.        try {
31.            socket = new Socket("127.0.0.1", port);
32.            in = new BufferedReader(new InputStreamReader(
33.                socket.getInputStream()));
34.            out = new PrintWriter(socket.getOutputStream(), true);
35.            out.println("QUERY TIME ORDER");
36.            System.out.println("Send order 2 server succeed.");
37.            String resp = in.readLine();
38.            System.out.println("Now is : " + resp);
39.        } catch (Exception e) {
40.            //不需要处理
41.        } finally {
42.            if (out != null) {
43.                out.close();
44.                out = null;
45.            }
46.
47.            if (in != null) {
48.                try {
49.                    in.close();
50.                } catch (IOException e) {
51.                    e.printStackTrace();
52.                }
53.                in = null;
54.            }
55.            if (socket != null) {
56.                try {
57.                    socket.close();
58.                } catch (IOException e) {
59.                    e.printStackTrace();
60.                }
61.                socket = null;
62.            }
63.        }
64.    }
65. }
```

第 35 行客户端通过 PrintWriter 向服务端发送"QUERY TIME ORDER"指令，然后通过 BufferedReader 的 readLine 读取响应并打印。

分别执行服务端和客户端，执行结果如下。

服务端执行结果如图 2-3 所示。

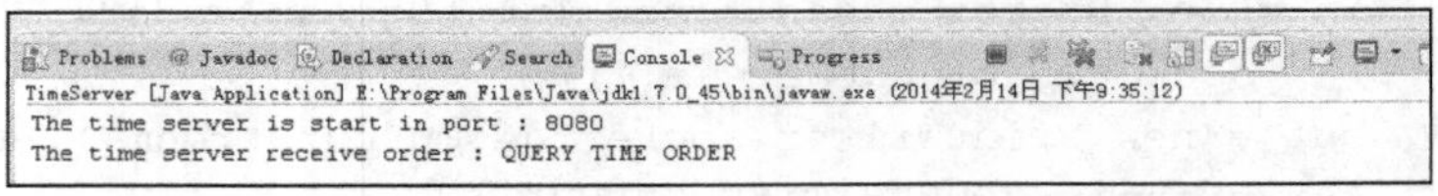

图 2-3　同步阻塞 I/O 时间服务器服务端运行结果

客户端执行结果如图 2-4 所示。

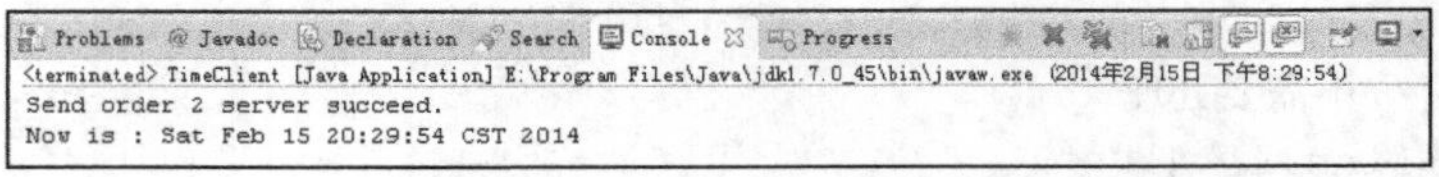

图 2-4　同步阻塞 IO 时间服务器客户端运行结果

到此为止，同步阻塞式 I/O 开发的时间服务器程序已经讲解完毕。我们发现，BIO 主要的问题在于每当有一个新的客户端请求接入时，服务端必须创建一个新的线程处理新接入的客户端链路，一个线程只能处理一个客户端连接。在高性能服务器应用领域，往往需要面向成千上万个客户端的并发连接，这种模型显然无法满足高性能、高并发接入的场景。

为了改进一线程一连接模型，后来又演进出了一种通过线程池或者消息队列实现 1 个或者多个线程处理 N 个客户端的模型，由于它的底层通信机制依然使用同步阻塞 I/O，所以被称为"伪异步"。下面的章节我们就对伪异步代码进行分析，看看伪异步是否能够满足我们对高性能、高并发接入的诉求。

2.2　伪异步 I/O 编程

为了解决同步阻塞 I/O 面临的一个链路需要一个线程处理的问题，后来有人对它的线程模型进行了优化——后端通过一个线程池来处理多个客户端的请求接入，形成客户端个数 M：线程池最大线程数 N 的比例关系，其中 M 可以远远大于 N。通过线程池可以灵活地调配线程资源，设置线程的最大值，防止由于海量并发接入导致线程耗尽。

下面，我们结合连接模型图和源码，对伪异步 I/O 进行分析，看它是否能够解决同步阻塞 I/O 面临的问题。

2.2.1 伪异步 I/O 模型图

采用线程池和任务队列可以实现一种叫做伪异步的 I/O 通信框架,它的模型图如图 2-5 所示。

当有新的客户端接入时,将客户端的 Socket 封装成一个 Task(该任务实现 java.lang. Runnable 接口)投递到后端的线程池中进行处理,JDK 的线程池维护一个消息队列和 N 个活跃线程,对消息队列中的任务进行处理。由于线程池可以设置消息队列的大小和最大线程数,因此,它的资源占用是可控的,无论多少个客户端并发访问,都不会导致资源的耗尽和宕机。

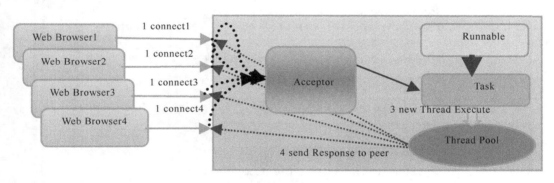

图 2-5　伪异步 I/O 服务端通信模型($M:N$)

下面的小节,我们依然采用时间服务器程序,将其改造成伪异步 I/O 时间服务器,然后通过对代码进行分析,找出其弊端。

2.2.2 伪异步 I/O 创建的 TimeServer 源码分析

我们对服务端代码进行一些改造,代码如下。

代码清单 2-4　伪异步 I/O 的 TimeServer

```
13. public class TimeServer {
14.
15.     /**
16.      * @param args
17.      * @throws IOException
18.      */
```

```
19.    public static void main(String[] args) throws IOException {
20.        int port = 8080;
21.        if (args != null && args.length > 0) {
22.            try {
23.                port = Integer.valueOf(args[0]);
24.            } catch (NumberFormatException e) {
25.                // 采用默认值
26.            }
27.        }
28.        ServerSocket server = null;
29.        try {
30.            server = new ServerSocket(port);
31.            System.out.println("The time server is start in port : " + port);
32.            Socket socket = null;
33.            TimeServerHandlerExecutePool singleExecutor = new TimeServerHandlerExecutePool(
34.                50, 10000);// 创建I/O任务线程池
35.            while (true) {
36.                socket = server.accept();
37.                singleExecutor.execute(new TimeServerHandler(socket));
38.            }
39.        } finally {
40.            if (server != null) {
41.                System.out.println("The time server close");
42.                server.close();
43.                server = null;
44.            }
45.        }
46.    }
47. }
```

伪异步I/O的主函数代码发生了变化,我们首先创建一个时间服务器处理类的线程池,当接收到新的客户端连接时,将请求Socket封装成一个Task,然后调用线程池的execute方法执行,从而避免了每个请求接入都创建一个新的线程。

代码清单2-5 伪异步I/O的TimeServerHandlerExecutePool

```
12. public class TimeServerHandlerExecutePool {
13.
14.     private ExecutorService executor;
15.
16.     public TimeServerHandlerExecutePool(int maxPoolSize, int queueSize) {
```

```
17.        executor = new ThreadPoolExecutor(Runtime.getRuntime()
18.            .availableProcessors(), maxPoolSize, 120L, TimeUnit.SECONDS,
19.            new ArrayBlockingQueue<java.lang.Runnable>(queueSize));
20.    }
21.    public void execute(java.lang.Runnable task) {
22.        executor.execute(task);
23.    }
24. }
```

由于线程池和消息队列都是有界的，因此，无论客户端并发连接数多大，它都不会导致线程个数过于膨胀或者内存溢出，相比于传统的一连接一线程模型，是一种改良。

由于客户端代码并没有改变，因此，我们直接运行服务端和客户端，执行结果如下。

服务端运行结果如图 2-6 所示。

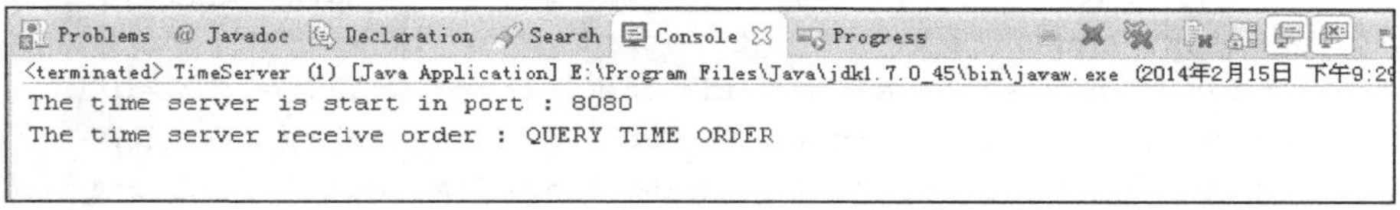

图 2-6　伪异步 I/O 时间服务器服务端运行结果

客户端运行结果如图 2-7 所示。

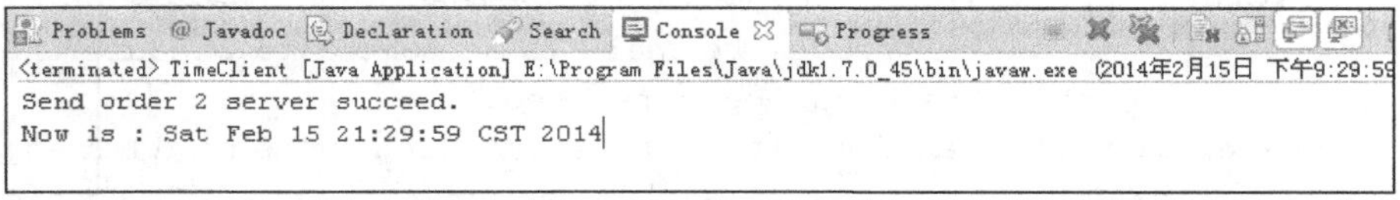

图 2-7　伪异步 I/O 时间服务器客户端运行结果

伪异步 I/O 通信框架采用了线程池实现，因此避免了为每个请求都创建一个独立线程造成的线程资源耗尽问题。但是由于它底层的通信依然采用同步阻塞模型，因此无法从根本上解决问题。下个小节我们对伪异步 I/O 进行深入分析，找到它的弊端，然后看看 NIO 是如何从根本上解决这个问题的。

2.2.3　伪异步 I/O 弊端分析

要对伪异步 I/O 的弊端进行深入分析，首先我们看两个 Java 同步 I/O 的 API 说明，随后结合代码进行详细分析。

代码清单 2-6　Java 输入流 InputStream

```
/**
 * Reads some number of bytes from the input stream and stores them into
 * the buffer array <code>b</code>. The number of bytes actually read is
 * returned as an integer. This method blocks until input data is
 * available, end of file is detected, or an exception is thrown.
 *
 * <p> If the length of <code>b</code> is zero, then no bytes are read and
 * <code>0</code> is returned; otherwise, there is an attempt to read at
 * least one byte. If no byte is available because the stream is at the
 * end of the file, the value <code>-1</code> is returned; otherwise, at
 * least one byte is read and stored into <code>b</code>.
 *
 * <p> The first byte read is stored into element <code>b[0]</code>, the
 * next one into <code>b[1]</code>, and so on. The number of bytes read is,
 * at most, equal to the length of <code>b</code>. Let <i>k</i> be the
 * number of bytes actually read; these bytes will be stored in elements
 * <code>b[0]</code> through <code>b[</code><i>k</i><code>-1]</code>,
 * leaving elements <code>b[</code><i>k</i><code>]</code> through
 * <code>b[b.length-1]</code> unaffected.
 *
 * @param      b   the buffer into which the data is read.
 * @return     the total number of bytes read into the buffer, or
 *             <code>-1</code> if there is no more data because the end of
 *             the stream has been reached.
 * @exception  IOException  If the first byte cannot be read for any reason
 * other than the end of the file, if the input stream has been closed, or
 * if some other I/O error occurs.
 * @exception  NullPointerException  if <code>b</code> is <code>null</code>.
 */
public int read(byte b[]) throws IOException {
    return read(b, 0, b.length);
}
```

请注意加粗斜体字部分的 API 说明，当对 Socket 的输入流进行读取操作的时候，它会一直阻塞下去，直到发生如下三种事件。

◎ 有数据可读；

◎ 可用数据已经读取完毕；

◎ 发生空指针或者 I/O 异常。

这意味着当对方发送请求或者应答消息比较缓慢,或者网络传输较慢时,读取输入流一方的通信线程将被长时间阻塞,如果对方要 60s 才能够将数据发送完成,读取一方的 I/O 线程也将会被同步阻塞 60s,在此期间,其他接入消息只能在消息队列中排队。

下面我们接着对输出流进行分析,还是看 JDK I/O 类库输出流的 API 文档,然后结合文档说明进行故障分析。

代码清单 2-7　Java 输入流 OutputStream

```
public void write(byte b[]) throws IOException
*Writes an array of bytes. This method will block until the bytes are *actually
written.
Parameters:
b - the data to be written
Throws: IOException
If an I/O error has occurred.
```

当调用 OutputStream 的 write 方法写输出流的时候,它将会被阻塞,直到所有要发送的字节全部写入完毕,或者发生异常。学习过 TCP/IP 相关知识的人都知道,当消息的接收方处理缓慢的时候,将不能及时地从 TCP 缓冲区读取数据,这将会导致发送方的 TCP window size 不断减小,直到为 0,双方处于 Keep-Alive 状态,消息发送方将不能再向 TCP 缓冲区写入消息,这时如果采用的是同步阻塞 I/O,write 操作将会被无限期阻塞,直到 TCP window size 大于 0 或者发生 I/O 异常。

通过对输入和输出流的 API 文档进行分析,我们了解到读和写操作都是同步阻塞的,阻塞的时间取决于对方 I/O 线程的处理速度和网络 I/O 的传输速度。本质上来讲,我们无法保证生产环境的网络状况和对端的应用程序能足够快,如果我们的应用程序依赖对方的处理速度,它的可靠性就非常差。也许在实验室进行的性能测试结果令人满意,但是一旦上线运行,面对恶劣的网络环境和良莠不齐的第三方系统,问题就会如火山一样喷发。

伪异步 I/O 实际上仅仅是对之前 I/O 线程模型的一个简单优化,它无法从根本上解决同步 I/O 导致的通信线程阻塞问题。下面我们就简单分析下通信对方返回应答时间过长会引起的级联故障。

(1)服务端处理缓慢,返回应答消息耗费 60s,平时只需要 10ms。

(2)采用伪异步 I/O 的线程正在读取故障服务节点的响应,由于读取输入流是阻塞的,它将会被同步阻塞 60s。

（3）假如所有的可用线程都被故障服务器阻塞，那后续所有的 I/O 消息都将在队列中排队。

（4）由于线程池采用阻塞队列实现，当队列积满之后，后续入队列的操作将被阻塞。

（5）由于前端只有一个 Accptor 线程接收客户端接入，它被阻塞在线程池的同步阻塞队列之后，新的客户端请求消息将被拒绝，客户端会发生大量的连接超时。

（6）由于几乎所有的连接都超时，调用者会认为系统已经崩溃，无法接收新的请求消息。

如何破解这个难题？下节的 NIO 将给出答案。

2.3　NIO 编程

在介绍 NIO 编程之前，我们首先需要澄清一个概念：NIO 到底是什么的简称？有人称之为 New I/O，原因在于它相对于之前的 I/O 类库是新增的。这是它的官方叫法。但是，由于之前老的 I/O 类库是阻塞 I/O，New I/O 类库的目标就是要让 Java 支持非阻塞 I/O，所以，更多的人喜欢称之为非阻塞 I/O（Non-block I/O）。由于非阻塞 I/O 更能够体现 NIO 的特点，所以本书使用的 NIO 都指的是非阻塞 I/O。

与 Socket 类和 ServerSocket 类相对应，NIO 也提供了 SocketChannel 和 ServerSocketChannel 两种不同的套接字通道实现。这两种新增的通道都支持阻塞和非阻塞两种模式。阻塞模式使用非常简单，但是性能和可靠性都不好，非阻塞模式则正好相反。开发人员可以根据自己的需要来选择合适的模式。一般来说，低负载、低并发的应用程序可以选择同步阻塞 I/O 以降低编程复杂度；对于高负载、高并发的网络应用，需要使用 NIO 的非阻塞模式进行开发。

下面的小节首先介绍 NIO 编程中的一些基本概念，然后通过 NIO 服务端的序列图和源码讲解，让大家快速地熟悉 NIO 编程的关键步骤和 API 的使用。如果你已经熟悉了 NIO 编程，可以跳过 2.3 节直接学习后面的章节。

2.3.1　NIO 类库简介

新的输入/输出（NIO）库是在 JDK 1.4 中引入的。NIO 弥补了原来同步阻塞 I/O 的不足,它在标准 Java 代码中提供了高速的、面向块的 I/O。通过定义包含数据的类,以及通过以块的形式处理这些数据,NIO 不用使用本机代码就可以利用低级优化,这是原来的 I/O 包所无法做到的。下面我们对 NIO 的一些概念和功能做下简单介绍,以便大家能够快速地了解 NIO 类库和相关概念。

1. 缓冲区 Buffer

我们首先介绍缓冲区(Buffer)的概念。Buffer 是一个对象,它包含一些要写入或者要读出的数据。在 NIO 类库中加入 Buffer 对象,体现了新库与原 I/O 的一个重要区别。在面向流的 I/O 中,可以将数据直接写入或者将数据直接读到 Stream 对象中。

在 NIO 库中,所有数据都是用缓冲区处理的。在读取数据时,它是直接读到缓冲区中的;在写入数据时,写入到缓冲区中。任何时候访问 NIO 中的数据,都是通过缓冲区进行操作。

缓冲区实质上是一个数组。通常它是一个字节数组(ByteBuffer),也可以使用其他种类的数组。但是一个缓冲区不仅仅是一个数组,缓冲区提供了对数据的结构化访问以及维护读写位置(limit)等信息。

最常用的缓冲区是 ByteBuffer,一个 ByteBuffer 提供了一组功能用于操作 byte 数组。除了 ByteBuffer,还有其他的一些缓冲区,事实上,每一种 Java 基本类型(除了 Boolean 类型)都对应有一种缓冲区,具体如下。

- ◎ ByteBuffer:字节缓冲区
- ◎ CharBuffer:字符缓冲区
- ◎ ShortBuffer:短整型缓冲区
- ◎ IntBuffer:整形缓冲区
- ◎ LongBuffer:长整形缓冲区
- ◎ FloatBuffer:浮点型缓冲区
- ◎ DoubleBuffer:双精度浮点型缓冲区

缓冲区的类图继承关系如图 2-8 所示。

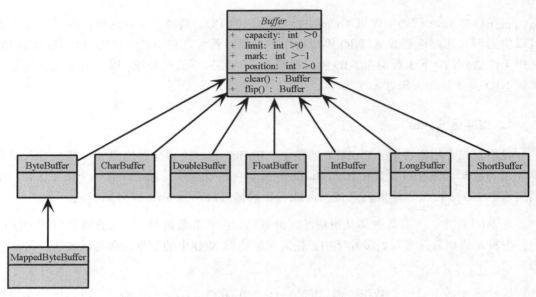

图 2-8 Buffer 继承关系图

每一个 Buffer 类都是 Buffer 接口的一个子实例。除了 ByteBuffer，每一个 Buffer 类都有完全一样的操作，只是它们所处理的数据类型不一样。因为大多数标准 I/O 操作都使用 ByteBuffer，所以它在具有一般缓冲区的操作之外还提供了一些特有的操作，以方便网络读写。

2. 通道 Channel

Channel 是一个通道，它就像自来水管一样，网络数据通过 Channel 读取和写入。通道与流的不同之处在于通道是双向的，流只是在一个方向上移动（一个流必须是 InputStream 或者 OutputStream 的子类），而通道可以用于读、写或者二者同时进行。

因为 Channel 是全双工的，所以它可以比流更好地映射底层操作系统的 API。特别是在 UNIX 网络编程模型中，底层操作系统的通道都是全双工的，同时支持读写操作。

Channel 的类图继承关系如图 2-9 所示。

自顶向下看，前三层主要是 Channel 接口，用于定义它的功能，后面是一些具体的功能类（抽象类）。从类图可以看出，实际上 Channel 可以分为两大类：用于网络读写的 SelectableChannel 和用于文件操作的 FileChannel。

本书涉及的 ServerSocketChannel 和 SocketChannel 都是 SelectableChannel 的子类，它们的具体用法将在后续的代码中体现。

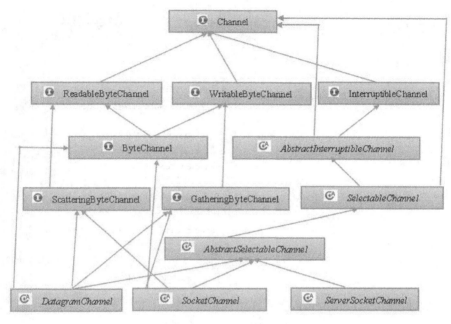

图 2-9　Channel 继承关系类图

3. 多路复用器 Selector

在本节中，我们将探索多路复用器 Selector，它是 Java NIO 编程的基础，熟练地掌握 Selector 对于 NIO 编程至关重要。多路复用器提供选择已经就绪的任务的能力。简单来讲，Selector 会不断地轮询注册在其上的 Channel，如果某个 Channel 上面发生读或者写事件，这个 Channel 就处于就绪状态，会被 Selector 轮询出来，然后通过 SelectionKey 可以获取就绪 Channel 的集合，进行后续的 I/O 操作。

一个多路复用器 Selector 可以同时轮询多个 Channel，由于 JDK 使用了 epoll()代替传统的 select 实现，所以它并没有最大连接句柄 1024/2048 的限制。这也就意味着只需要一个线程负责 Selector 的轮询，就可以接入成千上万的客户端，这确实是个非常巨大的进步。

下面，我们通过 NIO 编程的序列图和源码分析来熟悉相关的概念，以便巩固前面所学的 NIO 基础知识。

2.3.2 NIO 服务端序列图

NIO 服务端通信序列图如图 2-10 所示。

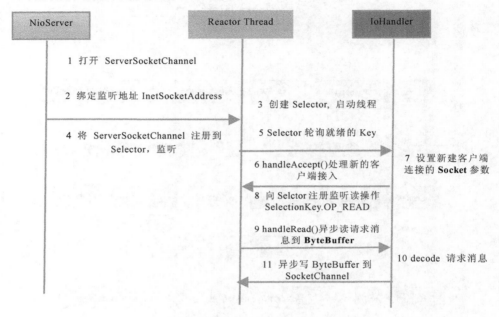

图 2-10 NIO 服务端通信序列图

下面，我们对 NIO 服务端的主要创建过程进行讲解和说明，作为 NIO 的基础入门，这里将忽略掉一些在生产环境中部署所需要的特性和功能。

步骤一：打开 ServerSocketChannel，用于监听客户端的连接，它是所有客户端连接的父管道，示例代码如下。

```
ServerSocketChannel acceptorSvr = ServerSocketChannel.open();
```

步骤二：绑定监听端口，设置连接为非阻塞模式，示例代码如下。

```
acceptorSvr.socket().bind(new
InetSocketAddress(InetAddress.getByName("IP"), port));
acceptorSvr.configureBlocking(false);
```

步骤三：创建 Reactor 线程，创建多路复用器并启动线程，示例代码如下。

```
Selector selector = Selector.open();
New Thread(new ReactorTask()).start();
```

步骤四：将 ServerSocketChannel 注册到 Reactor 线程的多路复用器 Selector 上，监听 ACCEPT 事件，示例代码如下。

```
SelectionKey key = acceptorSvr.register( selector, SelectionKey.OP_ACCEPT, ioHandler);
```

步骤五：多路复用器在线程 run 方法的无限循环体内轮询准备就绪的 Key，示例代码如下。

```
int num = selector.select();
Set selectedKeys = selector.selectedKeys();
Iterator it = selectedKeys.iterator();
while (it.hasNext()) {
    SelectionKey key = (SelectionKey)it.next();
    // ... deal with I/O event ...
}
```

步骤六：多路复用器监听到有新的客户端接入，处理新的接入请求，完成 TCP 三次握手，建立物理链路，示例代码如下。

```
SocketChannel channel = svrChannel.accept();
```

步骤七：设置客户端链路为非阻塞模式，示例代码如下。

```
channel.configureBlocking(false);
channel.socket().setReuseAddress(true);
......
```

步骤八：将新接入的客户端连接注册到 Reactor 线程的多路复用器上，监听读操作，读取客户端发送的网络消息，示例代码如下。

```
SelectionKey key = socketChannel.register( selector, SelectionKey.OP_READ, ioHandler);
```

步骤九：异步读取客户端请求消息到缓冲区，示例代码如下。

```
int readNumber = channel.read(receivedBuffer);
```

步骤十：对 ByteBuffer 进行编解码，如果有半包消息指针 reset，继续读取后续的报文，将解码成功的消息封装成 Task，投递到业务线程池中，进行业务逻辑编排，示例代码如下。

```
Object message = null;
while(buffer.hasRemain())
```

```
    {
        byteBuffer.mark();
        Object message = decode(byteBuffer);
        if (message == null)
        {
           byteBuffer.reset();
           break;
        }
        messageList.add(message );
    }
    if (!byteBuffer.hasRemain())
    byteBuffer.clear();
    else
       byteBuffer.compact();
    if (messageList != null & !messageList.isEmpty())
    {
    for(Object messageE : messageList)
       handlerTask(messageE);
    }
```

步骤十一:将 POJO 对象 encode 成 ByteBuffer,调用 SocketChannel 的异步 write 接口,将消息异步发送给客户端,示例代码如下。

```
socketChannel.write(buffer);
```

注意:如果发送区 TCP 缓冲区满,会导致写半包,此时,需要注册监听写操作位,循环写,直到整包消息写入 TCP 缓冲区。对于这些内容此处暂不赘述,后续 Netty 源码分析章节会详细分析 Netty 的处理策略。

在了解创建 NIO 服务端的基本步骤之后,下面我们将前面的时间服务器程序通过 NIO 重写一遍,让大家能够学习到完整版的 NIO 服务端创建。

2.3.3 NIO 创建的 TimeServer 源码分析

我们将在 TimeServer 例程中给出完整的 NIO 创建的时间服务器源码。

代码清单 2-8　NIO 时间服务器　TimeServer

```
9.    public class TimeServer {
10.
11.       /**
```

```
12.     * @param args
13.     * @throws IOException
14.     */
15.    public static void main(String[] args) throws IOException {
16.        int port = 8080;
17.        if (args != null && args.length > 0) {
18.            try {
19.                port = Integer.valueOf(args[0]);
20.            } catch (NumberFormatException e) {
21.                // 采用默认值
22.            }
23.        }
24.        MultiplexerTimeServer timeServer = new MultiplexerTimeServer(port);
25.        New Thread(timeServer, "NIO-MultiplexerTimeServer-001").start();
26.    }
27. }
```

下面对 NIO 创建的 TimeServer 进行简单分析。第 16~23 行跟之前的一样,设置监听端口。第 24~25 行创建了一个被称为 MultiplexerTimeServer 的多路复用类,它是个一个独立的线程,负责轮询多路复用器 Selctor,可以处理多个客户端的并发接入。现在我们继续看 MultiplexerTimeServer 的源码。

代码清单 2-8　NIO 时间服务器　MultiplexerTimeServer

```
17. public class MultiplexerTimeServer implements Runnable {
18.
19.     private Selector selector;
20.
21.     private ServerSocketChannel servChannel;
22.
23.     private volatile boolean stop;
24.
25.     /**
26.      * 初始化多路复用器,绑定监听端口
27.      *
28.      * @param port
29.      */
30.     public MultiplexerTimeServer(int port) {
31.         try {
32.             selector = Selector.open();
33.             servChannel = ServerSocketChannel.open();
```

```
34.         servChannel.configureBlocking(false);
35.         servChannel.socket().bind(new InetSocketAddress(port), 1024);
36.         servChannel.register(selector, SelectionKey.OP_ACCEPT);
37.         System.out.println("The time server is start in port : " + port);
38.      } catch (IOException e) {
39.         e.printStackTrace();
40.         System.exit(1);
41.      }
42.   }
43.
44.   public void stop() {
45.    this.stop = true;
46.   }
47.
48.   /*
49.    * (non-Javadoc)
50.    *
51.    * @see java.lang.Runnable#run()
52.    */
53.   @Override
54.   public void run() {
55.    while (!stop) {
56.       try {
57.        selector.select(1000);
58.        Set<SelectionKey> selectedKeys = selector.selectedKeys();
59.        Iterator<SelectionKey> it = selectedKeys.iterator();
60.        SelectionKey key = null;
61.        while (it.hasNext()) {
62.           key = it.next();
63.           it.remove();
64.           try {
65.            handleInput(key);
66.           } catch (Exception e) {
67.            if (key != null) {
68.               key.cancel();
69.               if (key.channel() != null)
70.                key.channel().close();
71.            }
72.           }
73.        }
74.       } catch (Throwable t) {
```

```
75.            t.printStackTrace();
76.        }
77.    }
78.
79.    // 多路复用器关闭后，所有注册在上面的Channel和Pipe等资源都会被自动去注册
并关闭，所以不需要重复释放资源
80.    if (selector != null)
81.        try {
82.            selector.close();
83.        } catch (IOException e) {
84.            e.printStackTrace();
85.        }
86.    }
87.
88.    private void handleInput(SelectionKey key) throws IOException {
89.
90.        if (key.isValid()) {
91.            // 处理新接入的请求消息
92.            if (key.isAcceptable()) {
93.                // Accept the new connection
94.                ServerSocketChannel ssc = (ServerSocketChannel) key.channel();
95.                SocketChannel sc = ssc.accept();
96.                sc.configureBlocking(false);
97.                // Add the new connection to the selector
98.                sc.register(selector, SelectionKey.OP_READ);
99.            }
100.            if (key.isReadable()) {
101.                // Read the data
102.                SocketChannel sc = (SocketChannel) key.channel();
103.                ByteBuffer readBuffer = ByteBuffer.allocate(1024);
104.                int readBytes = sc.read(readBuffer);
105.                if (readBytes > 0) {
106.                    readBuffer.flip();
107.                    byte[] bytes = new byte[readBuffer.remaining()];
108.                    readBuffer.get(bytes);
109.                    String body = new String(bytes, "UTF-8");
110.                    System.out.println("The time server receive order : "
111.                        + body);
112.                    String currentTime = "QUERY TIME ORDER"
113.                        .equalsIgnoreCase(body) ? new java.util.Date(
114.                        System.currentTimeMillis()).toString()
115.                        : "BAD ORDER";
```

```
116.                    doWrite(sc, currentTime);
117.                } else if (readBytes < 0) {
118.                    // 对端链路关闭
119.                    key.cancel();
120.                    sc.close();
121.                } else
122.                    ; // 读到 0 字节,忽略
123.            }
124.        }
125.    }
126.
127.    private void doWrite(SocketChannel channel, String response)
128.            throws IOException {
129.        if (response != null && response.trim().length() > 0) {
130.            byte[] bytes = response.getBytes();
131.            ByteBuffer writeBuffer = ByteBuffer.allocate(bytes.length);
132.            writeBuffer.put(bytes);
133.            writeBuffer.flip();
134.            channel.write(writeBuffer);
135.        }
136.    }
137. }
```

由于这个类相比于传统的 Socket 编程会稍微复杂一些,在此展开进行详细分析,从如下几个关键步骤来讲解多路复用处理类。

(1)30~42 行为构造方法,在构造方法中进行资源初始化。创建多路复用器 Selector、ServerSocketChannel,对 Channel 和 TCP 参数进行配置。例如,将 ServerSocketChannel 设置为异步非阻塞模式,它的 backlog 设为 1024。系统资源初始化成功后,将 ServerSocketChannel 注册到 Selector,监听 SelectionKey.OP_ACCEPT 操作位。如果资源初始化失败(例如端口被占用),则退出。

(2)55~77 行是在线程的 run 方法的 while 循环体中循环遍历 selector,它的休眠时间为 1s。无论是否有读写等事件发生,selector 每隔 1s 都被唤醒一次。selector 也提供了一个无参的 select 方法:当有处于就绪状态的 Channel 时,selector 将返回该 Channel 的 SelectionKey 集合。通过对就绪状态的 Channel 集合进行迭代,可以进行网络的异步读写操作。

(3)92~99 行处理新接入的客户端请求消息,根据 SelectionKey 的操作位进行判断即可获知网络事件的类型,通过 ServerSocketChannel 的 accept 接收客户端的连接请求并创

建 SocketChannel 实例。完成上述操作后，相当于完成了 TCP 的三次握手，TCP 物理链路正式建立。注意，我们需要将新创建的 SocketChannel 设置为异步非阻塞，同时也可以对其 TCP 参数进行设置，例如 TCP 接收和发送缓冲区的大小等。但作为入门的例子，以上例程没有进行额外的参数设置。

（4）100～125 行用于读取客户端的请求消息。首先创建一个 ByteBuffer，由于我们事先无法得知客户端发送的码流大小，作为例程，我们开辟一个 1MB 的缓冲区。然后调用 SocketChannel 的 read 方法读取请求码流。注意，由于我们已经将 SocketChannel 设置为异步非阻塞模式，因此它的 read 是非阻塞的。使用返回值进行判断，看读取到的字节数，返回值有以下三种可能的结果。

◎ 返回值大于 0：读到了字节，对字节进行编解码；

◎ 返回值等于 0：没有读取到字节，属于正常场景，忽略；

◎ 返回值为-1：链路已经关闭，需要关闭 SocketChannel，释放资源。

当读取到码流以后，进行解码。首先对 readBuffer 进行 flip 操作，它的作用是将缓冲区当前的 limit 设置为 position，position 设置为 0，用于后续对缓冲区的读取操作。然后根据缓冲区可读的字节个数创建字节数组，调用 ByteBuffer 的 get 操作将缓冲区可读的字节数组复制到新创建的字节数组中，最后调用字符串的构造函数创建请求消息体并打印。如果请求指令是"QUERY TIME ORDER"，则把服务器的当前时间编码后返回给客户端。下面我们看看异步发送应答消息给客户端的情况。

（5）127～135 行将应答消息异步发送给客户端。我们看下关键代码，首先将字符串编码成字节数组，根据字节数组的容量创建 ByteBuffer，调用 ByteBuffer 的 put 操作将字节数组复制到缓冲区中，然后对缓冲区进行 flip 操作，最后调用 SocketChannel 的 write 方法将缓冲区中的字节数组发送出去。需要指出的是，由于 SocketChannel 是异步非阻塞的，它并不保证一次能够把需要发送的字节数组发送完，此时会出现"写半包"问题。我们需要注册写操作，不断轮询 Selector 将没有发送完的 ByteBuffer 发送完毕，然后可以通过 ByteBuffer 的 hasRemain()方法判断消息是否发送完成。此处仅仅是个简单的入门级例程，没有演示如何处理"写半包"场景，后续的章节会有详细说明。

使用 NIO 创建 TimeServer 服务器完成之后，我们继续学习如何创建 NIO 客户端。首先还是通过时序图了解关键步骤和过程，然后结合代码进行详细分析。

2.3.4 NIO 客户端序列图

NIO 客户端创建序列图如图 2-11 所示。

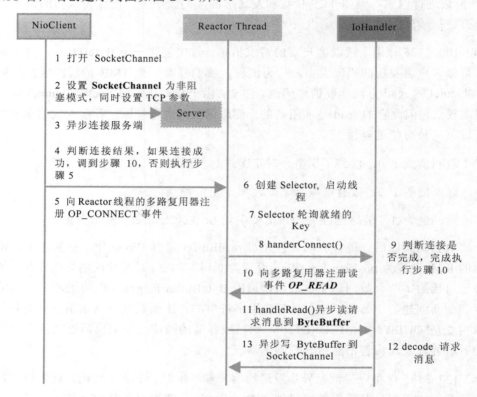

图 2-11 NIO 客户端创建序列图

步骤一：打开 SocketChannel，绑定客户端本地地址（可选，默认系统会随机分配一个可用的本地地址），示例代码如下。

```
SocketChannel clientChannel = SocketChannel.open();
```

步骤二：设置 SocketChannel 为非阻塞模式，同时设置客户端连接的 TCP 参数，示例代码如下。

```
clientChannel.configureBlocking(false);
socket.setReuseAddress(true);
socket.setReceiveBufferSize(BUFFER_SIZE);
socket.setSendBufferSize(BUFFER_SIZE);
```

步骤三：异步连接服务端，示例代码如下。

```
boolean connected=clientChannel.connect(new InetSocketAddress("ip",port));
```

步骤四：判断是否连接成功，如果连接成功，则直接注册读状态位到多路复用器中，如果当前没有连接成功（异步连接，返回 false，说明客户端已经发送 sync 包，服务端没有返回 ack 包，物理链路还没有建立），示例代码如下。

```
if (connected)
{
    clientChannel.register( selector, SelectionKey.OP_READ, ioHandler);
}
else
{
    clientChannel.register( selector, SelectionKey.OP_CONNECT, ioHandler);
}
```

步骤五：向 Reactor 线程的多路复用器注册 OP_CONNECT 状态位，监听服务端的 TCP ACK 应答，示例代码如下。

```
clientChannel.register( selector, SelectionKey.OP_CONNECT, ioHandler);
```

步骤六：创建 Reactor 线程，创建多路复用器并启动线程，代码如下。

```
Selector selector = Selector.open();
New Thread(new ReactorTask()).start();
```

步骤七：多路复用器在线程 run 方法的无限循环体内轮询准备就绪的 Key，代码如下。

```
int num = selector.select();
Set selectedKeys = selector.selectedKeys();
Iterator it = selectedKeys.iterator();
while (it.hasNext()) {
    SelectionKey key = (SelectionKey)it.next();
    // ... deal with I/O event ...
}
```

步骤八：接收 connect 事件进行处理，示例代码如下。

```
if (key.isConnectable())
  //handlerConnect();
```

步骤九：判断连接结果，如果连接成功，注册读事件到多路复用器，示例代码如下。

```
if (channel.finishConnect())
    registerRead();
```

步骤十：注册读事件到多路复用器，示例代码如下。

```
clientChannel.register( selector, SelectionKey.OP_READ, ioHandler);
```

步骤十一：异步读客户端请求消息到缓冲区，示例代码如下。

```
int readNumber = channel.read(receivedBuffer);
```

步骤十二：对 ByteBuffer 进行编解码，如果有半包消息接收缓冲区 Reset，继续读取后续的报文，将解码成功的消息封装成 Task，投递到业务线程池中，进行业务逻辑编排。示例代码如下。

```
Object message = null;
while(buffer.hasRemain())
{
    byteBuffer.mark();
    Object message = decode(byteBuffer);
    if (message == null)
    {
        byteBuffer.reset();
        break;
    }
    messageList.add(message );
}
if (!byteBuffer.hasRemain())
byteBuffer.clear();
else
    byteBuffer.compact();
if (messageList != null & !messageList.isEmpty())
{
for(Object messageE : messageList)
    handlerTask(messageE);
}
```

步骤十三：将 POJO 对象 encode 成 ByteBuffer，调用 SocketChannel 的异步 write 接口，将消息异步发送给客户端。示例代码如下。

```
socketChannel.write(buffer);
```

通过序列图和关键代码的解说，相信大家对创建 NIO 客户端程序已经有了一个初步的了解。下面，就跟随着我们的脚步，继续看看如何使用 NIO 改造之前的时间服务器客户端 TimeClient 吧。

2.3.5 NIO 创建的 TimeClient 源码分析

我们首先还是看下如何对 TimeClient 进行改造。

代码清单 2-9　NIO 时间服务器客户端　TimeClient

```
16.     if (args != null && args.length > 0) {
17.         try {
18.             port = Integer.valueOf(args[0]);
19.         } catch (NumberFormatException e) {
20.             // 采用默认值
21.         }
22.     }
23.     new Thread(new TimeClientHandle("127.0.0.1", port), "TimeClient-001")
24.             .start();
25. }
26. }
```

与之前唯一不同的地方在于通过创建 TimeClientHandle 线程来处理异步连接和读写操作，由于 TimeClient 非常简单且变更不大，这里重点分析 TimeClientHandle，代码如下。

代码清单 2-10　NIO 时间服务器客户端　TimeClientHandle

```
1.  package com.phei.netty.nio;
2.  import java.io.IOException;
3.  import java.net.InetSocketAddress;
4.  import java.nio.ByteBuffer;
5.  import java.nio.channels.SelectionKey;
6.  import java.nio.channels.Selector;
7.  import java.nio.channels.SocketChannel;
8.  import java.util.Iterator;
9.  import java.util.Set;
10.
11. /**
12.  * @author Administrator
13.  * @date 2014年2月16日
```

```
14.    * @version 1.0
15.    */
16.   public class TimeClientHandle implements Runnable {
17.       private String host;
18.       private int port;
19.       private Selector selector;
20.       private SocketChannel socketChannel;
21.       private volatile boolean stop;
22.
23.       public TimeClientHandle(String host, int port) {
24.        this.host = host == null ? "127.0.0.1" : host;
25.        this.port = port;
26.        try {
27.            selector = Selector.open();
28.            socketChannel = SocketChannel.open();
29.            socketChannel.configureBlocking(false);
30.        } catch (IOException e) {
31.            e.printStackTrace();
32.            System.exit(1);
33.        }
34.       }
35.
36.       /*
37.        * (non-Javadoc)
38.        *
39.        * @see java.lang.Runnable#run()
40.        */
41.       @Override
42.       public void run() {
43.        try {
44.            doConnect();
45.        } catch (IOException e) {
46.            e.printStackTrace();
47.            System.exit(1);
48.        }
49.        while (!stop) {
50.            try {
51.             selector.select(1000);
52.             Set<SelectionKey> selectedKeys = selector.selectedKeys();
53.             Iterator<SelectionKey> it = selectedKeys.iterator();
54.             SelectionKey key = null;
```

```
55.        while (it.hasNext()) {
56.            key = it.next();
57.            it.remove();
58.            try {
59.                handleInput(key);
60.            } catch (Exception e) {
61.                if (key != null) {
62.                    key.cancel();
63.                    if (key.channel() != null)
64.                        key.channel().close();
65.                }
66.            }
67.        }
68.    } catch (Exception e) {
69.        e.printStackTrace();
70.        System.exit(1);
71.    }
72. }
73.
74.    // 多路复用器关闭后，所有注册在上面的 Channel 和 Pipe 等资源都会被自动去注册并关闭，所以不需要重复释放资源
75.    if (selector != null)
76.        try {
77.            selector.close();
78.        } catch (IOException e) {
79.            e.printStackTrace();
80.        }
81. }
82.
83. private void handleInput(SelectionKey key) throws IOException {
84.
85.    if (key.isValid()) {
86.        // 判断是否连接成功
87.        SocketChannel sc = (SocketChannel) key.channel();
88.        if (key.isConnectable()) {
89.            if (sc.finishConnect()) {
90.                sc.register(selector, SelectionKey.OP_READ);
91.                doWrite(sc);
92.            } else
93.                System.exit(1);// 连接失败，进程退出
94.        }
```

```
95.         if (key.isReadable()) {
96.             ByteBuffer readBuffer = ByteBuffer.allocate(1024);
97.             int readBytes = sc.read(readBuffer);
98.             if (readBytes > 0) {
99.                 readBuffer.flip();
100.                 byte[] bytes = new byte[readBuffer.remaining()];
101.                 readBuffer.get(bytes);
102.                 String body = new String(bytes, "UTF-8");
103.                 System.out.println("Now is : " + body);
104.                 this.stop = true;
105.             } else if (readBytes < 0) {
106.                 // 对端链路关闭
107.                 key.cancel();
108.                 sc.close();
109.             } else
110.                 ; // 读到0字节,忽略
111.             }
112.         }
113.
114.     }
115.
116.     private void doConnect() throws IOException {
117.         // 如果直接连接成功,则注册到多路复用器上,发送请求消息,读应答
118.         if(socketChannel.connect(new InetSocketAddress(host, port))) {
119.             socketChannel.register(selector, SelectionKey.OP_READ);
120.             doWrite(socketChannel);
121.         } else
122.             socketChannel.register(selector, SelectionKey.OP_CONNECT);
123.     }
124.
125.     private void doWrite(SocketChannel sc) throws IOException {
126.         byte[] req = "QUERY TIME ORDER".getBytes();
127.         ByteBuffer writeBuffer = ByteBuffer.allocate(req.length);
128.         writeBuffer.put(req);
129.         writeBuffer.flip();
130.         sc.write(writeBuffer);
131.         if (!writeBuffer.hasRemaining())
132.             System.out.println("Send order 2 server succeed.");
133.     }
134. }
```

与服务端类似，接下来我们通过对关键步骤的源码进行分析和解读，让大家深入了解如何创建 NIO 客户端以及如何使用 NIO 的 API。

（1）23～34 行构造函数用于初始化 NIO 的多路复用器和 SocketChannel 对象。需要注意的是，创建 SocketChannel 之后，需要将其设置为异步非阻塞模式。就像在 2.3.3 小节中所讲的，我们可以设置 SocketChannel 的 TCP 参数，例如接收和发送的 TCP 缓冲区大小。

（2）43～48 行用于发送连接请求，作为示例，连接是成功的，所以不需要做重连操作，因此将其放到循环之前。下面我们具体看看 doConnect 的实现，代码跳到第 116～123 行，首先对 SocketChannel 的 connect()操作进行判断。如果连接成功，则将 SocketChannel 注册到多路复用器 Selector 上，注册 SelectionKey.OP_READ；如果没有直接连接成功，则说明服务端没有返回 TCP 握手应答消息，但这并不代表连接失败。我们需要将 SocketChannel 注册到多路复用器 Selector 上，注册 SelectionKey.OP_CONNECT，当服务端返回 TCP syn-ack 消息后，Selector 就能够轮询到这个 SocketChannel 处于连接就绪状态。

（3）49～72 行在循环体中轮询多路复用器 Selector。当有就绪的 Channel 时，执行第 59 行的 handleInput(key)方法。下面我们就对 handleInput 方法进行分析。

（4）跳到第 83 行，我们首先对 SelectionKey 进行判断，看它处于什么状态。如果是处于连接状态，说明服务端已经返回 ACK 应答消息。这时我们需要对连接结果进行判断，调用 SocketChannel 的 finishConnect()方法。如果返回值为 true，说明客户端连接成功；如果返回值为 false 或者直接抛出 IOException，说明连接失败。在本例程中，返回值为 true，说明连接成功。将 SocketChannel 注册到多路复用器上，注册 SelectionKey.OP_READ 操作位，监听网络读操作，然后发送请求消息给服务端。

下面我们对 doWrite(sc)进行分析。代码跳到第 125 行，我们构造请求消息体，然后对其编码，写入到发送缓冲区中，最后调用 SocketChannel 的 write 方法进行发送。由于发送是异步的，所以会存在"半包写"问题，此处不再赘述。最后通过 hasRemaining()方法对发送结果进行判断，如果缓冲区中的消息全部发送完成，打印"Send order 2 server succeed."

（5）返回代码第 95 行，我们继续分析客户端是如何读取时间服务器应答消息的。如果客户端接收到了服务端的应答消息，则 SocketChannel 是可读的，由于无法事先判断应答码流的大小，我们就预分配 1MB 的接收缓冲区用于读取应答消息，调用 SocketChannel 的 read()方法进行异步读取操作。由于是异步操作，所以必须对读取的结果进行判断，这部分的处理逻辑已经在 2.3.3 小节详细介绍过，此处不再赘述。如果读取到了消息，则对消息进行解码，最后打印结果。执行完成后将 stop 置为 true，线程退出循环。

（6）线程退出循环后，我们需要对连接资源进行释放，以实现“优雅退出”。75～80 行用于多路复用器的资源释放，由于多路复用器上可能注册成千上万的 Channel 或者 pipe，如果一一对这些资源进行释放显然不合适。因此，JDK 底层会自动释放所有跟此多路复用器关联的资源，JDK 的 API DOC 如图 2-12 所示。

到此为止，我们已经通过 NIO 对时间服务器完成了改造，并对源码进行了分析和解读，下面分别执行时间服务器的服务端和客户端，看看执行结果。

void java.nio.channels.Selector.close() throws IOException

Closes this selector.

If a thread is currently blocked in one of this selector's selection methods then it is interrupted as if by invoking the selector's wakeup method.

Any uncancelled keys still associated with this selector are invalidated, their channels are deregistered, and any other resources associated with this selector are released.

If this selector is already closed then invoking this method has no effect.

After a selector is closed, any further attempt to use it, except by invoking this method or the wakeup method, will cause a ClosedSelectorException to be thrown.

Specified by: close() in Closeable
Throws:
IOException - If an I/O error occurs

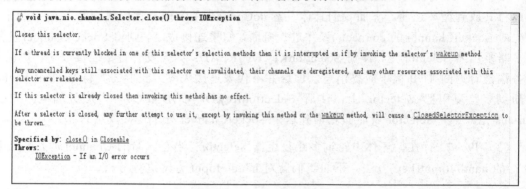

图 2-12　多路复用器 Selector 的资源释放

服务端执行结果如图 2-13 所示。

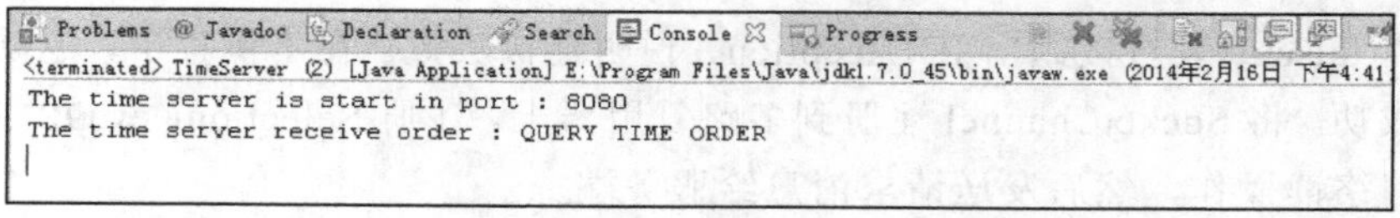

图 2-13　NIO 时间服务器服务端执行结果

客户端执行结果如图 2-14 所示。

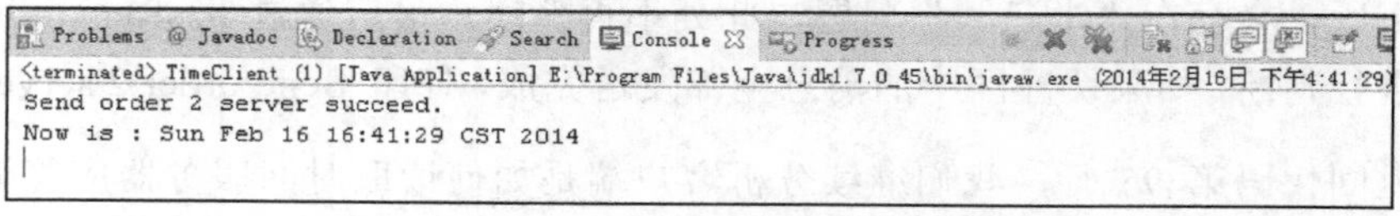

图 2-14　NIO 时间服务器客户端执行结果

通过源码对比分析，我们发现 NIO 编程的难度确实比同步阻塞 BIO 的大很多，我们的 NIO 例程并没有考虑“半包读”和“半包写”，如果加上这些，代码将会更加复杂。NIO

代码既然这么复杂，为什么它的应用却越来越广泛呢？使用 NIO 编程的优点总结如下。

（1）客户端发起的连接操作是异步的，可以通过在多路复用器注册 OP_CONNECT 等待后续结果，不需要像之前的客户端那样被同步阻塞。

（2）SocketChannel 的读写操作都是异步的，如果没有可读写的数据它不会同步等待，直接返回，这样 I/O 通信线程就可以处理其他的链路，不需要同步等待这个链路可用。

（3）线程模型的优化：由于 JDK 的 Selector 在 Linux 等主流操作系统上通过 epoll 实现，它没有连接句柄数的限制（只受限于操作系统的最大句柄数或者对单个进程的句柄限制），这意味着一个 Selector 线程可以同时处理成千上万个客户端连接，而且性能不会随着客户端的增加而线性下降。因此，它非常适合做高性能、高负载的网络服务器。

JDK1.7 升级了 NIO 类库，升级后的 NIO 类库被称为 NIO 2.0。引人注目的是，Java 正式提供了异步文件 I/O 操作，同时提供了与 UNIX 网络编程事件驱动 I/O 对应的 AIO。下面的 2.4 节我们将学习如何利用 NIO2.0 编写 AIO 程序，依旧以时间服务器为例进行讲解。

2.4 AIO 编程

NIO 2.0 引入了新的异步通道的概念，并提供了异步文件通道和异步套接字通道的实现。异步通道提供以下两种方式获取获取操作结果。

- 通过 java.util.concurrent.Future 类来表示异步操作的结果；
- 在执行异步操作的时候传入一个 java.nio.channels。

CompletionHandler 接口的实现类作为操作完成的回调。

NIO 2.0 的异步套接字通道是真正的异步非阻塞 I/O，对应于 UNIX 网络编程中的事件驱动 I/O（AIO）。它不需要通过多路复用器（Selector）对注册的通道进行轮询操作即可实现异步读写，从而简化了 NIO 的编程模型。

下面通过代码来熟悉 NIO2.0 AIO 的相关类库，仍旧以时间服务器为例程进行讲解。

2.4.1 AIO 创建的 TimeServer 源码分析

首先看下时间服务器的主函数。

代码清单2-11　AIO 时间服务器服务端　TimeClientHandle

```
10. public class TimeServer {
11.
12.     /**
13.      * @param args
14.      * @throws IOException
15.      */
16.     public static void main(String[] args) throws IOException {
17.         int port = 8080;
18.         if (args != null && args.length > 0) {
19.             try {
20.                 port = Integer.valueOf(args[0]);
21.             } catch (NumberFormatException e) {
22.                 // 采用默认值
23.             }
24.         }
25.         AsyncTimeServerHandler timeServer=new AsyncTimeServerHandler(port);
26.         new Thread(timeServer, "AIO-AsyncTimeServerHandler-001").start();
27.     }
28. }
```

我们直接从第 25 行开始看，首先创建异步的时间服务器处理类，然后启动线程将 AsyncTimeServerHandler 拉起，代码如下。

代码清单2-12　AIO 时间服务器服务端

```
13. public class AsyncTimeServerHandler implements Runnable {
14.
15.     private int port;
16.
17.     CountDownLatch latch;
18.     AsynchronousServerSocketChannel asynchronousServerSocketChannel;
19.
20.     public AsyncTimeServerHandler(int port) {
21.         this.port = port;
22.         try {
23.             asynchronousServerSocketChannel = AsynchronousServerSocketChannel
24.                     .open();
```

```
25.            asynchronousServerSocketChannel.bind(new
InetSocketAddress(port));
26.            System.out.println("The time server is start in port : " + port);
27.        } catch (IOException e) {
28.            e.printStackTrace();
29.        }
30.    }
31.
32.    /*
33.     * (non-Javadoc)
34.     *
35.     * @see java.lang.Runnable#run()
36.     */
37.    @Override
38.    public void run() {
39.
40.        latch = new CountDownLatch(1);
41.        doAccept();
42.        try {
43.            latch.await();
44.        } catch (InterruptedException e) {
45.            e.printStackTrace();
46.        }
47.    }
48.
49.    public void doAccept() {
50.        asynchronousServerSocketChannel.accept(this,
51.            new AcceptCompletionHandler());
52.    }
```

我们重点对 AsyncTimeServerHandler 进行分析。首先看第 20~27 行，在构造方法中，首先创建一个异步的服务端通道 AsynchronousServerSocketChannel，然后调用它的 bind 方法绑定监听端口。如果端口合法且没被占用，则绑定成功，打印启动成功提示到控制台。

在线程的 run 方法中，第 40 行初始化 CountDownLatch 对象，它的作用是在完成一组正在执行的操作之前，允许当前的线程一直阻塞。在本例程中，我们让线程在此阻塞，防止服务端执行完成退出。在实际项目应用中，不需要启动独立的线程来处理 AsynchronousServerSocketChannel，这里仅仅是个 demo 演示。

第 41 行用于接收客户端的连接，由于是异步操作，我们可以传递一个 CompletionHandler <AsynchronousSocketChannel,? super A>类型的 handler 实例接收 accept 操作成功的通知消息。在本例程中我们通过 AcceptCompletionHandler 实例作为 handler 来接收通知消息，下面继续对 AcceptCompletionHandler 进行分析。

代码清单 2-13　AIO 时间服务器服务端　AcceptCompletionHandler

```
14.
15.     @Override
16.     public void completed(AsynchronousSocketChannel result,
17.         AsyncTimeServerHandler attachment) {
18.         attachment.asynchronousServerSocketChannel.accept(attachment,
this);
19.         ByteBuffer buffer = ByteBuffer.allocate(1024);
20.         result.read(buffer, buffer, new ReadCompletionHandler(result));
21.     }
22.
23.     @Override
24.     public void failed(Throwable exc,AsyncTimeServerHandler attachment) {
25.         exc.printStackTrace();
26.         attachment.latch.countDown();
27.     }
28. }
```

CompletionHandler 有两个方法，分别如下。

◎　public void completed(AsynchronousSocketChannel result, AsyncTimeServerHandler attachment)；

◎　public void failed(Throwable exc, AsyncTimeServerHandler attachment)。

下面分别对这两个接口的实现进行分析。首先看 completed 接口的实现，代码第 18～20 行，我们从 attachment 获取成员变量 AsynchronousServerSocketChannel，然后继续调用它的 accept 方法。有的读者可能会心存疑惑：既然已经接收客户端成功了，为什么还要再次调用 accept 方法呢？原因是这样的：调用 AsynchronousServerSocketChannel 的 accept 方法后，如果有新的客户端连接接入，系统将回调我们传入的 CompletionHandler 实例的 completed 方法，表示新的客户端已经接入成功。因为一个 AsynchronousServerSocket Channel 可以接收成千上万个客户端，所以需要继续调用它的 accept 方法，接收其他的客户端连接，最终形成一个循环。每当接收一个客户读连接成功之后，再异步接收新的客户端连接。

链路建立成功之后，服务端需要接收客户端的请求消息，在代码第 19 行创建新的 ByteBuffer，预分配 1MB 的缓冲区。第 20 行通过调用 AsynchronousSocketChannel 的 read

方法进行异步读操作。下面我们看看异步 read 方法的参数。

- ByteBuffer dst：接收缓冲区，用于从异步 Channel 中读取数据包；
- A attachment：异步 Channel 携带的附件，通知回调的时候作为入参使用；
- CompletionHandler<Integer,? super A>：接收通知回调的业务 Handler，在本例程中为 ReadCompletionHandler。

下面我们继续对 ReadCompletionHandler 进行分析。

代码清单 2-14 AIO 时间服务器服务端 ReadCompletionHandler

```
8.
9.  /**
10.  * @author lilinfeng
11.  * @date 2014年2月16日
12.  * @version 1.0
13.  */
14. public class ReadCompletionHandler implements
15.     CompletionHandler<Integer, ByteBuffer> {
16.
17.     private AsynchronousSocketChannel channel;
18.
19.     public ReadCompletionHandler(AsynchronousSocketChannel channel) {
20.      if (this.channel == null)
21.          this.channel = channel;
22.      }
23.
24.     @Override
25.     public void completed(Integer result, ByteBuffer attachment) {
26.      attachment.flip();
27.      byte[] body = new byte[attachment.remaining()];
28.      attachment.get(body);
29.      try {
30.          String req = new String(body, "UTF-8");
31.          System.out.println("The time server receive order : " + req);
32.          String currentTime = "QUERY TIME ORDER".equalsIgnoreCase(req) ? new java.util.Date(
33.              System.currentTimeMillis()).toString() : "BAD ORDER";
34.          doWrite(currentTime);
35.      } catch (UnsupportedEncodingException e) {
36.          e.printStackTrace();
```

```
37.        }
38.      }
39.
40.    private void doWrite(String currentTime) {
41.      if (currentTime != null && currentTime.trim().length() > 0) {
42.          byte[] bytes = (currentTime).getBytes();
43.          ByteBuffer writeBuffer = ByteBuffer.allocate(bytes.length);
44.          writeBuffer.put(bytes);
45.          writeBuffer.flip();
46.          channel.write(writeBuffer, writeBuffer,
47.             new CompletionHandler<Integer, ByteBuffer>() {
48.               @Override
49.               public void completed(Integer result, ByteBuffer buffer) {
50.                   // 如果没有发送完成,继续发送
51.                   if (buffer.hasRemaining())
52.                     channel.write(buffer, buffer, this);
53.               }
54.
55.               @Override
56.               public void failed(Throwable exc, ByteBuffer attachment) {
57.                  try {
58.                    channel.close();
59.                  } catch (IOException e) {
60.                    // ingnore on close
61.                  }
62.               }
63.          });
64.        }
65.      }
66.
67.    @Override
68.    public void failed(Throwable exc, ByteBuffer attachment) {
69.      try {
70.         this.channel.close();
71.      } catch (IOException e) {
72.         e.printStackTrace();
73.      }
74.    }
75. }
```

首先看构造方法。我们将 AsynchronousSocketChannel 通过参数传递到 ReadCompletion Handler 中，当作成员变量来使用，主要用于读取半包消息和发送应答。本例程不对半包读写进行具体说明，对此感兴趣的读者可以关注后续章节对 Netty 半包处理的专题介绍。

继续看代码，第 25～38 行是读取到消息后的处理。首先对 attachment 进行 flip 操作，为后续从缓冲区读取数据做准备。根据缓冲区的可读字节数创建 byte 数组，然后通过 new String 方法创建请求消息，对请求消息进行判断，如果是"QUERY TIME ORDER"则获取当前系统服务器的时间，调用 doWrite 方法发送给客户端。下面我们对 doWrite 方法进行详细分析。

跳到代码第 41 行，首先对当前时间进行合法性校验，如果合法，调用字符串的解码方法将应答消息编码成字节数组，然后将它复制到发送缓冲区 writeBuffer 中，最后调用 AsynchronousSocketChannel 的异步 write 方法。正如前面介绍的异步 read 方法一样，它也有三个与 read 方法相同的参数，在本例程中我们直接实现 write 方法的异步回调接口 CompletionHandler。代码跳到第 51 行，对发送的 writeBuffer 进行判断，如果还有剩余的字节可写，说明没有发送完成，需要继续发送，直到发送成功。

最后，我们关注下 failed 方法，它的实现很简单，就是当发生异常的时候，对异常 Throwable 进行判断：如果是 I/O 异常，就关闭链路，释放资源；如果是其他异常，按照业务自己的逻辑进行处理。本例程作为简单的 demo，没有对异常进行分类判断，只要发生了读写异常，就关闭链路，释放资源。

异步非阻塞 I/O 版本的时间服务器服务端已经介绍完毕，下面我们继续看客户端的实现。

2.4.2　AIO 创建的 TimeClient 源码分析

首先看下客户端主函数的实现。

代码清单 2-15　AIO 时间服务器客户端　TimeClient

```
16.         try {
17.             port = Integer.valueOf(args[0]);
18.         } catch (NumberFormatException e) {
19.             // 采用默认值
20.         }
21.     }
22.     new Thread(new AsyncTimeClientHandler("127.0.0.1", port),
```

```
23.             "AIO-AsyncTimeClientHandler-001").start();
24.     }
25. }
```

第 22 行通过一个独立的 I/O 线程创建异步时间服务器客户端 Handler。在实际项目中，我们不需要独立的线程创建异步连接对象，因为底层都是通过 JDK 的系统回调实现的，在后面运行时间服务器程序的时候，我们会抓取线程调用堆栈给大家展示。

继续看代码，AsyncTimeClientHandler 的实现类源码如下。

代码清单 2-16　AIO 时间服务器客户端 AsyncTimeClientHandler

```
1.  package com.phei.netty.aio;
2.
3.  import java.io.IOException;
4.  import java.io.UnsupportedEncodingException;
5.  import java.net.InetSocketAddress;
6.  import java.nio.ByteBuffer;
7.  import java.nio.channels.AsynchronousSocketChannel;
8.  import java.nio.channels.CompletionHandler;
9.  import java.util.concurrent.CountDownLatch;
10.
11. /**
12.  * @author Administrator
13.  * @date 2014年2月16日
14.  * @version 1.0
15.  */
16. public class AsyncTimeClientHandler implements
17.     CompletionHandler<Void, AsyncTimeClientHandler>, Runnable {
18.
19.     private AsynchronousSocketChannel client;
20.     private String host;
21.     private int port;
22.     private CountDownLatch latch;
23.
24.     public AsyncTimeClientHandler(String host, int port) {
25.      this.host = host;
26.      this.port = port;
27.      try {
28.         client = AsynchronousSocketChannel.open();
29.      } catch (IOException e) {
```

```
30.            e.printStackTrace();
31.         }
32.      }
33.
34.     @Override
35.     public void run() {
36.         latch = new CountDownLatch(1);
37.         client.connect(new InetSocketAddress(host, port), this, this);
38.         try {
39.             latch.await();
40.         } catch (InterruptedException e1) {
41.             e1.printStackTrace();
42.         }
43.         try {
44.             client.close();
45.         } catch (IOException e) {
46.             e.printStackTrace();
47.         }
48.     }
49.
50.     @Override
51.     public void completed(Void result, AsyncTimeClientHandler attachment) {
52.         byte[] req = "QUERY TIME ORDER".getBytes();
53.         ByteBuffer writeBuffer = ByteBuffer.allocate(req.length);
54.         writeBuffer.put(req);
55.         writeBuffer.flip();
56.         client.write(writeBuffer, writeBuffer,
57.             new CompletionHandler<Integer, ByteBuffer>() {
58.                 @Override
59.                 public void completed(Integer result, ByteBuffer buffer) {
60.                     if (buffer.hasRemaining()) {
61.                         client.write(buffer, buffer, this);
62.                     } else {
63.                         ByteBuffer readBuffer = ByteBuffer.allocate(1024);
64.                         client.read(
65.                             readBuffer,
66.                             readBuffer,
67.                             new CompletionHandler<Integer, ByteBuffer>() {
68.                                 @Override
69.                                 public void completed(Integer result,
```

```
70.                    ByteBuffer buffer) {
71.                        buffer.flip();
72.                        byte[] bytes = new byte[buffer
73.                            .remaining()];
74.                        buffer.get(bytes);
75.                        String body;
76.                        try {
77.                            body = new String(bytes,
78.                                "UTF-8");
79.                            System.out.println("Now is : "
80.                                + body);
81.                            latch.countDown();
82.                        } catch (UnsupportedEncodingException e) {
83.                            e.printStackTrace();
84.                        }
85.                    }
86.
87.                    @Override
88.                    public void failed(Throwable exc,
89.                        ByteBuffer attachment) {
90.                        try {
91.                            client.close();
92.                            latch.countDown();
93.                        } catch (IOException e) {
94.                            // ingnore on close
95.                        }
96.                    }
97.                });
98.            }
99.        }
100.
101.            @Override
102.            public void failed(Throwable exc,ByteBuffer attachment) {
103.                try {
104.                    client.close();
105.                    latch.countDown();
106.                } catch (IOException e) {
107.                    // ingnore on close
108.                }
109.            }
110.        });
```

```
111.            }
112.
113.            @Override
114.            public void failed(Throwable exc, AsyncTimeClientHandler
attachment) {
115.                exc.printStackTrace();
116.                try {
117.                    client.close();
118.                    latch.countDown();
119.                } catch (IOException e) {
120.                    e.printStackTrace();
121.                }
122.            }
123.        }
```

由于在 AsyncTimeClientHandler 中大量使用了内部匿名类，所以代码看起来稍微有些复杂，下面我们就对主要代码进行详细讲解。

第 24~32 行是构造方法，首先通过 AsynchronousSocketChannel 的 open 方法创建一个新的 AsynchronousSocketChannel 对象。然后跳到第 36 行，创建 CountDownLatch 进行等待，防止异步操作没有执行完成线程就退出。第 37 行通过 connect 方法发起异步操作，它有两个参数，分别如下。

◎ A attachment：AsynchronousSocketChannel 的附件，用于回调通知时作为入参被传递，调用者可以自定义；

◎ CompletionHandler<Void,? super A> handler：异步操作回调通知接口，由调用者实现。

在本例程中，这两个参数都使用 AsyncTimeClientHandler 类本身，因为它实现了 CompletionHandler 接口。

接下来我们看异步连接成功之后的方法回调——completed 方法。代码第 52 行，我们创建请求消息体，对其进行编码，然后复制到发送缓冲区 writeBuffer 中，调用 AsynchronousSocketChannel 的 write 方法进行异步写。与服务端类似，我们可以实现 CompletionHandler <Integer, ByteBuffer>接口用于写操作完成后的回调。代码第 60~62 行，如果发送缓冲区中仍有尚未发送的字节，将继续异步发送，如果已经发送完成，则执行异步读取操作。

代码第 64~97 行是客户端异步读取时间服务器服务端应答消息的处理逻辑。代码第 64 行调用 AsynchronousSocketChannel 的 read 方法异步读取服务端的响应消息。由于 read

操作是异步的，所以我们通过内部匿名类实现 CompletionHandler<Integer, ByteBuffer>接口，当读取完成被 JDK 回调时，构造应答消息。第 71～78 行从 CompletionHandler 的 ByteBuffer 中读取应答消息，然后打印结果。

第 102～111 行，当读取发生异常时，关闭链路，同时调用 CountDownLatch 的 countDown 方法让 AsyncTimeClientHandler 线程执行完毕，客户端退出执行。

需要指出的是，正如之前的 NIO 例程，我们并没有完整的处理网络的半包读写，在对例程进行功能测试的时候没有问题。但是，如果对代码稍加改造，进行压力或者性能测试，就会发现输出结果存在问题。

由于半包读写会作为专门的小节在 Netty 的应用和源码分析章节进行详细讲解，在 NIO 的入门章节就不详细展开介绍了，以便读者能够将注意力集中在 NIO 的入门知识上来。

在下面的小节中我们会运行 AIO 版本的时间服务器程序，并通过打印线程堆栈的方式看下 JDK 回调异步 Channel CompletionHandler 的调用情况。

2.4.3 AIO 版本时间服务器运行结果

执行 TimeServer，运行结果如图 2-15 所示。

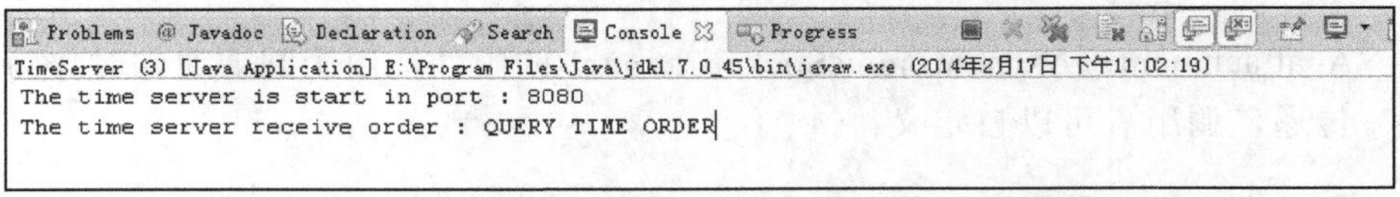

图 2-15　AIO 时间服务器服务端运行结果

执行 TimeClient，运行结果如图 2-16 所示。

图 2-16　AIO 时间服务器客户端运行结果

下面继续看下 JDK 异步回调 CompletionHandler 的线程执行堆栈。

```
"AIO-AsyncTimeClientHandler-001" prio=6 tid=0x16c64400 nid=0xa28 waiting on condition [0x1719f000]
   java.lang.Thread.State: WAITING (parking)
        at sun.misc.Unsafe.park(Native Method)
        - parking to wait for  <0x08061b30> (a java.util.concurrent.CountDownLatch$Sync)
        at java.util.concurrent.locks.LockSupport.park(LockSupport.java:186)
        at java.util.concurrent.locks.AbstractQueuedSynchronizer.parkAndCheckInterrupt(AbstractQueuedSynchronizer.java:834)
        at java.util.concurrent.locks.AbstractQueuedSynchronizer.doAcquireSharedInterruptibly(AbstractQueuedSynchronizer.java:994)
        at java.util.concurrent.locks.AbstractQueuedSynchronizer.acquireSharedInterruptibly(AbstractQueuedSynchronizer.java:1303)
        at java.util.concurrent.CountDownLatch.await(CountDownLatch.java:236)
        at com.phei.netty.aio.AsyncTimeClientHandler.run(AsyncTimeClientHandler.java:55)
        at java.lang.Thread.run(Thread.java:744)

   Locked ownable synchronizers:
        - None

"Thread-2" daemon prio=6 tid=0x16c62000 nid=0xe70 at breakpoint[0x1714f000]
   java.lang.Thread.State: RUNNABLE
        at com.phei.netty.aio.AsyncTimeClientHandler$1.completed(AsyncTimeClientHandler.java:80)
        at com.phei.netty.aio.AsyncTimeClientHandler$1.completed(AsyncTimeClientHandler.java:1)
        at sun.nio.ch.Invoker.invokeUnchecked(Invoker.java:126)
        at sun.nio.ch.Invoker.invokeUnchecked(Invoker.java:289)
        at sun.nio.ch.WindowsAsynchronousSocketChannelImpl$WriteTask.completed(WindowsAsynchronousSocketChannelImpl.java:815)
        at sun.nio.ch.Iocp$EventHandlerTask.run(Iocp.java:397)
        at sun.nio.ch.AsynchronousChannelGroupImpl$1.run(AsynchronousChannelGroupImpl.java:112)
        at java.util.concurrent.ThreadPoolExecutor.runWorker(ThreadPoolExecutor.java:1145)
        at java.util.concurrent.ThreadPoolExecutor$Worker.run(ThreadPoolExecutor.java:615)
```

图 2-17　AIO 时间服务器异步回调线程堆栈

从"Thread-2"线程堆栈中可以发现，JDK 底层通过线程池 ThreadPoolExecutor 来执行回调通知，异步回调通知类由 sun.nio.ch.AsynchronousChannelGroupImpl 实现，它经过层层调用，最终回调 com.phei.netty.aio.AsyncTimeClientHandler$1.completed 方法，完成回调通知。由此我们也可以得出结论：异步 Socket Channel 是被动执行对象，我们不需要像 NIO 编程那样创建一个独立的 I/O 线程来处理读写操作。对于 AsynchronousServerSocketChannel 和 AsynchronousSocketChannel，它们都由 JDK 底层的线程池负责回调并驱动读写操作。正因为如此，基于 NIO 2.0 新的异步非阻塞 Channel 进行编程比 NIO 编程更为简单。

本小节我们讲解了 JDK 1.7 提供的新异步非阻塞 I/O（AIO）的用法，由于国内商用的主流 Java 版本仍然是 JDK 1.6，因此，本小节不再详细介绍 NIO 2.0 其他新增的特性，如果大家对 NIO 2.0 的异步文件操作等特性感兴趣，可以选择阅读 JDK 1.7 的相关书籍或者查看甲骨文发布的 JDK 1.7 白皮书。

下个小节我们对本章列举的各种 I/O 进行概念澄清和比较，让大家从整体上掌握这些 I/O 模型的差异，以便在未来的工作中能够根据产品的实际情况选择合适的 I/O 模型。

2.5 4种I/O的对比

2.5.1 概念澄清

为了防止由于对一些技术概念和术语的理解或者叫法不一致而引起歧义，本小节特意对本书中的专业术语或者技术用语做下声明：如果它们与其他一些技术书籍的称呼不一致，请以本小节的解释为准。

1. 异步非阻塞 I/O

很多人喜欢将JDK 1.4提供的NIO框架称为异步非阻塞I/O，但是，如果严格按照UNIX网络编程模型和JDK的实现进行区分，实际上它只能被称为非阻塞I/O，不能叫异步非阻塞I/O。在早期的JDK 1.4和1.5 update10版本之前，JDK的Selector基于select/poll模型实现，它是基于I/O复用技术的非阻塞I/O，不是异步I/O。在JDK 1.5 update10和Linux core2.6以上版本，Sun优化了Selctor的实现，它在底层使用epoll替换了select/poll，上层的API并没有变化，可以认为是JDK NIO的一次性能优化，但是它仍旧没有改变I/O的模型。相关优化的官方说明如图2-18所示。

```
Changes in 1.5.0_10
The full internal version number for this update release is 1.5.0_10-b03 (where "b" means "build").
The external version number is 5.0u10.

Support for epoll
The Linux downloads of this update release include an implementation of
java.nio.channels.spi.SelectorProvider that is based on the epoll I/O event notification facility. The
epoll facility is available in the Linux 2.6 kernel, and is more scalable than the traditional poll system
call. This epoll-based implementation may improve the performance of server applications that use
the New I/O API and that register hundreds of channels with a selector. For more information, refer to
the epoll(4) and poll(2) man pages.

The epoll-based implementation of SelectorProvider is not selected by default. To select it, specify a
property value from the command line as follows:

java -Djava.nio.channels.spi.SelectorProvider=sun.nio.ch.EPollSelectorProvider ...
```

图2-18 JDK1.5_update10 支持epoll

由JDK1.7提供的NIO 2.0新增了异步的套接字通道，它是真正的异步I/O，在异步I/O操作的时候可以传递信号变量，当操作完成之后会回调相关的方法，异步I/O也被称为AIO。

NIO类库支持非阻塞读和写操作，相比于之前的同步阻塞读和写，它是异步的，因此很多人习惯于称NIO为异步非阻塞I/O，包括很多介绍NIO编程的书籍也沿用了这个说法。为了符合大家的习惯，本书也会将NIO称为异步非阻塞I/O或者非阻塞I/O，请大家理解，不要过分纠结在一些技术术语的咬文嚼字上。

2. 多路复用器 Selector

几乎所有的中文技术书籍都将 Selector 翻译为选择器，但是实际上我认为这样的翻译并不恰当，选择器仅仅是字面上的意思，体现不出 Selector 的功能和特点。

在前面的章节我们介绍过 Java NIO 的实现关键是多路复用 I/O 技术，多路复用的核心就是通过 Selector 来轮询注册在其上的 Channel，当发现某个或者多个 Channel 处于就绪状态后，从阻塞状态返回就绪的 Channel 的选择键集合，进行 I/O 操作。由于多路复用器是 NIO 实现非阻塞 I/O 的关键，它又是主要通过 Selector 实现的，所以本书将 Selector 翻译为多路复用器，与其他技术书籍所说的选择器是同一个东西，请大家了解。

3. 伪异步 I/O

伪异步 I/O 的概念完全来源于实践。在 JDK NIO 编程没有流行之前，为了解决 Tomcat 通信线程同步 I/O 导致业务线程被挂住的问题，大家想到了一个办法：在通信线程和业务线程之间做个缓冲区，这个缓冲区用于隔离 I/O 线程和业务线程间的直接访问，这样业务线程就不会被 I/O 线程阻塞。而对于后端的业务侧来说，将消息或者 Task 放到线程池后就返回了，它不再直接访问 I/O 线程或者进行 I/O 读写，这样也就不会被同步阻塞。类似的设计还包括前端启动一组线程，将接收的客户端封装成 Task，放到后端的线程池执行，用于解决一连接一线程问题。像这样通过线程池做缓冲区的做法，本书中习惯于称它为伪异步 I/O，而官方并没有伪异步 I/O 这种说法，请大家注意。

下面的小节我们对几种常见的 I/O 进行对比，以便大家能够理解它们的差异。

2.5.2 不同 I/O 模型对比

不同的 I/O 模型由于线程模型、API 等差别很大，所以用法的差异也非常大。由于之前的几个小节已经集中对这几种 I/O 的 API 和用法进行了说明，本小节会重点对它们进行功能对比。如表 2-1 所示。

表 2-1 几种 I/O 模型的功能和特性对比

	同步阻塞 I/O（BIO）	伪异步 I/O	非阻塞 I/O（NIO）	异步 I/O（AIO）
客户端个数：I/O 线程	1:1	M:N（其中 M 可以大于 N）	M:1（1 个 I/O 线程处理多个客户端连接）	M:0（不需要启动额外的 I/O 线程，被动回调）
I/O 类型（阻塞）	阻塞 I/O	阻塞 I/O	非阻塞 I/O	非阻塞 I/O

续表

	同步阻塞 I/O（BIO）	伪异步 I/O	非阻塞 I/O（NIO）	异步 I/O（AIO）
I/O 类型（同步）	同步 I/O	同步 I/O	同步 I/O（I/O 多路复用）	异步 I/O
API 使用难度	简单	简单	非常复杂	复杂
调试难度	简单	简单	复杂	复杂
可靠性	非常差	差	高	高
吞吐量	低	中	高	高

尽管本书是专门介绍 NIO 框架 Netty 的，但是，并不意味着所有的 Java 网络编程都必须要选择 NIO 和 Netty，具体选择什么样的 I/O 模型或者 NIO 框架，完全基于业务的实际应用场景和性能诉求，如果客户端并发连接数不多，周边对接的网元不多，服务器的负载也不重，那就完全没必要选择 NIO 做服务端；如果是相反情况，那就要考虑选择合适的 NIO 框架进行开发。

对比完 Java 的几种主流 I/O 模型之后，我们继续看下为什么要选择 Netty 进行 NIO 开发，而不是直接使用 JDK 的 NIO 原生类库。

2.6 选择 Netty 的理由

在开始本节之前，我先讲一个亲身经历的故事：曾经有两个项目组同时用到了 NIO 编程技术，一个项目组选择自己开发 NIO 服务端，直接使用 JDK 原生的 API，结果两个多月过去了，他们的 NIO 服务端始终无法稳定，问题频出。由于 NIO 通信是它们的核心组件之一，因此项目的进度受到了严重的影响。另一个项目组直接使用 Netty 作为 NIO 服务端，业务的定制开发工作量非常小，测试表明，功能和性能都完全达标，项目组几乎没有在 NIO 服务端上花费额外的时间和精力，项目进展也非常顺利。

这两个项目组的不同遭遇告诉我们：开发出高质量的 NIO 程序并不是一件简单的事情，除去 NIO 固有的复杂性和 BUG 不谈，作为一个 NIO 服务端，需要能够处理网络的闪断、客户端的重复接入、客户端的安全认证、消息的编解码、半包读写等情况，如果你没有足够的 NIO 编程经验积累，一个 NIO 框架的稳定往往需要半年甚至更长的时间。更为糟糕的是，一旦在生产环境中发生问题，往往会导致跨节点的服务调用中断，严重的可能会导致整个集群环境都不可用，需要重启服务器，这种非正常停机会带来巨大的损失。

从可维护性角度看，由于 NIO 采用了异步非阻塞编程模型，而且是一个 I/O 线程处理多条链路,它的调试和跟踪非常麻烦,特别是生产环境中的问题,我们无法进行有效的调试和跟踪,往往只能靠一些日志来辅助分析,定位难度很大。

2.6.1 不选择 Java 原生 NIO 编程的原因

现在我们总结一下为什么不建议开发者直接使用 JDK 的 NIO 类库进行开发,具体原因如下。

(1) NIO 的类库和 API 繁杂,使用麻烦,你需要熟练掌握 Selector、ServerSocketChannel、SocketChannel、ByteBuffer 等。

(2) 需要具备其他的额外技能做铺垫,例如熟悉 Java 多线程编程。这是因为 NIO 编程涉及到 Reactor 模式,你必须对多线程和网络编程非常熟悉,才能编写出高质量的 NIO 程序。

(3) 可靠性能力补齐,工作量和难度都非常大。例如客户端面临断连重连、网络闪断、半包读写、失败缓存、网络拥塞和异常码流的处理等问题,NIO 编程的特点是功能开发相对容易,但是可靠性能力补齐的工作量和难度都非常大。

(4) JDK NIO 的 BUG,例如臭名昭著的 epoll bug,它会导致 Selector 空轮询,最终导致 CPU 100%。官方声称在 JDK 1.6 版本的 update18 修复了该问题,但是直到 JDK 1.7 版本该问题仍旧存在,只不过该 BUG 发生概率降低了一些而已,它并没有得到根本性解决。该 BUG 以及与该 BUG 相关的问题单可以参见以下链接内容。

◎ http://bugs.java.com/bugdatabase/view_bug.do?bug_id=6403933

◎ http://bugs.java.com/bugdatabase/view_bug.do?bug_id=2147719

异常堆栈如下。

```
java.lang.Thread.State: RUNNABLE
    at sun.nio.ch.EPollArrayWrapper.epollWait(Native Method)
    at sun.nio.ch.EPollArrayWrapper.poll(EPollArrayWrapper.java:210)
    at sun.nio.ch.EPollSelectorImpl.doSelect(EPollSelectorImpl.java:65)
    at sun.nio.ch.SelectorImpl.lockAndDoSelect(SelectorImpl.java:69)
    - locked <0x0000000750928190> (a sun.nio.ch.Util$2)
    - locked <0x00000007509281a8> (a java.util.Collections$UnmodifiableSet)
    - locked <0x0000000750946098> (a sun.nio.ch.EPollSelectorImpl)
    at sun.nio.ch.SelectorImpl.select(SelectorImpl.java:80)
```

```
        at      net.spy.memcached.MemcachedConnection.handleIO(Memcached
Connection.java:217)
        at  net.spy.memcached.MemcachedConnection.run(MemcachedConnection.
java:836)
```

由于上述原因,在大多数场景下,不建议大家直接使用 JDK 的 NIO 类库,除非你精通 NIO 编程或者有特殊的需求。在绝大多数的业务场景中,我们可以使用 NIO 框架 Netty 来进行 NIO 编程,它既可以作为客户端也可以作为服务端,同时支持 UDP 和异步文件传输,功能非常强大。

下个小节我们就看看为什么选择 Netty 作为基础通信框架。

2.6.2 为什么选择 Netty

Netty 是业界最流行的 NIO 框架之一,它的健壮性、功能、性能、可定制性和可扩展性在同类框架中都是首屈一指的,它已经得到成百上千的商用项目验证,例如 Hadoop 的 RPC 框架 Avro 就使用了 Netty 作为底层通信框架,其他还有业界主流的 RPC 框架,也使用 Netty 来构建高性能的异步通信能力。

通过对 Netty 的分析,我们将它的优点总结如下:

◎ API 使用简单,开发门槛低;
◎ 功能强大,预置了多种编解码功能,支持多种主流协议;
◎ 定制能力强,可以通过 ChannelHandler 对通信框架进行灵活地扩展;
◎ 性能高,通过与其他业界主流的 NIO 框架对比,Netty 的综合性能最优;
◎ 成熟、稳定,Netty 修复了已经发现的所有 JDK NIO BUG,业务开发人员不需要再为 NIO 的 BUG 而烦恼;
◎ 社区活跃,版本迭代周期短,发现的 BUG 可以被及时修复,同时,更多的新功能会加入;
◎ 经历了大规模的商业应用考验,质量得到验证。Netty 在互联网、大数据、网络游戏、企业应用、电信软件等众多行业已经得到了成功商用,证明它已经完全能够满足不同行业的商业应用了。

正是因为这些优点,Netty 逐渐成为了 Java NIO 编程的首选框架。

2.7 总结

本章通过一个简单的 demo 开发,即时间服务器程序,让大家熟悉传统的同步阻塞 I/O、伪异步 I/O、非阻塞 I/O（NIO）和异步 I/O（AIO）的编程和使用差异,然后对比了各自的优缺点,并给出了使用建议。

最后,我们详细介绍了为什么不建议读者朋友们直接使用 JDK 的 NIO 原生类库进行异步 I/O 的开发,同时对 Netty 的优点进行分析和总结,给出使用 Netty 进行 NIO 开发的理由。

相信学完本章之后,大家对 Java 的网络编程已经有了初步的认识,从下一个章节开始,我们正式进入 Netty 的世界,学习基于 Netty 的网络开发。

入门篇

Netty NIO 开发指南

第 3 章　Netty 入门应用

第 4 章　TCP 粘包/拆包问题的解决之道

第 5 章　分隔符和定长解码器的应用

第 3 章

Netty 入门应用

作为 Netty 的第一个应用程序，我们依然以第 2 章的时间服务器为例进行开发，通过 Netty 版本的时间服务器的开发，让初学者尽快学到如何搭建 Netty 开发环境和运行 Netty 应用程序。

如果你已经熟悉 Netty 的基础应用，可以跳过本章，继续后面知识的学习。

本章主要内容包括：

- ◎ Netty 开发环境的搭建
- ◎ 服务端程序 TimeServer 开发
- ◎ 客户端程序 TimeClient 开发
- ◎ 时间服务器的运行和调试

3.1 Netty 开发环境的搭建

首先假设你已经在本机安装了 JDK1.7，配置了 JDK 的环境变量 path，同时下载并正确启动了 IDE 工具 Eclipse。如果你是个 Java 初学者，从来没有在本机搭建过 Java 开发环境，建议你先选择一本 Java 基础入门的书籍或者课程进行学习。

假如你习惯于使用其他 IDE 工具进行 Java 开发，例如 NetBeans IDE，也可以运行本节的入门例程。但是，你需要根据自己实际使用的 IDE 进行对应的配置修改和调整，本书统一使用 eclipse-jee-kepler-SR1-win32 作为 Java 开发工具。

下面我们开始学习如何搭建 Netty 的开发环境。

3.1.1 下载 Netty 的软件包

访问 Netty 的官网 http://netty.io/，从【Downloads】标签页选择下载 5.0.0.Alpha1 安装包，安装包不大，8.95MB 左右，下载之后的安装包如图 3-1 所示。

图 3-1　Netty 5.0 压缩包

通过解压缩工具打开压缩包，目录如图 3-2 所示。

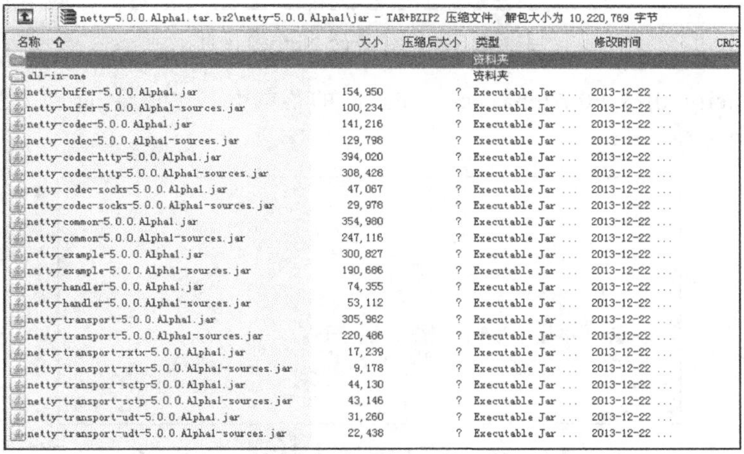

图 3-2　Netty 5.0 压缩包内部目录

这时会发现里面包含了各个模块的 .jar 包和源码，由于我们直接以二进制类库的方式使用 Netty，所以只需要获取 netty-all-5.0.0.Alpha1.jar 即可。

3.1.2 搭建 Netty 应用工程

使用 Eclipse 创建普通的 Java 工程，同时创建 Java 源文件的 package，如图 3-3 所示。

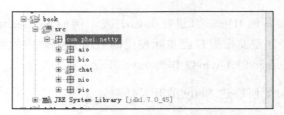

图 3-3　Netty 应用工程

创建第三方类库存放文件夹 lib，同时将 netty-all-5.0.0.Alpha1.jar 复制到 lib 目录下，如图 3-4 所示。

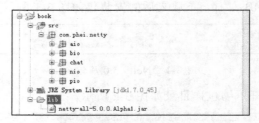

图 3-4　配置引用的 Netty Jar 包

右键单击 netty-all-5.0.0.Alpha1.jar，在弹出的菜单中，选择将.jar 包添加到 Build Path 中，操作如图 3-5 所示。

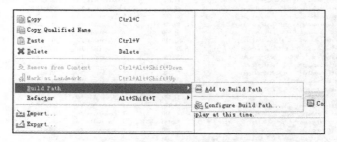

图 3-5　将 Netty Jar 包添加到 ClassPath 中

到此，Netty 应用开发环境已经搭建完成，下面的小节将演示如何基于 Netty 开发时间服务器程序。

3.2　Netty 服务端开发

作为第一个 Netty 的应用例程，为了让读者能够将精力集中在 Netty 的使用上，我们依然以第 2 章的时间服务器为例进行源码开发和代码讲解。

在开始使用 Netty 开发 TimeServer 之前，先回顾一下使用 NIO 进行服务端开发的步骤。

（1）创建 ServerSocketChannel，配置它为非阻塞模式；

（2）绑定监听，配置 TCP 参数，例如 backlog 大小；

（3）创建一个独立的 I/O 线程，用于轮询多路复用器 Selector；

（4）创建 Selector，将之前创建的 ServerSocketChannel 注册到 Selector 上，监听 SelectionKey.ACCEPT；

（5）启动 I/O 线程，在循环体中执行 Selector.select() 方法，轮询就绪的 Channel；

（6）当轮询到了处于就绪状态的 Channel 时，需要对其进行判断，如果是 OP_ACCEPT 状态，说明是新的客户端接入，则调用 ServerSocketChannel.accept() 方法接受新的客户端；

（7）设置新接入的客户端链路 SocketChannel 为非阻塞模式，配置其他的一些 TCP 参数；

（8）将 SocketChannel 注册到 Selector，监听 OP_READ 操作位；

（9）如果轮询的 Channel 为 OP_READ，则说明 SocketChannel 中有新的就绪的数据包需要读取，则构造 ByteBuffer 对象，读取数据包；

（10）如果轮询的 Channel 为 OP_WRITE，说明还有数据没有发送完成，需要继续发送。

一个简单的 NIO 服务端程序，如果我们直接使用 JDK 的 NIO 类库进行开发，竟然需要经过烦琐的十多步操作才能完成最基本的消息读取和发送，这也是我们要选择 Netty 等 NIO 框架的原因了，下面我们看看使用 Netty 是如何轻松搞定服务端开发的。

代码清单 3-1　Netty 时间服务器服务端 TimeServer

```
16. public class TimeServer {
17.
18.     public void bind(int port) throws Exception {
19.         // 配置服务端的NIO线程组
20.         EventLoopGroup bossGroup = new NioEventLoopGroup();
21.         EventLoopGroup workerGroup = new NioEventLoopGroup();
22.         try {
23.             ServerBootstrap b = new ServerBootstrap();
24.             b.group(bossGroup, workerGroup)
25.              .channel(NioServerSocketChannel.class)
26.              .option(ChannelOption.SO_BACKLOG, 1024)
```

```
27.              .childHandler(new ChildChannelHandler());
28.         // 绑定端口，同步等待成功
29.         ChannelFuture f = b.bind(port).sync();
30.
31.         // 等待服务端监听端口关闭
32.         f.channel().closeFuture().sync();
33.     } finally {
34.         // 优雅退出，释放线程池资源
35.         bossGroup.shutdownGracefully();
36.         workerGroup.shutdownGracefully();
37.     }
38. }
39.
40. private class ChildChannelHandler extends ChannelInitializer<SocketChannel> {
41.     @Override
42.     protected void initChannel(SocketChannel arg0) throws Exception {
43.         arg0.pipeline().addLast(new TimeServerHandler());
44.     }
45.
46. }
47.
48. /**
49.  * @param args
50.  * @throws Exception
51.  */
52. public static void main(String[] args) throws Exception {
53.     int port = 8080;
54.     if (args != null && args.length > 0) {
55.         try {
56.             port = Integer.valueOf(args[0]);
57.         } catch (NumberFormatException e) {
58.             // 采用默认值
59.         }
60.     }
61.     new TimeServer().bind(port);
62. }
63. }
```

由于本章的重点是讲解 Netty 的应用开发，所以对于一些 Netty 的类库和用法仅仅做基础性的讲解，我们从黑盒的角度理解这些概念即可。后续源码分析章节会专门对 Netty

核心的类库和功能进行分析，感兴趣的同学可以跳到源码分析章节进行后续的学习。

我们从 bind 方法开始学习，在代码第 20~21 行创建了两个 NioEventLoopGroup 实例。NioEventLoopGroup 是个线程组，它包含了一组 NIO 线程，专门用于网络事件的处理，实际上它们就是 Reactor 线程组。这里创建两个的原因是一个用于服务端接受客户端的连接，另一个用于进行 SocketChannel 的网络读写。第 23 行创建 ServerBootstrap 对象，它是 Netty 用于启动 NIO 服务端的辅助启动类，目的是降低服务端的开发复杂度。第 24 行调用 ServerBootstrap 的 group 方法，将两个 NIO 线程组当作入参传递到 ServerBootstrap 中。接着设置创建的 Channel 为 NioServerSocketChannel，它的功能对应于 JDK NIO 类库中的 ServerSocketChannel 类。然后配置 NioServerSocketChannel 的 TCP 参数，此处将它的 backlog 设置为 1024，最后绑定 I/O 事件的处理类 ChildChannelHandler，它的作用类似于 Reactor 模式中的 Handler 类，主要用于处理网络 I/O 事件，例如记录日志、对消息进行编解码等。

服务端启动辅助类配置完成之后，调用它的 bind 方法绑定监听端口，随后，调用它的同步阻塞方法 sync 等待绑定操作完成。完成之后 Netty 会返回一个 ChannelFuture，它的功能类似于 JDK 的 java.util.concurrent.Future，主要用于异步操作的通知回调。

第 32 行使用 f.channel().closeFuture().sync()方法进行阻塞，等待服务端链路关闭之后 main 函数才退出。

第 34~36 行调用 NIO 线程组的 shutdownGracefully 进行优雅退出，它会释放跟 shutdownGracefully 相关联的资源。

下面看看 TimeServerHandler 类是如何实现的。

代码清单 3-2　Netty 时间服务器服务端 TimeServerHandler

```
12. public class TimeServerHandler extends ChannelHandlerAdapter {
13.
14.     @Override
15.     public void channelRead(ChannelHandlerContext ctx, Object msg)
16.         throws Exception {
17.     ByteBuf buf = (ByteBuf) msg;
18.     byte[] req = new byte[buf.readableBytes()];
19.     buf.readBytes(req);
20.     String body = new String(req, "UTF-8");
21.     System.out.println("The time server receive order : " + body);
22.     String currentTime = "QUERY TIME ORDER".equalsIgnoreCase(body) ? new java.util.Date(
23.         System.currentTimeMillis()).toString() : "BAD ORDER";
```

```
24.        ByteBuf resp = Unpooled.copiedBuffer(currentTime.getBytes());
25.        ctx.write(resp);
26.    }
27.
28.    @Override
29.    public void channelReadComplete(ChannelHandlerContext ctx) throws Exception {
30.        ctx.flush();
31.    }
32.
33.    @Override
34.    public void exceptionCaught(ChannelHandlerContext ctx, Throwable cause) {
35.        ctx.close();
36.    }
37. }
```

TimeServerHandler 继承自 ChannelHandlerAdapter，它用于对网络事件进行读写操作，通常我们只需要关注 channelRead 和 exceptionCaught 方法。下面对这两个方法进行简单说明。

第 17 行做类型转换，将 msg 转换成 Netty 的 ByteBuf 对象。ByteBuf 类似于 JDK 中的 java.nio.ByteBuffer 对象，不过它提供了更加强大和灵活的功能。通过 ByteBuf 的 readableBytes 方法可以获取缓冲区可读的字节数，根据可读的字节数创建 byte 数组，通过 ByteBuf 的 readBytes 方法将缓冲区中的字节数组复制到新建的 byte 数组中，最后通过 new String 构造函数获取请求消息。这时对请求消息进行判断，如果是"QUERY TIME ORDER"则创建应答消息，通过 ChannelHandlerContext 的 write 方法异步发送应答消息给客户端。

第 30 行我们发现还调用了 ChannelHandlerContext 的 flush 方法，它的作用是将消息发送队列中的消息写入到 SocketChannel 中发送给对方。从性能角度考虑，为了防止频繁地唤醒 Selector 进行消息发送，Netty 的 write 方法并不直接将消息写入 SocketChannel 中，调用 write 方法只是把待发送的消息放到发送缓冲数组中，再通过调用 flush 方法，将发送缓冲区中的消息全部写到 SocketChannel 中。

第 35 行，当发生异常时，关闭 ChannelHandlerContext，释放和 ChannelHandlerContext 相关联的句柄等资源。

通过对代码进行统计分析可以看出，不到 30 行的业务逻辑代码，即完成了 NIO 服务端的开发，相比于传统基于 JDK NIO 原生类库的服务端，代码量大大减少，开发难度也降低了很多。

下面我们继续学习客户端的开发,并使用 Netty 改造 TimeClient。

3.3 Netty 客户端开发

Netty 客户端的开发相比于服务端更简单,下面我们就看下客户端的代码如何实现。

代码清单 3-3 Netty 时间服务器客户端 TimeClient

```
16. public class TimeClient {
17.
18.     public void connect(int port, String host) throws Exception {
19.         // 配置客户端NIO线程组
20.         EventLoopGroup group = new NioEventLoopGroup();
21.         try {
22.             Bootstrap b = new Bootstrap();
23.             b.group(group).channel(NioSocketChannel.class)
24.                 .option(ChannelOption.TCP_NODELAY, true)
25.                 .handler(new ChannelInitializer<SocketChannel>() {
26.                     @Override
27.                     public void initChannel(SocketChannel ch)
28.                         throws Exception {
29.                         ch.pipeline().addLast(new TimeClientHandler());
30.                     }
31.                 });
32.
33.             // 发起异步连接操作
34.             ChannelFuture f = b.connect(host, port).sync();
35.
36.             // 等待客户端链路关闭
37.             f.channel().closeFuture().sync();
38.         } finally {
39.             // 优雅退出,释放NIO线程组
40.             group.shutdownGracefully();
41.         }
42.     }
43.
44.     /**
45.      * @param args
46.      * @throws Exception
47.      */
```

```
48.    public static void main(String[] args) throws Exception {
49.        int port = 8080;
50.        if (args != null && args.length > 0) {
51.            try {
52.                port = Integer.valueOf(args[0]);
53.            } catch (NumberFormatException e) {
54.                // 采用默认值
55.            }
56.        }
57.        new TimeClient().connect(port, "127.0.0.1");
58.    }
59. }
```

我们从 connect 方法讲起,在第 20 行首先创建客户端处理 I/O 读写的 NioEventLoopGroup 线程组,然后继续创建客户端辅助启动类 Bootstrap,随后需要对其进行配置。与服务端不同的是,它的 Channel 需要设置为 NioSocketChannel,然后为其添加 Handler。此处为了简单直接创建匿名内部类,实现 initChannel 方法,其作用是当创建 NioSocketChannel 成功之后,在进行初始化时,将它的 ChannelHandler 设置到 ChannelPipeline 中,用于处理网络 I/O 事件。

客户端启动辅助类设置完成之后,调用 connect 方法发起异步连接,然后调用同步方法等待连接成功。

最后,当客户端连接关闭之后,客户端主函数退出,退出之前释放 NIO 线程组的资源。

下面我们继续看 TimeClientHandler 的代码如何实现。

代码清单 3-4　Netty 时间服务器客户端 TimeClientHandler

```
14. public class TimeClientHandler extends ChannelHandlerAdapter {
15.
16.     private static final Logger logger = Logger
17.             .getLogger(TimeClientHandler.class.getName());
18.
19.     private final ByteBuf firstMessage;
20.
21.     /**
22.      * Creates a client-side handler.
23.      */
24.     public TimeClientHandler() {
25.         byte[] req = "QUERY TIME ORDER".getBytes();
26.         firstMessage = Unpooled.buffer(req.length);
```

```
27.        firstMessage.writeBytes(req);
28.
29.    }
30.
31.    @Override
32.    public void channelActive(ChannelHandlerContext ctx) {
33.     ctx.writeAndFlush(firstMessage);
34.    }
35.
36.    @Override
37.    public void channelRead(ChannelHandlerContext ctx, Object msg)
38.        throws Exception {
39.     ByteBuf buf = (ByteBuf) msg;
40.     byte[] req = new byte[buf.readableBytes()];
41.     buf.readBytes(req);
42.     String body = new String(req, "UTF-8");
43.     System.out.println("Now is : " + body);
44.    }
45.
46.    @Override
47.    public void exceptionCaught(ChannelHandlerContext ctx, Throwable cause) {
48.     // 释放资源
49.     logger.warning("Unexpected exception from downstream : "
50.         + cause.getMessage());
51.     ctx.close();
52.    }
53. }
```

这里重点关注三个方法：channelActive、channelRead 和 exceptionCaught。当客户端和服务端 TCP 链路建立成功之后，Netty 的 NIO 线程会调用 channelActive 方法，发送查询时间的指令给服务端，调用 ChannelHandlerContext 的 writeAndFlush 方法将请求消息发送给服务端。

当服务端返回应答消息时，channelRead 方法被调用，第 39~43 行从 Netty 的 ByteBuf 中读取并打印应答消息。

第 47~52 行，当发生异常时，打印异常日志，释放客户端资源。

3.4 运行和调试

3.4.1 服务端和客户端的运行

在 Eclipse 开发环境中运行和调试 Java 程序非常简单，下面我们看下如何运行 TimeServer：将光标定位到 TimeServer 类中，单击右键，在弹出菜单中选择 Run As→Java Application，或者直接使用快捷键 Alt + Shift + X 执行，如图 3-6 所示。

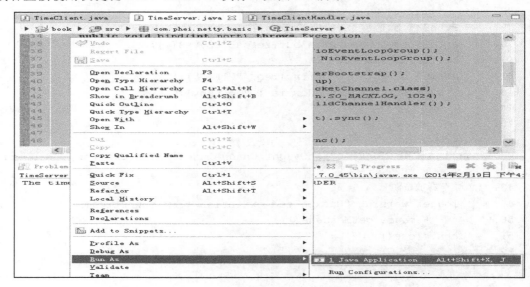

图 3-6 运行 TimeServer

客户端的执行类似，可以看到以下执行结果。

服务端运行结果如图 3-7 所示。

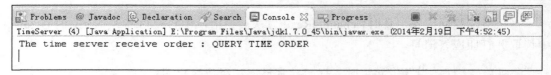

图 3-7 TimeServer 运行结果

客户端运行结果如图 3-8 所示。

图 3-8 TimeClient 运行结果

运行结果正确。可以发现，通过 Netty 开发的 NIO 服务端和客户端非常简单，短短几十行代码，就能完成之前 NIO 程序需要几百行才能完成的功能。基于 Netty 的应用开发不但 API 使用简单、开发模式固定，而且扩展性和定制性非常好，后面，我们会通过更多应用来介绍 Netty 的强大功能。

需要指出的是，本例程依然没有考虑读半包的处理，对于功能演示或者测试，上述程序没有问题，但是稍加改造进行性能或者压力测试，它就不能正确地工作了。在下一个章节我们会给出能够正确处理半包消息的应用实例。

3.4.2 打包和部署

基于 Netty 开发的都是非 Web 的 Java 应用，它的打包形态非常简单，就是一个普通的 .jar 包，通常情况下，在正式的商业开发中，我们会使用三种打包方式。

（1）Eclipse 提供的导出功能。它可以将指定的 Java 工程或者源码包、代码输出成指定的 .jar 包，它属于手工操作，在项目模块较多时非常不方便，所以一般不使用这种方式；

（2）使用 ant 脚本对工程进行打包。将 Netty 的应用程序打包成指定的 .jar 包，一般会输出一个软件安装包：xxxx_install.gz；

（3）使用 Maven 进行工程构建。它可以对模块间的依赖进行管理，支持版本的自动化测试、编译和构建，是目前主流的项目管理工具。

3.5 总结

本章节讲解了 Netty 的入门应用，通过使用 Netty 重构时间服务器程序，可以发现相比于传统的 NIO 程序，Netty 的代码更加简洁、开发难度更低，扩展性也更好，非常适合作为基础通信框架被用户集成和使用。

在介绍 Netty 服务端和客户端时，简单地对代码进行了讲解，由于后续会有专门章节对 Netty 进行源码分析，所以在 Netty 应用部分我们不进行详细的源码解读和分析。

第 4 章会讲解一个稍微复杂的应用，它利用 Netty 提供的默认编解码功能解决了之前没有解决的读半包问题。事实上，对于读半包问题，Netty 提供了很多种好的解决方案。下面一起学习一下如何利用 Netty 默认的编解码功能解决半包读取问题。

第 4 章

TCP 粘包/拆包问题的解决之道

熟悉 TCP 编程的读者可能都知道,无论是服务端还是客户端,当我们读取或者发送消息的时候,都需要考虑 TCP 底层的粘包/拆包机制。本章开始我们先简单介绍 TCP 粘包/拆包的基础知识,然后模拟一个没有考虑 TCP 粘包/拆包导致功能异常的案例,最后,通过正确例程来探讨 Netty 是如何解决这个问题的。

如果你已经熟悉了 TCP 粘包和拆包的相关知识,建议直接跳到代码讲解小节,看 Netty 是如何解决这个问题的。

本章主要内容包括:

◎ TCP 粘包/拆包的基础知识

◎ 没考虑 TCP 粘包/拆包的问题案例

◎ 使用 Netty 解决读半包问题

4.1 TCP 粘包/拆包

TCP 是个 "流" 协议,所谓流,就是没有界限的一串数据。大家可以想想河里的流水,它们是连成一片的,其间并没有分界线。TCP 底层并不了解上层业务数据的具体含义,它会根据 TCP 缓冲区的实际情况进行包的划分,所以在业务上认为,一个完整的包可能会被 TCP 拆分成多个包进行发送，也有可能把多个小的包封装成一个大的数据包发送，这就是所谓的 TCP 粘包和拆包问题。

4.1.1 TCP 粘包/拆包问题说明

我们可以通过图解对 TCP 粘包和拆包问题进行说明，粘包问题示例如图 4-1 所示。

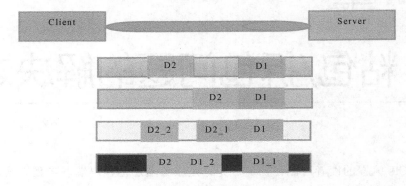

图 4-1 TCP 粘包/拆包问题

假设客户端分别发送了两个数据包 D1 和 D2 给服务端，由于服务端一次读取到的字节数是不确定的，故可能存在以下 4 种情况。

（1）服务端分两次读取到了两个独立的数据包，分别是 D1 和 D2，没有粘包和拆包；

（2）服务端一次接收到了两个数据包，D1 和 D2 粘合在一起，被称为 TCP 粘包；

（3）服务端分两次读取到了两个数据包，第一次读取到了完整的 D1 包和 D2 包的部分内容，第二次读取到了 D2 包的剩余内容，这被称为 TCP 拆包；

（4）服务端分两次读取到了两个数据包，第一次读取到了 D1 包的部分内容 D1_1，第二次读取到了 D1 包的剩余内容 D1_2 和 D2 包的整包。

如果此时服务端 TCP 接收滑窗非常小，而数据包 D1 和 D2 比较大，很有可能会发生第 5 种可能，即服务端分多次才能将 D1 和 D2 包接收完全，期间发生多次拆包。

4.1.2 TCP 粘包/拆包发生的原因

问题产生的原因有三个，分别如下。

（1）应用程序 write 写入的字节大小大于套接口发送缓冲区大小；

（2）进行 MSS 大小的 TCP 分段；

（3）以太网帧的 payload 大于 MTU 进行 IP 分片。

图解如图 4-2 所示。

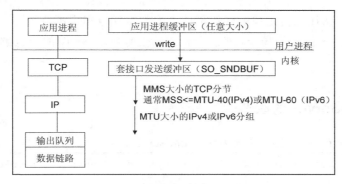

图 4-2　TCP 粘包/拆包问题原因

4.1.3　粘包问题的解决策略

由于底层的 TCP 无法理解上层的业务数据，所以在底层是无法保证数据包不被拆分和重组的，这个问题只能通过上层的应用协议栈设计来解决，根据业界的主流协议的解决方案，可以归纳如下。

（1）消息定长，例如每个报文的大小为固定长度 200 字节，如果不够，空位补空格；

（2）在包尾增加回车换行符进行分割，例如 FTP 协议；

（3）将消息分为消息头和消息体，消息头中包含表示消息总长度（或者消息体长度）的字段，通常设计思路为消息头的第一个字段使用 int32 来表示消息的总长度；

（4）更复杂的应用层协议。

介绍完了 TCP 粘包/拆包的基础知识，下面我们就通过实际例程来看看如何使用 Netty 提供的半包解码器来解决 TCP 粘包/拆包问题。

4.2 未考虑 TCP 粘包导致功能异常案例

在前面的时间服务器例程中,我们多次强调并没有考虑读半包问题,这在功能测试时往往没有问题,但是一旦压力上来,或者发送大报文之后,就会存在粘包/拆包问题。如果代码没有考虑,往往就会出现解码错位或者错误,导致程序不能正常工作。下面我们以 3.1 节的代码为例,模拟故障场景,然后看看如何正确使用 Netty 的半包解码器来解决 TCP 粘包/拆包问题。

4.2.1 TimeServer 的改造

代码清单 4-1 Netty 时间服务器服务端 TimeServerHandler

```
12. public class TimeServerHandler extends ChannelHandlerAdapter {
13.
14.     private int counter;
15.
16.     @Override
17.     public void channelRead(ChannelHandlerContext ctx, Object msg)
18.        throws Exception {
19. ByteBuf buf = (ByteBuf) msg;
20. byte[] req = new byte[buf.readableBytes()];
21. buf.readBytes(req);
22. String body = new String(req, "UTF-8").substring(0, req.length
23.     - System.getProperty("line.separator").length());
24. System.out.println("The time server receive order : " + body
25.     + " ; the counter is : " + ++counter);
26. String currentTime = "QUERY TIME ORDER".equalsIgnoreCase(body) ? new java.util.Date(
27.         System.currentTimeMillis()).toString() : "BAD ORDER";
28. currentTime = currentTime + System.getProperty("line.separator");
29. ByteBuf resp = Unpooled.copiedBuffer(currentTime.getBytes());
30. ctx.writeAndFlush(resp);
31.     }
32.
33.     @Override
34.     public void exceptionCaught(ChannelHandlerContext ctx, Throwable cause) {
```

```
35.        ctx.close();
36.    }
37. }
```

每读到一条消息后,就计一次数,然后发送应答消息给客户端。按照设计,服务端接收到的消息总数应该跟客户端发送的消息总数相同,而且请求消息删除回车换行符后应该为"QUERY TIME ORDER"。下面我们继续看下客户端的改造。

4.2.2 TimeClient 的改造

代码清单 4-2 Netty 时间服务器客户端 TimeClientHandler

```
14. public class TimeClientHandler extends ChannelHandlerAdapter {
15.
16.     private static final Logger logger = Logger
17.         .getLogger(TimeClientHandler.class.getName());
18.
19.     private int counter;
20.
21.     private byte[] req;
22.
23.     /**
24.      * Creates a client-side handler.
25.      */
26.     public TimeClientHandler() {
27. req = ("QUERY TIME ORDER" + System.getProperty("line.separator"))
28.         .getBytes();
29.     }
30.
31.     @Override
32.     public void channelActive(ChannelHandlerContext ctx) {
33. ByteBuf message = null;
34. for (int i = 0; i < 100; i++) {
35.     message = Unpooled.buffer(req.length);
36.     message.writeBytes(req);
37.     ctx.writeAndFlush(message);
38. }
39.     }
40.
41.     @Override
```

```
42.     public void channelRead(ChannelHandlerContext ctx, Object msg)
43.         throws Exception {
44. ByteBuf buf = (ByteBuf) msg;
45. byte[] req = new byte[buf.readableBytes()];
46. buf.readBytes(req);
47. String body = new String(req, "UTF-8");
48. System.out.println("Now is : " + body + " ; the counter is : "
49.     + ++counter);
50.     }
51.
52.     @Override
53.     public void exceptionCaught(ChannelHandlerContext ctx, Throwable cause) {
54. // 释放资源
55. logger.warning("Unexpected exception from downstream : "
56.     + cause.getMessage());
57. ctx.close();
58.     }
59. }
```

主要的修改点就是代码第 33~38 行，客户端跟服务端链路建立成功之后，循环发送 100 条消息，每发送一条就刷新一次，保证每条消息都会被写入 Channel 中。按照我们的设计，服务端应该接收到 100 条查询时间指令的请求消息。

第 48~49 行，客户端每接收到服务端一条应答消息之后，就打印一次计数器。按照设计初衷，客户端应该打印 100 次服务端的系统时间。

下面的小节就来看下运行结果是否符合设计初衷。

4.2.3 运行结果

分别执行服务端和客户端，运行结果如下。

服务端运行结果如下。

```
The time server receive order : QUERY TIME ORDER
此处省略 55 行 QUERY TIME ORDER ......
QUERY TIME ORD ; the counter is : 1
The time server receive order :
此处省略 42 行 QUERY TIME ORDER ......
```

```
QUERY TIME ORDER ; the counter is : 2
```

服务端运行结果表明它只接收到了两条消息，第一条包含 57 条"QUERY TIME ORDER"指令，第二条包含了 43 条"QUERY TIME ORDER"指令，总数正好是 100 条。我们期待的是收到 100 条消息，每条包含一条"QUERY TIME ORDER"指令。这说明发生了 TCP 粘包。

客户端运行结果如下。

```
Now is : BAD ORDER
BAD ORDER
 ; the counter is : 1
```

按照设计初衷，客户端应该收到 100 条当前系统时间的消息，但实际上只收到了一条。这不难理解，因为服务端只收到了 2 条请求消息，所以实际服务端只发送了 2 条应答，由于请求消息不满足查询条件，所以返回了 2 条"BAD ORDER"应答消息。但是实际上客户端只收到了一条包含 2 条"BAD ORDER"指令的消息，说明服务端返回的应答消息也发生了粘包。

由于上面的例程没有考虑 TCP 的粘包/拆包，所以当发生 TCP 粘包时，我们的程序就不能正常工作。

下面的章节将演示如何通过 Netty 的 LineBasedFrameDecoder 和 StringDecoder 来解决 TCP 粘包问题。

4.3　利用 LineBasedFrameDecoder 解决 TCP 粘包问题

为了解决 TCP 粘包/拆包导致的半包读写问题，Netty 默认提供了多种编解码器用于处理半包，只要能熟练掌握这些类库的使用，TCP 粘包问题从此会变得非常容易，你甚至不需要关心它们，这也是其他 NIO 框架和 JDK 原生的 NIO API 所无法匹敌的。

下面我们就以修正时间服务器为目标进行开发和讲解，通过对实际代码的讲解让大家能够尽快熟悉和掌握半包解码器的使用。

4.3.1 支持 TCP 粘包的 TimeServer

直接看代码，然后对 LineBasedFrameDecoder 和 StringDecoder 的 API 进行说明。

代码清单 4-3　Netty 时间服务器服务端 TimeServer

```
18. public class TimeServer {
19.
20.     public void bind(int port) throws Exception {
21. // 配置服务端的NIO线程组
22. EventLoopGroup bossGroup = new NioEventLoopGroup();
23. EventLoopGroup workerGroup = new NioEventLoopGroup();
24. try {
25.     ServerBootstrap b = new ServerBootstrap();
26.     b.group(bossGroup, workerGroup)
27.      .channel(NioServerSocketChannel.class)
28.      .option(ChannelOption.SO_BACKLOG, 1024)
29.      .childHandler(new ChildChannelHandler());
30. // 绑定端口，同步等待成功
31.     ChannelFuture f = b.bind(port).sync();
32.
33. // 等待服务端监听端口关闭
34.     f.channel().closeFuture().sync();
35. } finally {
36. // 优雅退出，释放线程池资源
37.     bossGroup.shutdownGracefully();
38.     workerGroup.shutdownGracefully();
39. }
40.     }
41.
42.     private class ChildChannelHandler extends ChannelInitializer<SocketChannel> {
43. @Override
44. protected void initChannel(SocketChannel arg0) throws Exception {
45.     arg0.pipeline().addLast(new LineBasedFrameDecoder(1024));
46.     arg0.pipeline().addLast(new StringDecoder());
47.     arg0.pipeline().addLast(new TimeServerHandler());
48. }
49.     }
50.
51.     /**
52.      * @param args
```

```
53.     * @throws Exception
54.     */
55.    public static void main(String[] args) throws Exception {
56. int port = 8080;
57. if (args != null && args.length > 0) {
58.     try {
59.      port = Integer.valueOf(args[0]);
60.     } catch (NumberFormatException e) {
61.     // 采用默认值
62.     }
63. }
64. new TimeServer().bind(port);
65.    }
66. }
```

重点看 45～47 行，在原来的 TimeServerHandler 之前新增了两个解码器：LineBasedFrameDecoder 和 StringDecoder。这两个类的功能后续会进行介绍，下面继续看 TimeServerHandler 的代码修改。

代码清单 4-4　Netty 时间服务器服务端 TimeServerHandler

```
12. public class TimeServerHandler extends ChannelHandlerAdapter {
13.
14.     private int counter;
15.
16.     @Override
17.     public void channelRead(ChannelHandlerContext ctx, Object msg)
18.         throws Exception {
19. String body = (String) msg;
20. System.out.println("The time server receive order : " + body
21.     + " ; the counter is : " + ++counter);
22. String currentTime = "QUERY TIME ORDER".equalsIgnoreCase(body) ? new java.util.Date(
23.         System.currentTimeMillis()).toString() : "BAD ORDER";
24. currentTime = currentTime + System.getProperty("line.separator");
25. ByteBuf resp = Unpooled.copiedBuffer(currentTime.getBytes());
26. ctx.writeAndFlush(resp);
27.     }
28.
29.     @Override
30.     public void exceptionCaught(ChannelHandlerContext ctx, Throwable cause) {
31. ctx.close();
```

```
32.     }
33. }
```

直接看 19～21 行，可以发现接收到的 msg 就是删除回车换行符后的请求消息，不需要额外考虑处理读半包问题，也不需要对请求消息进行编码，代码非常简洁。读者可能会质疑这样是否可行，不着急，我们先继续看看客户端的类似改造，然后运行程序看执行结果，最后再揭开其中的奥秘。

4.3.2 支持 TCP 粘包的 TimeClient

支持 TCP 粘包的客户端修改起来也非常简单，代码如下。

代码清单 4-5　Netty 时间服务器客户端 TimeClient

```
18. public class TimeClient {
19.
20.     public void connect(int port, String host) throws Exception {
21.         // 配置客户端NIO线程组
22.         EventLoopGroup group = new NioEventLoopGroup();
23.         try {
24.             Bootstrap b = new Bootstrap();
25.             b.group(group).channel(NioSocketChannel.class)
26.                 .option(ChannelOption.TCP_NODELAY, true)
27.                 .handler(new ChannelInitializer<SocketChannel>() {
28.                     @Override
29.                     public void initChannel(SocketChannel ch)
30.                         throws Exception {
31.                         ch.pipeline().addLast(
32.                             new LineBasedFrameDecoder(1024));
33.                         ch.pipeline().addLast(new StringDecoder());
34.                         ch.pipeline().addLast(new TimeClientHandler());
35.                     }
36.                 });
37.
38.             // 发起异步连接操作
39.             ChannelFuture f = b.connect(host, port).sync();
40.
41.             // 等待客户端链路关闭
42.             f.channel().closeFuture().sync();
43.         } finally {
```

```
44.        // 优雅退出，释放NIO线程组
45.        group.shutdownGracefully();
46.    }
47.    }
48.
49.    /**
50.     * @param args
51.     * @throws Exception
52.     */
53.    public static void main(String[] args) throws Exception {
54.    int port = 8080;
55.    if (args != null && args.length > 0) {
56.        try {
57.            port = Integer.valueOf(args[0]);
58.        } catch (NumberFormatException e) {
59.            // 采用默认值
60.        }
61.    }
62.    new TimeClient().connect(port, "127.0.0.1");
63.        }
64.    }
```

31～34 行与服务端类似，直接在 TimeClientHandler 之前新增 LineBasedFrameDecoder 和 StringDecoder 解码器，下面我们继续看 TimeClientHandler 的代码修改。

代码清单 4-6 Netty 时间服务器客户端 TimeClientHandler

```
14. public class TimeClientHandler extends ChannelHandlerAdapter {
15.
16.     private static final Logger logger = Logger
17.         .getLogger(TimeClientHandler.class.getName());
18.
19.     private int counter;
20.
21.     private byte[] req;
22.
23.     /**
24.      * Creates a client-side handler.
25.      */
26.     public TimeClientHandler() {
27.     req = ("QUERY TIME ORDER" + System.getProperty("line.separator"))
28.         .getBytes();
```

```
29.     }
30.
31.     @Override
32.     public void channelActive(ChannelHandlerContext ctx) {
33. ByteBuf message = null;
34. for (int i = 0; i < 100; i++) {
35.     message = Unpooled.buffer(req.length);
36.     message.writeBytes(req);
37.     ctx.writeAndFlush(message);
38. }
39.     }
40.
41.     @Override
42.     public void channelRead(ChannelHandlerContext ctx, Object msg)
43.     throws Exception {
44. String body = (String) msg;
45. System.out.println("Now is : " + body + " ; the counter is : "
46.     + ++counter);
47.     }
48.
49.     @Override
50.     public void exceptionCaught(ChannelHandlerContext ctx, Throwable cause) {
51. // 释放资源
52. logger.warning("Unexpected exception from downstream : "
53.     + cause.getMessage());
54. ctx.close();
55.     }
56. }
```

第 44～46 行拿到的 msg 已经是解码成字符串之后的应答消息了，相比于之前的代码简洁了很多。

下个小节我们运行重构后的时间服务器服务端和客户端，看看它能否像设计预期那样正常工作。

4.3.3 运行支持 TCP 粘包的时间服务器程序

分别运行 TimeServer 和 TimeClient，执行结果如下。

服务端执行结果如下。

```
The time server receive order : QUERY TIME ORDER ; the counter is : 1
//此处省略2-99行 sThe time server receive order ......
The time server receive order : QUERY TIME ORDER ; the counter is : 100
```

客户端运行结果如下。

```
Now is : Thu Feb 20 00:00:14 CST 2014 ; the counter is : 1
//此处省略2-99行 Now is : Thu Feb 20 00:00:14 CST 2014 ......
Now is : Thu Feb 20 00:00:14 CST 2014 ; the counter is : 100
```

程序的运行结果完全符合预期,说明通过使用 LineBasedFrameDecoder 和 StringDecoder 成功解决了 TCP 粘包导致的读半包问题。对于使用者来说,只要将支持半包解码的 Handler 添加到 ChannelPipeline 中即可,不需要写额外的代码,用户使用起来非常简单。

下个小节,我们就对添加 LineBasedFrameDecoder 和 StringDecoder 之后就能解决 TCP 粘包导致的读半包或者多包问题的原因进行分析。

4.3.4　LineBasedFrameDecoder 和 StringDecoder 的原理分析

LineBasedFrameDecoder 的工作原理是它依次遍历 ByteBuf 中的可读字节,判断看是否有 "\n" 或者 "\r\n",如果有,就以此位置为结束位置,从可读索引到结束位置区间的字节就组成了一行。它是以换行符为结束标志的解码器,支持携带结束符或者不携带结束符两种解码方式,同时支持配置单行的最大长度。如果连续读取到最大长度后仍然没有发现换行符,就会抛出异常,同时忽略掉之前读到的异常码流。

StringDecoder 的功能非常简单,就是将接收到的对象转换成字符串,然后继续调用后面的 Handler。LineBasedFrameDecoder + StringDecoder 组合就是按行切换的文本解码器,它被设计用来支持 TCP 的粘包和拆包。

可能读者会提出新的疑问:如果发送的消息不是以换行符结束的,该怎么办呢？或者没有回车换行符,靠消息头中的长度字段来分包怎么办？是不是需要自己写半包解码器？答案是否定的,Netty 提供了多种支持 TCP 粘包/拆包的解码器,用来满足用户的不同诉求。

第 5 章我们将学习分隔符解码器,由于它在实际项目中应用非常广泛,所以单独用一章对其用法和原理进行讲解。

4.4 总结

本章首先对 TCP 的粘包和拆包进行了讲解，给出了解决这个问题的通用做法。然后我们对第 3 章的时间服务器进行改造和测试，首先验证没有考虑 TCP 粘包/拆包导致的问题。随后给出了解决方案，即利用 LineBasedFrameDecoder + StringDecoder 来解决 TCP 的粘包/拆包问题。

第 5 章

分隔符和定长解码器的应用

TCP 以流的方式进行数据传输，上层的应用协议为了对消息进行区分，往往采用如下 4 种方式。

（1）消息长度固定，累计读取到长度总和为定长 LEN 的报文后，就认为读取到了一个完整的消息；将计数器置位，重新开始读取下一个数据报；

（2）将回车换行符作为消息结束符，例如 FTP 协议，这种方式在文本协议中应用比较广泛；

（3）将特殊的分隔符作为消息的结束标志，回车换行符就是一种特殊的结束分隔符；

（4）通过在消息头中定义长度字段来标识消息的总长度。

Netty 对上面 4 种应用做了统一的抽象，提供了 4 种解码器来解决对应的问题，使用起来非常方便。有了这些解码器，用户不需要自己对读取的报文进行人工解码，也不需要考虑 TCP 的粘包和拆包。

第 4 章我们介绍了如何利用 LineBasedFrameDecoder 解决 TCP 的粘包问题，本章我们继续学习另外两种实用的解码器——DelimiterBasedFrameDecoder 和 FixedLengthFrameDecoder，前者可以自动完成以分隔符做结束标志的消息的解码，后者可以自动完成对定长消息的解码，它们都能解决 TCP 粘包/拆包导致的读半包问题。

本章主要内容包括：

◎ DelimiterBasedFrameDecoder 服务端开发

◎ DelimiterBasedFrameDecoder 客户端开发

◎ 运行 DelimiterBasedFrameDecoder 服务端和客户端

◎ FixedLengthFrameDecoder 服务端开发

◎ 通过 telnet 命令行调试 FixedLengthFrameDecoder 服务端

5.1 DelimiterBasedFrameDecoder 应用开发

通过对 DelimiterBasedFrameDecoder 的使用，我们可以自动完成以分隔符作为码流结束标识的消息的解码，下面通过一个演示程序来学习下如何使用 DelimiterBased FrameDecoder 进行开发。

演示程序以经典的 Echo 服务为例。EchoServer 接收到 EchoClient 的请求消息后，将其打印出来，然后将原始消息返回给客户端，消息以"$_"作为分隔符。

5.1.1 DelimiterBasedFrameDecoder 服务端开发

下面我们直接看 EchoServer 的源代码。

代码清单 5-1　EchoServer 服务端 EchoServer

```
22. public class EchoServer {
23.     public void bind(int port) throws Exception {
24.         // 配置服务端的NIO线程组
25.         EventLoopGroup bossGroup = new NioEventLoopGroup();
26.         EventLoopGroup workerGroup = new NioEventLoopGroup();
27.         try {
28.             ServerBootstrap b = new ServerBootstrap();
29.             b.group(bossGroup, workerGroup)
30.              .channel(NioServerSocketChannel.class)
31.              .option(ChannelOption.SO_BACKLOG, 100)
32.              .handler(new LoggingHandler(LogLevel.INFO))
33.              .childHandler(new ChannelInitializer<SocketChannel>() {
```

```
34.            @Override
35.            public void initChannel(SocketChannel ch)
36.                throws Exception {
37.                ByteBuf delimiter = Unpooled.copiedBuffer("$_"
38.                    .getBytes());
39.                ch.pipeline().addLast(
40.                    new DelimiterBasedFrameDecoder(1024,
41.                        delimiter));
42.                ch.pipeline().addLast(new StringDecoder());
43.                ch.pipeline().addLast(new EchoServerHandler());
44.            }
45.            });
46.
47.            // 绑定端口，同步等待成功
48.            ChannelFuture f = b.bind(port).sync();
49.
50.            // 等待服务端监听端口关闭
51.            f.channel().closeFuture().sync();
52.        } finally {
53.            // 优雅退出，释放线程池资源
54.            bossGroup.shutdownGracefully();
55.            workerGroup.shutdownGracefully();
56.        }
57.    }
58.
59.    public static void main(String[] args) throws Exception {
60.        int port = 8080;
61.        if (args != null && args.length > 0) {
62.            try {
63.                port = Integer.valueOf(args[0]);
64.            } catch (NumberFormatException e) {
65.                // 采用默认值
66.            }
67.        }
68.        new EchoServer().bind(port);
69.    }
70. }
```

我们重点看 37～41 行，首先创建分隔符缓冲对象 ByteBuf，本例程中使用 "$_" 作为分隔符。第 40 行，创建 DelimiterBasedFrameDecoder 对象，将其加入到 ChannelPipeline 中。DelimiterBasedFrameDecoder 有多个构造方法，这里我们传递两个参数：第一个 1024

表示单条消息的最大长度,当达到该长度后仍然没有查找到分隔符,就抛出 TooLongFrame Exception 异常,防止由于异常码流缺失分隔符导致的内存溢出,这是 Netty 解码器的可靠性保护;第二个参数就是分隔符缓冲对象。

下面继续看 EchoServerHandler 的实现。

代码清单 5-2　EchoServer 服务端 EchoServerHandler

```
13.  @Sharable
14.  public class EchoServerHandler extends ChannelHandlerAdapter {
15.
16.      int counter = 0;
17.
18.      @Override
19.      public void channelRead(ChannelHandlerContext ctx, Object msg)
20.          throws Exception {
21.      String body = (String) msg;
22.      System.out.println("This is " + ++counter + " times receive client : ["
23.          + body + "]");
24.      body += "$_";
25.      ByteBuf echo = Unpooled.copiedBuffer(body.getBytes());
26.      ctx.writeAndFlush(echo);
27.      }
28.
29.      @Override
30.      public void exceptionCaught(ChannelHandlerContext ctx, Throwable cause) {
31.      cause.printStackTrace();
32.      ctx.close();// 发生异常,关闭链路
33.      }
34.  }
```

第 21～23 行直接将接收的消息打印出来,由于 DelimiterBasedFrameDecoder 自动对请求消息进行了解码,后续的 ChannelHandler 接收到的 msg 对象就是个完整的消息包;第二个 ChannelHandler 是 StringDecoder,它将 ByteBuf 解码成字符串对象;第三个 EchoServerHandler 接收到的 msg 消息就是解码后的字符串对象。

由于我们设置 DelimiterBasedFrameDecoder 过滤掉了分隔符,所以,返回给客户端时需要在请求消息尾部拼接分隔符"$_",最后创建 ByteBuf,将原始消息重新返回给客户端。

下面我们继续看下客户端的实现。

5.1.2 DelimiterBasedFrameDecoder 客户端开发

首先看下 EchoClient 的实现。

代码清单 5-3 EchoClient 客户端 EchoClient

```
20.  public class EchoClient {
21.
22.      public void connect(int port, String host) throws Exception {
23.          // 配置客户端NIO线程组
24.          EventLoopGroup group = new NioEventLoopGroup();
25.          try {
26.              Bootstrap b = new Bootstrap();
27.              b.group(group).channel(NioSocketChannel.class)
28.                  .option(ChannelOption.TCP_NODELAY, true)
29.                  .handler(new ChannelInitializer<SocketChannel>() {
30.                      @Override
31.                      public void initChannel(SocketChannel ch)
32.                          throws Exception {
33.                          ByteBuf delimiter = Unpooled.copiedBuffer("$_"
34.                              .getBytes());
35.                          ch.pipeline().addLast(
36.                              new DelimiterBasedFrameDecoder(1024,
37.                                  delimiter));
38.                          ch.pipeline().addLast(new StringDecoder());
39.                          ch.pipeline().addLast(new EchoClientHandler());
40.                      }
41.                  });
42.
43.              // 发起异步连接操作
44.              ChannelFuture f = b.connect(host, port).sync();
45.
46.              // 等待客户端链路关闭
47.              f.channel().closeFuture().sync();
48.          } finally {
49.              // 优雅退出，释放NIO线程组
50.              group.shutdownGracefully();
51.          }
52.      }
```

```
53.
54.    /**
55.     * @param args
56.     * @throws Exception
57.     */
58.    public static void main(String[] args) throws Exception {
59.        int port = 8080;
60.        if (args != null && args.length > 0) {
61.            try {
62.                port = Integer.valueOf(args[0]);
63.            } catch (NumberFormatException e) {
64.                // 采用默认值
65.            }
66.        }
67.        new EchoClient().connect(port, "127.0.0.1");
68.    }
69. }
```

与服务端类似，分别将 DelimiterBasedFrameDecoder 和 StringDecoder 添加到客户端 ChannelPipeline 中，最后添加客户端 I/O 事件处理类 EchoClientHandler，下面继续看 EchoClientHandler 的实现。

代码清单 5-4　EchoClient 客户端 EchoClientHandler

```
11. public class EchoClientHandler extends ChannelHandlerAdapter {
12.
13.     private int counter;
14.
15.     static final String ECHO_REQ = "Hi, Lilinfeng. Welcome to Netty.$_";
16.
17.     /**
18.      * Creates a client-side handler.
19.      */
20.     public EchoClientHandler() {
21.     }
22.
23.     @Override
24.     public void channelActive(ChannelHandlerContext ctx) {
25.         for (int i = 0; i < 10; i++) {
ctx.writeAndFlush(Unpooled.copiedBuffer(ECHO_REQ.getBytes()));
26.         }
27.     }
```

```
28.
29.      @Override
30.      public void channelRead(ChannelHandlerContext ctx, Object msg)
31.            throws Exception {
32.          System.out.println("This is " + ++counter + " times receive server : ["
33.              + msg + "]");
34.      }
35.
36.      @Override
37.      public void channelReadComplete(ChannelHandlerContext ctx) throws Exception {
38.          ctx.flush();
39.      }
40.
41.      @Override
42.      public void exceptionCaught(ChannelHandlerContext ctx, Throwable cause) {
43.          cause.printStackTrace();
44.          ctx.close();
45.      }
46. }
```

第 25～26 行在 TCP 链路建立成功之后循环发送请求消息给服务端，第 32～33 行打印接收到的服务端应答消息同时进行计数。

下个小节，运行上面开发的服务端和客户端，看看运行结果是否正确。

5.1.3 运行 DelimiterBasedFrameDecoder 服务端和客户端

服务端运行结果如下。

```
This is 1 times receive client : [Hi, Lilinfeng. Welcome to Netty.]
This is 2 times receive client : [Hi, Lilinfeng. Welcome to Netty.]
This is 3 times receive client : [Hi, Lilinfeng. Welcome to Netty.]
This is 4 times receive client : [Hi, Lilinfeng. Welcome to Netty.]
This is 5 times receive client : [Hi, Lilinfeng. Welcome to Netty.]
This is 6 times receive client : [Hi, Lilinfeng. Welcome to Netty.]
This is 7 times receive client : [Hi, Lilinfeng. Welcome to Netty.]
This is 8 times receive client : [Hi, Lilinfeng. Welcome to Netty.]
This is 9 times receive client : [Hi, Lilinfeng. Welcome to Netty.]
This is 10 times receive client : [Hi, Lilinfeng. Welcome to Netty.]
```

客户端运行结果如下。

```
This is 1 times receive server : [Hi, Lilinfeng. Welcome to Netty.]
This is 2 times receive server : [Hi, Lilinfeng. Welcome to Netty.]
This is 3 times receive server : [Hi, Lilinfeng. Welcome to Netty.]
This is 4 times receive server : [Hi, Lilinfeng. Welcome to Netty.]
This is 5 times receive server : [Hi, Lilinfeng. Welcome to Netty.]
This is 6 times receive server : [Hi, Lilinfeng. Welcome to Netty.]
This is 7 times receive server : [Hi, Lilinfeng. Welcome to Netty.]
This is 8 times receive server : [Hi, Lilinfeng. Welcome to Netty.]
This is 9 times receive server : [Hi, Lilinfeng. Welcome to Netty.]
This is 10 times receive server : [Hi, Lilinfeng. Welcome to Netty.]
```

服务端成功接收到了客户端发送的 10 条 "Hi, Lilinfeng. Welcome to Netty." 请求消息，客户端成功接收到了服务端返回的 10 条 "Hi, Lilinfeng. Welcome to Netty." 应答消息。测试结果表明使用 DelimiterBasedFrameDecoder 可以自动对采用分隔符做码流结束标识的消息进行解码。

本例程运行 10 次的原因是模拟 TCP 粘包/拆包，在笔者的机器上，连续发送 10 条 Echo 请求消息会发生粘包，如果没有 DelimiterBasedFrameDecoder 解码器的处理，服务端和客户端程序都将运行失败。下面我们将服务端的 DelimiterBasedFrameDecoder 注释掉，最终代码如图 5-1 所示。

```
public void initChannel(SocketChannel ch)
            throws Exception {
    ByteBuf delimiter = Unpooled.copiedBuffer("$_"
            .getBytes());
    // ch.pipeline().addLast(
    // new DelimiterBasedFrameDecoder(1024,
    // delimiter));
    ch.pipeline().addLast(new StringDecoder());
    ch.pipeline().addLast(new EchoServerHandler());
}
```

图 5-1　删除掉 DelimiterBasedFrameDecoder 后的服务端代码

服务端运行结果如下。

```
    This is 1 times receive client : [Hi, Lilinfeng. Welcome to Netty.$_Hi, Lilinfeng. Welcome to Netty.$_Hi, Lilinfeng. Welcome to Netty.$_Hi, Lilinfeng. Welcome to Netty.$_Hi, Lilinfeng. Welcome to Netty.$_Hi, Lilinfeng. Welcome to Netty.$_Hi, Lilinfeng. Welcome to Netty.$_Hi, Lilinfeng. Welcome to Netty.$_Hi, Lilinfeng. Welcome to Netty.$_Hi, Lilinfeng. Welcome to Netty.$_]
```

由于没有分隔符解码器，导致服务端一次读取了客户端发送的所有消息，这就是典型的没有考虑 TCP 粘包导致的问题。

5.2　FixedLengthFrameDecoder 应用开发

FixedLengthFrameDecoder 是固定长度解码器，它能够按照指定的长度对消息进行自动解码，开发者不需要考虑 TCP 的粘包/拆包问题，非常实用。下面我们通过一个应用实例对其用法进行讲解。

5.2.1　FixedLengthFrameDecoder 服务端开发

在服务端的 ChannelPipeline 中新增 FixedLengthFrameDecoder，长度设置为 20，然后再依次增加字符串解码器和 EchoServerHandler，代码如下。

代码清单 5-5　EchoServer 服务端　EchoServer

```
20. public class EchoServer {
21.     public void bind(int port) throws Exception {
22.         // 配置服务端的NIO线程组
23.         EventLoopGroup bossGroup = new NioEventLoopGroup();
24.         EventLoopGroup workerGroup = new NioEventLoopGroup();
25.         try {
26.             ServerBootstrap b = new ServerBootstrap();
27.             b.group(bossGroup, workerGroup)
28.                 .channel(NioServerSocketChannel.class)
29.                 .option(ChannelOption.SO_BACKLOG, 100)
30.                 .handler(new LoggingHandler(LogLevel.INFO))
31.                 .childHandler(new ChannelInitializer<SocketChannel>() {
32.                     @Override
33.                     public void initChannel(SocketChannel ch)
34.                         throws Exception {
35.                         ch.pipeline().addLast(
36.                             new FixedLengthFrameDecoder(20));
37.                         ch.pipeline().addLast(new StringDecoder());
38.                         ch.pipeline().addLast(new EchoServerHandler());
39.                     }
40.                 });
41.
42.             // 绑定端口，同步等待成功
43.             ChannelFuture f = b.bind(port).sync();
44.
```

```
45.         // 等待服务端监听端口关闭
46.         f.channel().closeFuture().sync();
47.     } finally {
48.         // 优雅退出，释放线程池资源
49.         bossGroup.shutdownGracefully();
50.         workerGroup.shutdownGracefully();
51.     }
52. }
53.
54. public static void main(String[] args) throws Exception {
55.     int port = 8080;
56.     if (args != null && args.length > 0) {
57.         try {
58.             port = Integer.valueOf(args[0]);
59.         } catch (NumberFormatException e) {
60.             // 采用默认值
61.         }
62.     }
63.     new EchoServer().bind(port);
64.   }
65. }
```

EchoServerHandler 的功能比较简单，直接将读取到的消息打印出来，代码如下。

代码清单 5-6　EchoServer 服务端　EchoServerHandler

```
11. @Sharable
12. public class EchoServerHandler extends ChannelHandlerAdapter {
13.
14.     @Override
15.     public void channelRead(ChannelHandlerContext ctx, Object msg)
16.         throws Exception {
17.      System.out.println("Receive client : [" + msg + "]");
18.     }
19.
20.     @Override
21.     public void exceptionCaught(ChannelHandlerContext ctx, Throwable cause) {
22.         cause.printStackTrace();
23.         ctx.close();// 发生异常，关闭链路
24.     }
25. }
```

利用 FixedLengthFrameDecoder 解码器,无论一次接收到多少数据报,它都会按照构造函数中设置的固定长度进行解码,如果是半包消息,FixedLengthFrameDecoder 会缓存半包消息并等待下个包到达后进行拼包,直到读取到一个完整的包。

下面的章节我们通过 telnet 命令行来测试 EchoServer 服务端,看它能否按照预期进行工作。

5.2.2 利用 telnet 命令行测试 EchoServer 服务端

由于客户端代码比较简单,所以这次我们通过 telnet 命令行对服务端进行测试。

测试场景:在 Windows 操作系统上打开 CMD 命令行窗口,通过 telnet 命令行连接服务端,在控制台输入如下内容。

```
Lilinfeng welcome to Netty at Nanjing
```

然后看服务端打印的内容,预期输出的请求消息为"Lilinfeng welcome to"。

下面我们就具体看下详细的测试步骤。

(1) 在【运行】菜单中输入 cmd 命令,打开命令行窗口,如图 5-2 所示。

图 5-2 通过 cmd 命令打开 CMD 窗口

(2) 在命令行中输入"telnet localhost 8080",通过 telnet 连接服务端,如图 5-3 所示。

图 5-3 通过 telnet 命令连接服务端

(3)通过 set localecho 命令打开本地回显功能，输入命令行内容，如图 5-4 所示。

图 5-4　输入 Lilinfeng welcome to Netty at Nanjing

(4) EchoServer 服务端运行结果如图 5-5 所示。

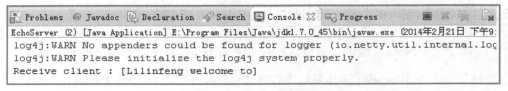

图 5-5　服务端运行结果

根据图 5-5 所示内容，服务端运行结果完全符合预期，FixedLengthFrameDecoder 解码器按照 20 个字节长度对请求消息进行截取，输出结果为"Lilinfeng welcome to"。

5.3　总结

本章我们学习了两个非常实用的解码器：DelimiterBasedFrameDecoder 和 FixedLengthFrameDecoder。

DelimiterBasedFrameDecoder 用于对使用分隔符结尾的消息进行自动解码，FixedLengthFrameDecoder 用于对固定长度的消息进行自动解码。有了上述两种解码器，再结合其他的解码器，如字符串解码器等，可以轻松地完成对很多消息的自动解码，而且不再需要考虑 TCP 粘包/拆包导致的读半包问题，极大地提升了开发效率。

应用 DelimiterBasedFrameDecoder 和 FixedLengthFrameDecoder 进行开发非常简单，在绝大数情况下，只要将 DelimiterBasedFrameDecoder 或 FixedLengthFrameDecoder 添加到对应 ChannelPipeline 的起始位即可。

熟悉了 Netty 的 NIO 基础应用开发之后，从第三部分开始，我们继续学习编解码技术。在了解编解码基础知识之后，继续学习 Netty 内置的编解码框架的使用，例如 Java 序列化、二进制编解码、谷歌的 protobuf 和 JBoss 的 Marshalling 序列化框架。

中级篇

Netty 编解码开发指南

第 6 章　编解码技术

第 7 章　MessagePack 编解码

第 8 章　Google Protobuf 编解码

第 9 章　JBoss Marshalling 编解码

第 6 章 编解码技术

基于 Java 提供的对象输入/输出流 ObjectInputStream 和 ObjectOutputStream，可以直接把 Java 对象作为可存储的字节数组写入文件，也可以传输到网络上。对程序员来说，基于 JDK 默认的序列化机制可以避免操作底层的字节数组，从而提升开发效率。

Java 序列化的目的主要有两个：

◎ 网络传输

◎ 对象持久化

由于本书主要介绍基于 Netty 的 NIO 网络开发，所以我们重点关注网络传输。当进行远程跨进程服务调用时，需要把被传输的 Java 对象编码为字节数组或者 ByteBuffer 对象。而当远程服务读取到 ByteBuffer 对象或者字节数组时，需要将其解码为发送时的 Java 对象。这被称为 Java 对象编解码技术。

Java 序列化仅仅是 Java 编解码技术的一种，由于它的种种缺陷，衍生出了多种编解码技术和框架，后续的章节我们会结合 Netty 介绍几种业界主流的编解码技术和框架，看看如何在 Netty 中应用这些编解码框架实现消息的高效序列化。

本章主要内容包括：

◎ Java 序列化的缺点

◎ 业界流行的几种编解码框架介绍

6.1 Java 序列化的缺点

Java 序列化从 JDK 1.1 版本就已经提供，它不需要添加额外的类库，只需实现 java.io.Serializable 并生成序列 ID 即可，因此，它从诞生之初就得到了广泛的应用。

但是在远程服务调用（RPC）时，很少直接使用 Java 序列化进行消息的编解码和传输，这又是什么原因呢？下面通过分析 Java 序列化的缺点来找出答案。

6.1.1 无法跨语言

无法跨语言，是 Java 序列化最致命的问题。对于跨进程的服务调用，服务提供者可能会使用 C++或者其他语言开发，当我们需要和异构语言进程交互时，Java 序列化就难以胜任。

由于 Java 序列化技术是 Java 语言内部的私有协议，其他语言并不支持，对于用户来说它完全是黑盒。对于 Java 序列化后的字节数组，别的语言无法进行反序列化，这就严重阻碍了它的应用。

事实上，目前几乎所有流行的 Java RCP 通信框架，都没有使用 Java 序列化作为编解码框架，原因就在于它无法跨语言，而这些 RPC 框架往往需要支持跨语言调用。

6.1.2 序列化后的码流太大

下面我们通过一个实例看下 Java 序列化后的字节数组大小。

代码清单 6-1　Java 序列化代码　POJO 对象类 UserInfo

```
10. public class UserInfo implements Serializable {
11.
12.     /**
13.      * 默认的序列号
14.      */
15.     private static final long serialVersionUID = 1L;
16.
17.     private String userName;
18.
```

```
19.    private int userID;
20.
21.    public UserInfo buildUserName(String userName) {
22.     this.userName = userName;
23.     return this;
24.    }
25.
26.    public UserInfo buildUserID(int userID) {
27.     this.userID = userID;
28.     return this;
29.    }
30.
31.    /**
32.     * @return the userName
33.     */
34.    public final String getUserName() {
35.     return userName;
36.    }
37.
38.    /**
39.     * @param userName
40.     *            the userName to set
41.     */
42.    public final void setUserName(String userName) {
43.     this.userName = userName;
44.    }
45.
46.    /**
47.     * @return the userID
48.     */
49.    public final int getUserID() {
50.     return userID;
51.    }
52.
53.    /**
54.     * @param userID
55.     *            the userID to set
56.     */
57.    public final void setUserID(int userID) {
58.     this.userID = userID;
59.    }
```

```
60.
61.    public byte[] codeC() {
62.        ByteBuffer buffer = ByteBuffer.allocate(1024);
63.        byte[] value = this.userName.getBytes();
64.        buffer.putInt(value.length);
65.        buffer.put(value);
66.        buffer.putInt(this.userID);
67.        buffer.flip();
68.        value = null;
69.        byte[] result = new byte[buffer.remaining()];
70.        buffer.get(result);
71.        return result;
72.    }
73. }
```

UserInfo 对象是个普通的 POJO 对象，它实现了 java.io.Serializable 接口，并且生成了一个默认的序列号 serialVersionUID = 1L。这说明 UserInfo 对象可以通过 JDK 默认的序列化机制进行序列化和反序列化。

第 61～72 行使用基于 ByteBuffer 的通用二进制编解码技术对 UserInfo 对象进行编码，编码结果仍然是 byte 数组，可以与传统的 JDK 序列化后的码流大小进行对比。

下面写一个测试程序，先调用两种编码接口对 POJO 对象编码，然后分别打印两者编码后的码流大小进行对比。

代码清单 6-2　Java 序列化代码　编码测试类 TestUserInfo

```
11. public class TestUserInfo {
12.
13.    /**
14.     * @param args
15.     * @throws IOException
16.     */
17.    public static void main(String[] args) throws IOException {
18.        UserInfo info = new UserInfo();
19.        info.buildUserID(100).buildUserName("Welcome to Netty");
20.        ByteArrayOutputStream bos = new ByteArrayOutputStream();
21.        ObjectOutputStream os = new ObjectOutputStream(bos);
22.        os.writeObject(info);
23.        os.flush();
24.        os.close();
25.        byte[] b = bos.toByteArray();
```

```
26.        System.out.println("The jdk serializable length is : " + b.length);
27.        bos.close();
28.        System.out.println("--------------------------------------");
29.        System.out.println("The byte array serializable length is : "
30.            + info.codeC().length);
31.    }
32.
33. }
```

测试结果如图 6-1 所示。

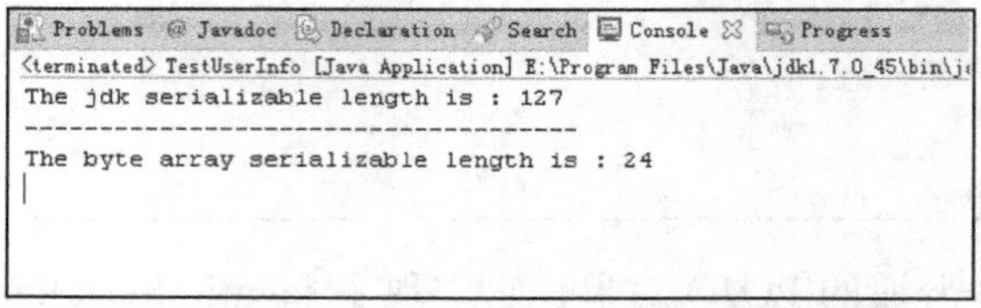

图 6-1 JDK 序列化机制和通用二进制编码测试结果

测试结果令人震惊，采用 JDK 序列化机制编码后的二进制数组大小竟然是二进制编码的 5.29 倍。

我们评判一个编解码框架的优劣时，往往会考虑以下几个因素。

◎ 是否支持跨语言，支持的语言种类是否丰富；
◎ 编码后的码流大小；
◎ 编解码的性能；
◎ 类库是否小巧，API 使用是否方便；
◎ 使用者需要手工开发的工作量和难度。

在同等情况下，编码后的字节数组越大，存储的时候就越占空间，存储的硬件成本就越高，并且在网络传输时更占带宽，导致系统的吞吐量降低。Java 序列化后的码流偏大也一直被业界所诟病，导致它的应用范围受到了很大限制。

6.1.3 序列化性能太低

下面我们从序列化的性能角度看下 JDK 的表现如何。将之前的例程代码稍做修改，改造成性能测试版本，如图 6-2 所示。

```
public byte[] codeC(ByteBuffer buffer) {
    buffer.clear();
    byte[] value = this.userName.getBytes();
    buffer.putInt(value.length);
    buffer.put(value);
    buffer.putInt(this.userID);
    buffer.flip();
    value = null;
    byte[] result = new byte[buffer.remaining()];
    buffer.get(result);
    return result;
}
```

图 6-2　UserInfo 性能测试版本修改

对 UserInfo 进行改造，新增上图所示的方法，再创建一个性能测试版本的 UserInfo 测试程序，代码如下。

代码清单 6-3　Java 序列化代码　编码性能测试类 PerformTestUserInfo

```
12. public class PerformTestUserInfo {
13.
14.     /**
15.      * @param args
16.      * @throws IOException
17.      */
18.     public static void main(String[] args) throws IOException {
19.         UserInfo info = new UserInfo();
20.         info.buildUserID(100).buildUserName("Welcome to Netty");
21.         int loop = 1000000;
22.         ByteArrayOutputStream bos = null;
23.         ObjectOutputStream os = null;
24.         long startTime = System.currentTimeMillis();
25.         for (int i = 0; i < loop; i++) {
26.             bos = new ByteArrayOutputStream();
27.             os = new ObjectOutputStream(bos);
28.             os.writeObject(info);
29.             os.flush();
30.             os.close();
31.             byte[] b = bos.toByteArray();
32.             bos.close();
33.         }
34.         long endTime = System.currentTimeMillis();
35.         System.out.println("The jdk serializable cost time is : "
36.             + (endTime - startTime) + " ms");
37.
38.         System.out.println("--------------------------------------");
```

```
39.
40.     ByteBuffer buffer = ByteBuffer.allocate(1024);
41.     startTime = System.currentTimeMillis();
42.     for (int i = 0; i < loop; i++) {
43.         byte[] b = info.codeC(buffer);
44.     }
45.     endTime = System.currentTimeMillis();
46.     System.out.println("The byte array serializable cost time is : "
47.         + (endTime - startTime) + " ms");
48.     }
49. }
```

对 Java 序列化和二进制编码分别进行性能测试，编码 100 万次，然后统计耗费的总时间，测试结果如图 6-3 所示。

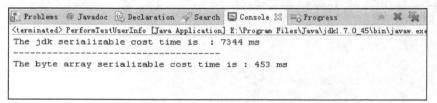

图 6-3　UserInfo 编码性能测试结果

这个结果也非常令人惊讶：Java 序列化的性能只有二进制编码的 6.17% 左右，可见 Java 原生序列化的性能实在太差。

下面我们结合编码速度，综合对比一下 Java 序列化和二进制编码的性能差异，如图 6-4 所示。

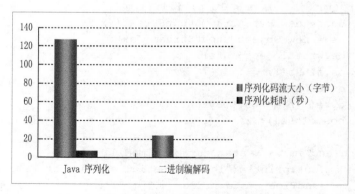

图 6-4　序列化性能对比图

从图 6-4 可以看出，无论是序列化后的码流大小，还是序列化的性能，JDK 默认的序列化机制表现得都很差。因此，我们通常不会选择 Java 序列化作为远程跨节点调用的编解码框架。

但是不使用 JDK 提供的默认序列化框架，自己开发编解码框架又是个非常复杂的工作，怎么办呢？不用着急，业界有很多优秀的编解码框架，它们在克服了 JDK 默认序列化框架缺点的基础上，还增加了很多亮点，下面让我们继续了解并学习业界流行的几款编解码框架。

6.2 业界主流的编解码框架

由于 Java 的编解码框架五花八门，穷举学习显然不是一个好的策略，本节挑选了一些业界主流的编解码框架和编解码技术进行介绍，希望读者在了解这些框架特性的基础上，做出合理的选择。

6.2.1 Google 的 Protobuf 介绍

Protobuf 全称 Google Protocol Buffers，它由谷歌开源而来，在谷歌内部久经考验。它将数据结构以.proto 文件进行描述，通过代码生成工具可以生成对应数据结构的 POJO 对象和 Protobuf 相关的方法和属性。

它的特点如下。

- 结构化数据存储格式（XML，JSON 等）；
- 高效的编解码性能；
- 语言无关、平台无关、扩展性好；
- 官方支持 Java、C++和 Python 三种语言。

首先我们来看下为什么不使用 XML，尽管 XML 的可读性和可扩展性非常好，也非常适合描述数据结构，但是 XML 解析的时间开销和 XML 为了可读性而牺牲的空间开销都非常大，因此不适合做高性能的通信协议。Protobuf 使用二进制编码，在空间和性能上具有更大的优势。

Protobuf 另一个比较吸引人的地方就是它的数据描述文件和代码生成机制,利用数据描述文件对数据结构进行说明的优点如下。

◎ 文本化的数据结构描述语言,可以实现语言和平台无关,特别适合异构系统间的集成;

◎ 通过标识字段的顺序,可以实现协议的前向兼容;

◎ 自动代码生成,不需要手工编写同样数据结构的 C++ 和 Java 版本;

◎ 方便后续的管理和维护。相比于代码,结构化的文档更容易管理和维护。

下面我们看下 Protobuf 编解码和其他几种序列化框架的性能对比数据,如图 6-5、图 6-6 所示。

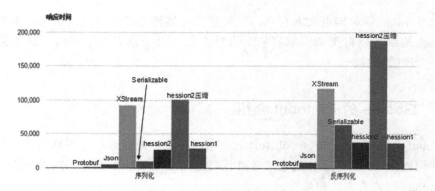

图 6-5　Protobuf 编解码和其他几种序列化框架的响应时间对比

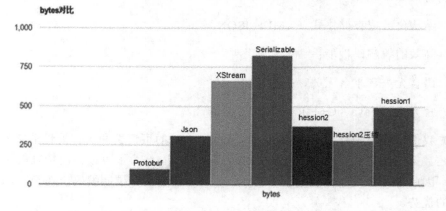

图 6-6　Protobuf 和其他几种序列化框架的字节数对比

从图 6-5 和图 6-6 两幅对比图可以发现，Protobuf 的编解码性能远远高于其他几种序列化框架的序列化和反序列化，这也是很多 RPC 框架选用 Protobuf 做编解码框架的原因。

6.2.2　Facebook 的 Thrift 介绍

Thrift 源于 Facebook，在 2007 年 Facebook 将 Thrift 作为一个开源项目提交给了 Apache 基金会。对于当时的 Facebook 来说，创造 Thrift 是为了解决 Facebook 各系统间大数据量的传输通信以及系统之间语言环境不同需要跨平台的特性，因此 Thrift 可以支持多种程序语言，如 C++、C#、Cocoa、Erlang、Haskell、Java、Ocami、Perl、PHP、Python、Ruby 和 Smalltalk。

在多种不同的语言之间通信，Thrift 可以作为高性能的通信中间件使用，它支持数据（对象）序列化和多种类型的 RPC 服务。Thrift 适用于静态的数据交换，需要先确定好它的数据结构，当数据结构发生变化时，必须重新编辑 IDL 文件，生成代码和编译，这一点跟其他 IDL 工具相比可以视为是 Thrift 的弱项。Thrift 适用于搭建大型数据交换及存储的通用工具，对于大型系统中的内部数据传输，相对于 JSON 和 XML 在性能和传输大小上都有明显的优势。

Thrift 主要由 5 部分组成。

（1）语言系统以及 IDL 编译器：负责由用户给定的 IDL 文件生成相应语言的接口代码；

（2）TProtocol：RPC 的协议层，可以选择多种不同的对象序列化方式，如 JSON 和 Binary；

（3）TTransport：RPC 的传输层，同样可以选择不同的传输层实现，如 socket、NIO、MemoryBuffer 等；

（4）TProcessor：作为协议层和用户提供的服务实现之间的纽带，负责调用服务实现的接口；

（5）TServer：聚合 TProtocol、TTransport 和 TProcessor 等对象。

我们重点关注的是编解码框架，与之对应的就是 TProtocol。由于 Thrift 的 RPC 服务调用和编解码框架绑定在一起，所以，通常我们使用 Thrift 的时候会采取 RPC 框架的方式。但是，它的 TProtocol 编解码框架还是可以以类库的方式独立使用的。

与 Protobuf 比较类似的是，Thrift 通过 IDL 描述接口和数据结构定义，它支持 8 种 Java 基本类型、Map、Set 和 List，支持可选和必选定义，功能非常强大。因为可以定义数据结构中字段的顺序，所以它也可以支持协议的前向兼容。

Thrift 支持三种比较典型的编解码方式。

◎ 通用的二进制编解码；

◎ 压缩二进制编解码；

◎ 优化的可选字段压缩编解码。

由于支持二进制压缩编解码，Thrift 的编解码性能表现也相当优异，远远超过 Java 序列化和 RMI 等，图 6-7 展示了同等测试条件下的编解码耗时信息。

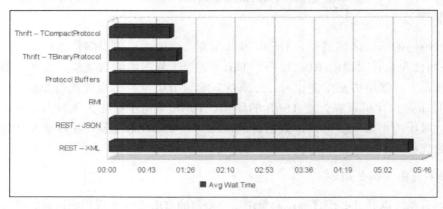

图 6-7　Thrift 性能测试对比图

6.2.3　JBoss Marshalling 介绍

JBoss Marshalling 是一个 Java 对象的序列化 API 包，修正了 JDK 自带的序列化包的很多问题，但又保持跟 java.io.Serializable 接口的兼容。同时，增加了一些可调的参数和附加的特性，并且这些参数和特性可通过工厂类进行配置。

相比于传统的 Java 序列化机制，它的优点如下。

◎ 可插拔的类解析器，提供更加便捷的类加载定制策略，通过一个接口即可实现定制；

◎ 可插拔的对象替换技术，不需要通过继承的方式；

◎ 可插拔的预定义类缓存表，可以减小序列化的字节数组长度，提升常用类型的对象序列化性能；

◎ 无须实现 java.io.Serializable 接口，即可实现 Java 序列化；

◎ 通过缓存技术提升对象的序列化性能。

相比于前面介绍的两种编解码框架，JBoss Marshalling 更多是在 JBoss 内部使用，应用范围有限。

JBoss Marshalling 的使用非常简单，后续在介绍 Netty 的 Marshalling 解码器时会给出例程。

6.3 总结

本章首先对 Java 的序列化技术进行了介绍，对 Java 序列化的缺点进行了总结说明，在此基础上引出了几款业界主流的编解码框架。由于编解码框架种类繁多，无法一一枚举，所以重点介绍了当前最流行的几种编解码框架。在后续的章节中我们会对这些编解码框架的使用进行说明，并给出具体的示例，同时，讲解如何在 Netty 中应用这些编解码框架。

第 7 章
MessagePack 编解码

MessagePack 是一个高效的二进制序列化框架,它像 JSON 一样支持不同语言间的数据交换,但是它的性能更快,序列化之后的码流也更小。

由于 MessagePack 在业界得到了非常广泛的应用,因此本章将介绍如何利用 Netty 的 CodeC 框架新增对 MessagePack 的支持。

本章主要内容包括:

◎ MessagePack 介绍
◎ MessagePack 编码器和解码器开发
◎ 粘包/半包支持

7.1 MessagePack 介绍

MessagePack 的特点如下:

◎ 编解码高效,性能高;
◎ 序列化之后的码流小;

◎ 支持跨语言。

7.1.1 MessagePack 多语言支持

衡量序列化框架通用性的一个重要指标就是对多语言的支持,因为数据交换的双方很难保证一定采用相同的语言开发,如果序列化框架和某种语言绑定,它就很难跨语言,例如 Java 的序列化机制。

MessagePack 提供了对多语言的支持,官方支持的语言如下:Java、Python、Ruby、Haskell、C#、OCaml、Lua、Go、C、C++等。更详细的支持列表可以参见它的官网:http://msgpack.org/。

7.1.2 MessagePack Java API 介绍

如果使用 Maven 工程开发,在你的 pom.xml 中配置 MessagePack 的坐标,如下。

```xml
<dependency>
   <groupId>org.msgpack</groupId>
   <artifactId>msgpack</artifactId>
   <version>${msgpack.version}</version>
</dependency>
```

它的 API 使用起来非常简单,编码和解码开发示例如下。

```java
// Create serialize objects.
List<String> src = new ArrayList<String>();
src.add("msgpack");
src.add("kumofs");
src.add("viver");
MessagePack msgpack = new MessagePack();
// Serialize
byte[] raw = msgpack.write(src);
// Deserialize directly using a template
List<String> dst1 = msgpack.read(raw, Templates.tList(Templates.TString));
System.out.println(dst1.get(0));
System.out.println(dst1.get(1));
System.out.println(dst1.get(2));
```

7.1.3 MessagePack 开发包下载

MessagePack 开发包的下载路径如下，用户可以根据实际情况选择合适的版本。

https://github.com/msgpack/msgpack-java/releases

7.2 MessagePack 编码器和解码器开发

利用 Netty 的编解码框架可以非常方便地集成第三方序列化框架，Netty 预集成了几种常用的编解码框架，用户也可以根据自己项目的实际情况集成其他编解码框架，或者进行自定义。

7.2.1 MessagePack 编码器开发

首先将 msgpack-0.6.6.jar 和 javassist-3.9.0.jar 加入到第三方类库中，然后新建 /io/netty/handler/codec/msgpack 目录，代码结构如图 7-1 所示。

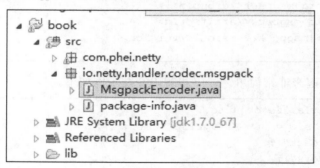

图 7-1 msgpack 编解码框架代码结构

接下来我们看下 msgpack 编码器的开发，代码如下。

代码清单 7-1 msgpack 编码器 MsgpackEncoder

```
public class MsgpackEncoder extends MessageToByteEncoder<Object> {
    @Override
    protected void encode(ChannelHandlerContext arg0, Object arg1, ByteBuf arg2)
            throws Exception {
```

```
        MessagePack msgpack = new MessagePack();
        // Serialize
        byte[] raw = msgpack.write(arg1);
        arg2.writeBytes(raw);
    }
}
```

MsgpackEncoder 继承 MessageToByteEncoder,它负责将 Object 类型的 POJO 对象编码为 byte 数组,然后写入到 ByteBuf 中。

7.2.2 MessagePack 解码器开发

完成编码器开发之后,下面我们继续看解码器如何编写,代码如下。

代码清单 7-2 msgpack 解码器 MsgpackDecoder

```
public class MsgpackDecoder extends MessageToMessageDecoder<ByteBuf>{

    @Override
    protected void decode(ChannelHandlerContext arg0, ByteBuf arg1,
            List<Object> arg2) throws Exception {
        final byte[] array;
        final int length = arg1.readableBytes();
        array = new byte[length];
        arg1.getBytes(arg1.readerIndex(), array, 0, length);
        MessagePack msgpack = new MessagePack();
        arg2.add(msgpack.read(array));
    }
}
```

首先从数据报 arg1 中获取需要解码的 byte 数组,然后调用 MessagePack 的 read 方法将其反序列化为 Object 对象,将解码后的对象加入到解码列表 arg2 中,这样就完成了 MessagePack 的解码操作。

7.2.3 功能测试

完成编解码器开发之后,我们以 Netty 原生的 Echo 程序为例,进行测试。对 Echo 进行简单改造,传输的对象由字符串修改为 POJO 对象,利用 MessagePack 对 POJO 对象进行序列化。

客户端代码清单如下。

代码清单 7-3 EchoClient

```java
public class EchoClient {
    private final String host;
    private final int port;
    private final int sendNumber;
    public EchoClient(String host, int port, int sendNumber) {
        this.host = host;
        this.port = port;
        this.sendNumber = sendNumber;
    }
    public void run() throws Exception {
        // Configure the client.
        EventLoopGroup group = new NioEventLoopGroup();
        try {
            Bootstrap b = new Bootstrap();
            b.group(group).channel(NioSocketChannel.class)
                .option(ChannelOption.TCP_NODELAY, true)
                .option(ChannelOption.CONNECT_TIMEOUT_MILLIS, 3000)
                .handler(new ChannelInitializer<SocketChannel>() {
                @Override
                public void initChannel(SocketChannel ch)
                    throws Exception {
                    ch.pipeline().addLast("msgpack decoder", new MsgpackDecoder());
                    ch.pipeline().addLast("msgpack encoder", new MsgpackEncoder());
                    ch.pipeline().addLast(
                        new EchoClientHandler(sendNumber));
                }
            });
        //代码省略......
    }
```

代码清单 7-4 EchoClientHandler

```java
public class EchoClientHandler extends ChannelHandlerAdapter {
    private final int sendNumber;
    public EchoClientHandler(int sendNumber) {
        this.sendNumber = sendNumber;
```

```java
    }
    @Override
    public void channelActive(ChannelHandlerContext ctx) {
        UserInfo [] infos = UserInfo();
        for(UserInfo infoE : infos)
        {
            ctx.write(infoE);
        }
        ctx.flush();
    }
    private UserInfo [] UserInfo()
    {
        UserInfo [] userInfos = new UserInfo[sendNumber];
        UserInfo userInfo = null;
        for (int i = 0; i < sendNumber; i++)
        {
            userInfo = new UserInfo();
            userInfo.setAge(i);
            userInfo.setName("ABCDEFG --->" + i);
            userInfos[i] = userInfo;
        }
        return userInfos;
    }
    @Override
    public void channelRead(ChannelHandlerContext ctx, Object msg)
            throws Exception {
        System.out.println("Client receive the msgpack message : " + msg);
        ctx.write(msg);
    }
    @Override
    public void channelReadComplete(ChannelHandlerContext ctx) throws Exception {
        ctx.flush();
    }
}
```

服务端代码与客户端类似，请参考书中附带的源码。

客户端运行结果如图 7-2 所示。

```
<terminated> EchoClient [Java Application] C:\Program Files\Java\jdk1.7.0_67\
Client receive the msgpack message : [0,"ABCDEFG --->0"]
Client receive the msgpack message : [0,"ABCDEFG --->0"]
Client receive the msgpack message : [0,"ABCDEFG --->0"]
Client receive the msgpack message : [0,"ABCDEFG --->0"]
Client receive the msgpack message : [0,"ABCDEFG --->0"]
Client receive the msgpack message : [0,"ABCDEFG --->0"]
```

图 7-2 客户端运行结果

服务端运行结果如图 7-3 所示。

```
<terminated> EchoServer [Java Application] C:\Program Files\Java\jdk1.7.0_67\
Server receive the msgpack message : [0,"ABCDEFG --->0"]
Server receive the msgpack message : [0,"ABCDEFG --->0"]
Server receive the msgpack message : [0,"ABCDEFG --->0"]
Server receive the msgpack message : [0,"ABCDEFG --->0"]
Server receive the msgpack message : [0,"ABCDEFG --->0"]
```

图 7-3 服务端运行结果

至此，MessagePack 编解码框架开发和测试完成。细心的读者可能发现了一个问题：测试代码中并没有考虑 TCP 粘包和半包的处理。下面我们模拟粘包的场景，看上述测试代码能否运行正确——链路建立成功之后，客户端批量发送 1000 条消息，运行结果如图 7-4 所示。

```
EchoServer [Java Application] C:\Program Files\Java\jdk1.7.0_67\bin\javaw.exe (2014年10月20日 上午11:32:11)
Server receive the msgpack message : [0,"ABCDEFG --->0"]
一月 20, 2014 11:32:56 上午 com.phei.netty.codec.msgpack.EchoServerHandler exceptionCaught
警告: Unexpected exception from downstream.
java.io.IOException: 远程主机强迫关闭了一个现有的连接。
    at sun.nio.ch.SocketDispatcher.read0(Native Method)
    at sun.nio.ch.SocketDispatcher.read(SocketDispatcher.java:43)
    at sun.nio.ch.IOUtil.readIntoNativeBuffer(IOUtil.java:223)
    at sun.nio.ch.IOUtil.read(IOUtil.java:192)
    at sun.nio.ch.SocketChannelImpl.read(SocketChannelImpl.java:379)
    at io.netty.buffer.UnpooledUnsafeDirectByteBuf.setBytes(UnpooledUnsafeDirectByteBuf.java:446)
    at io.netty.buffer.AbstractByteBuf.writeBytes(AbstractByteBuf.java:871)
    at io.netty.channel.socket.nio.NioSocketChannel.doReadBytes(NioSocketChannel.java:208)
    at io.netty.channel.nio.AbstractNioByteChannel$NioByteUnsafe.read(AbstractNioByteChannel.java:119)
    at io.netty.channel.nio.NioEventLoop.processSelectedKey(NioEventLoop.java:485)
    at io.netty.channel.nio.NioEventLoop.processSelectedKeysOptimized(NioEventLoop.java:452)
    at io.netty.channel.nio.NioEventLoop.run(NioEventLoop.java:346)
    at io.netty.util.concurrent.SingleThreadEventExecutor$5.run(SingleThreadEventExecutor.java:794)
    at java.lang.Thread.run(Thread.java:745)
Server receive the msgpack message : [0,"ABCDEFG --->0"]
Server receive the msgpack message : [16,"ABCDEFG --->16"]
Server receive the msgpack message : 66
```

图 7-4 粘包场景下测试结果

因为没有考虑粘包/半包的处理，所以我们开发的 MessagePack 编解码框架还不能正常工作，下面的 7.3 节将利用 Netty 的编解码特性，提供对粘包/半包场景的支持。

7.3 粘包/半包支持

在前面的章节中介绍到了粘包/半包的解决策略，其中最常用的就是在消息头中新增报文长度字段，利用该字段进行半包的编解码。

下面我们就利用 Netty 提供的 LengthFieldPrepender 和 LengthFieldBasedFrameDecoder，结合新开发的 MessagePack 编解码框架，实现对 TCP 粘包/半包的支持。

下面我们对客户端和服务端进行改造，代码如下。

代码清单 7-4　EchoClientV2 客户端代码片段

```
public void run() throws Exception {
    // Configure the client.
    EventLoopGroup group = new NioEventLoopGroup();
    try {
        Bootstrap b = new Bootstrap();
        b.group(group).channel(NioSocketChannel.class)
            .option(ChannelOption.TCP_NODELAY, true)
            .option(ChannelOption.CONNECT_TIMEOUT_MILLIS, 3000)
            .handler(new ChannelInitializer<SocketChannel>() {
            @Override
            public void initChannel(SocketChannel ch)
                throws Exception {
                ch.pipeline().addLast("frameDecoder", new LengthFieldBasedFrameDecoder(65535, 0, 2, 0, 2));
                ch.pipeline().addLast("msgpack decoder", new MsgpackDecoder());
                ch.pipeline().addLast("frameEncoder", new LengthFieldPrepender(2));
                ch.pipeline().addLast("msgpack encoder", new MsgpackEncoder());
                ch.pipeline().addLast(
                    new EchoClientHandler(sendNumber));
            }
        });
```

在 MessagePack 编码器之前增加 LengthFieldPrepender，它将在 ByteBuf 之前增加 2 个字节的消息长度字段，其原理如图 7-5 所示。

```
+-----------------+         +--------+-----------------+
| "HELLO, WORLD"  |   →     | 0x000C | "HELLO, WORLD"  |
+-----------------+         +--------+-----------------+
```

图 7-5　LengthFieldPrepender 原理示意图

在 MessagePack 解码器之前增加 LengthFieldBasedFrameDecoder，用于处理半包消息，

这样后面的 MsgpackDecoder 接收到的永远是整包消息。它的工作原理如图 7-6 所示。

```
BEFORE DECODE (14 bytes)            AFTER DECODE (12 bytes)
+--------+----------------+          +----------------+
| Length | Actual Content |----->|   Actual Content |
| 0x000C | "HELLO, WORLD" |          | "HELLO, WORLD" |
+--------+----------------+          +----------------+
```

图 7-6　LengthFieldBasedFrameDecoder 工作原理图

EchoServerV2 服务端代码改造如下：

代码清单 7-5　EchoServerV2 服务端代码片段

```java
public void run() throws Exception {
    // Configure the server.
    EventLoopGroup acceptorGroup = new NioEventLoopGroup();
    EventLoopGroup IOGroup = new NioEventLoopGroup();
    try {
        ServerBootstrap b = new ServerBootstrap();
        b.group(acceptorGroup, IOGroup)
            .channel(NioServerSocketChannel.class)
            .option(ChannelOption.SO_BACKLOG, 100)
            .handler(new LoggingHandler(LogLevel.INFO))
            .childHandler(new ChannelInitializer<SocketChannel>() {
                @Override
                public void initChannel(SocketChannel ch)
                    throws Exception {
                    ch.pipeline().addLast("frameDecoder", new LengthFieldBasedFrameDecoder(65535, 0, 2, 0, 2));
                    ch.pipeline().addLast("msgpack decoder", new MsgpackDecoder());
                    ch.pipeline().addLast("frameEncoder", new LengthFieldPrepender(2));
                    ch.pipeline().addLast("msgpack encoder", new MsgpackEncoder());
                    ch.pipeline().addLast(
                        new EchoServerHandler());
                }
            });
```

再次测试下，链路链接成功之后，客户端批量发送 1000 条消息，服务端接收到之后原路返回客户端，测试结果如下。

服务端运行结果如下。

```
Server receive the msgpack message : [0,"ABCDEFG --->0"]
Server receive the msgpack message : [1,"ABCDEFG --->1"]
Server receive the msgpack message : [2,"ABCDEFG --->2"]
Server receive the msgpack message : [3,"ABCDEFG --->3"]
Server receive the msgpack message : [4,"ABCDEFG --->4"]
Server receive the msgpack message : [5,"ABCDEFG --->5"]
Server receive the msgpack message : [6,"ABCDEFG --->6"]
Server receive the msgpack message : [7,"ABCDEFG --->7"]
Server receive the msgpack message : [8,"ABCDEFG --->8"]
//中间结果省略.....
Server receive the msgpack message : [8,"ABCDEFG --->1000"]
```

客户端运行结果如下。

```
Client receive the msgpack message : [0,"ABCDEFG --->0"]
Client receive the msgpack message : [1,"ABCDEFG --->1"]
Client receive the msgpack message : [2,"ABCDEFG --->2"]
Client receive the msgpack message : [3,"ABCDEFG --->3"]
Client receive the msgpack message : [4,"ABCDEFG --->4"]
Client receive the msgpack message : [5,"ABCDEFG --->5"]
Client receive the msgpack message : [6,"ABCDEFG --->6"]
Client receive the msgpack message : [7,"ABCDEFG --->7"]
Client receive the msgpack message : [8,"ABCDEFG --->8"]
//中间结果省略.....
Client receive the msgpack message : [8,"ABCDEFG --->1000"]
```

测试结果表明，利用 Netty 的半包编码和解码器 LengthFieldPrepender 和 LengthFieldBasedFrameDecoder，可以轻松地解决 TCP 粘包和半包问题。

7.4 总结

本章首先对 MessagePack 序列化框架进行了讲解，然后利用 Netty 提供的编解码框架扩展实现了对 MessagePack 编解码的支持。

最后结合 Netty 提供的半包编码和解码器，MessagePack 编解码框架实现了对 TCP 粘包和半包的支持。后续章节，我们还将专门对 LengthFieldPrepender 和 LengthFieldBasedFrameDecoder 的工作原理进行详细讲解。

第 8 章

Google Protobuf 编解码

Google 的 Protobuf 在业界非常流行，很多商业项目选择 Protobuf 作为编解码框架，这里一起回顾一下 Protobuf 的优点。

（1）在谷歌内部长期使用，产品成熟度高；

（2）跨语言、支持多种语言，包括 C++、Java 和 Python；

（3）编码后的消息更小，更加有利于存储和传输；

（4）编解码的性能非常高；

（5）支持不同协议版本的前向兼容；

（6）支持定义可选和必选字段。

本章主要内容包括：

- Protobuf 的入门
- 开发支持 Protobuf 的 Netty 服务端
- 开发支持 Protobuf 的 Netty 客户端
- 运行基于 Netty 开发的 Protobuf 例程

8.1 Protobuf 的入门

Protobuf 是一个灵活、高效、结构化的数据序列化框架，相比于 XML 等传统的序列化工具，它更小、更快、更简单。Protobuf 支持数据结构化一次可以到处使用，甚至跨语言使用，通过代码生成工具可以自动生成不同语言版本的源代码，甚至可以在使用不同版本的数据结构进程间进行数据传递，实现数据结构的前向兼容。

下面我们通过一个简单的例程来学习如何使用 Protobuf 对 POJO 对象进行编解码，然后以这个例程为基础，学习如何在 Netty 中对 POJO 对象进行 Protobuf 编解码，并在两个进程之间进行通信和数据交换。

8.1.1 Protobuf 开发环境搭建

首先下载 Protobuf 的最新 Windows 版本，网址如下：

http://code.google.com/p/protobuf/downloads/detail?name=protoc-2.5.0-win32.zip&can=2&q=

对下载的 protoc-2.5.0-win32.zip 进行解压，解压后的目录如图 8-1 所示。

图 8-1　Protobuf 解压后的目录

protoc.exe 工具主要根据 .proto 文件生成代码，下面我们以商品订购例程为例，定义 SubscribeReq.proto 和 SubscribeResp.proto，数据文件定义如下。

◎ SubscribeReq.proto，如图 8-2 所示。

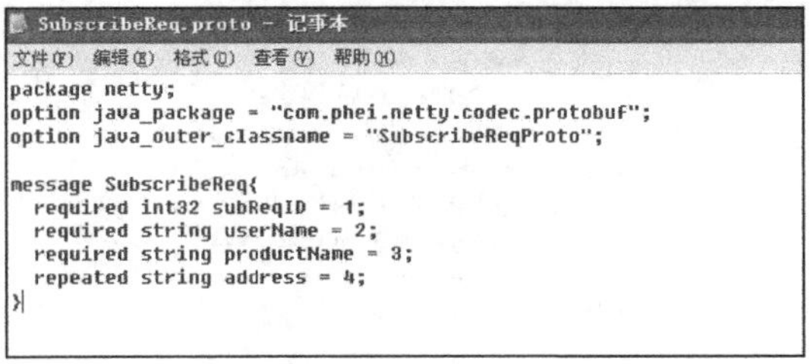

图 8-2　SubscribeReq.proto 文件定义

◎ SubscribeResp.proto，如图 8-3 所示。

```
package netty;
option java_package = "com.phei.netty.codec.protobuf";
option java_outer_classname = "SubscribeRespProto";
message SubscribeResp{
  required int32 subReqID = 1;
  required int32 respCode = 2;
  required string desc = 3;
}
```

图 8-3 SubscribeResp.proto 文件定义

通过 protoc.exe 命令行生成 Java 代码，如图 8-4 所示。

```
H:\java\nio\netty\protobuf\protoc-2.5.0-win32>
H:\java\nio\netty\protobuf\protoc-2.5.0-win32>protoc.exe --java_out=.\src .\nett
y\SubscribeReq.proto

H:\java\nio\netty\protobuf\protoc-2.5.0-win32>
```

图 8-4 通过 protoc.exe 工具生成源代码

将生成的 POJO 代码 SubscribeReqProto.java 和 SubscribeRespProto.java 复制到对应的 Eclipse 工程中，目录示例如图 8-5 所示。

图 8-5 将生成的 POJO 代码复制到源工程中

第 8 章　Google Protobuf 编解码

我们发现代码编译出错，原因是缺少 protobuf-java-2.5.0.jar 包，从 Google 官网下载后将其复制到 lib 目录后编译到引用类库中，如图 8-6 所示。

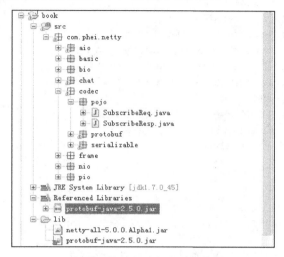

图 8-6　编译 Protobuf 工程

到此为止，Protobuf 开发环境已经搭建完毕，接下来将进行示例开发。

8.1.2　Protobuf 编解码开发

Protobuf 的类库使用比较简单，下面我们就通过对 SubscribeReqProto 进行编解码来介绍 Protobuf 的使用。

代码清单 8-1　Protobuf 入门　TestSubscribeReqProto

```
12. public class TestSubscribeReqProto {
13.
14.     private static byte[] encode(SubscribeReqProto.SubscribeReq req) {
15.      return req.toByteArray();
16.     }
17.
18.     private static SubscribeReqProto.SubscribeReq decode(byte[] body)
19.         throws InvalidProtocolBufferException {
20.      return SubscribeReqProto.SubscribeReq.parseFrom(body);
21.     }
22.
23.     private static SubscribeReqProto.SubscribeReq createSubscribeReq() {
```

```
24.     SubscribeReqProto.SubscribeReq.Builder builder = SubscribeReqProto.SubscribeReq
25.        .newBuilder();
26.     builder.setSubReqID(1);
27.     builder.setUserName("Lilinfeng");
28.     builder.setProductName("Netty Book");
29.     List<String> address = new ArrayList<>();
30.     address.add("NanJing YuHuaTai");
31.     address.add("BeiJing LiuLiChang");
32.     address.add("ShenZhen HongShuLin");
33.     builder.addAllAddress(address);
34.     return builder.build();
35.  }
36.
37.  /**
38.   * @param args
39.   * @throws InvalidProtocolBufferException
40.   */
41.  public static void main(String[] args)
42.        throws InvalidProtocolBufferException {
43.     SubscribeReqProto.SubscribeReq req = createSubscribeReq();
44.     System.out.println("Before encode : " + req.toString());
45.     SubscribeReqProto.SubscribeReq req2 = decode(encode(req));
46.     System.out.println("After decode : " + req.toString());
47.     System.out.println("Assert equal : --> " + req2.equals(req));
48.  }
49. }
```

我们首先看如何创建 SubscribeReqProto.SubscribeReq 实例，第 24 行通过 SubscribeReqProto.SubscribeReq 的静态方法 newBuilder 创建 SubscribeReqProto.SubscribeReq 的 Builder 实例，通过 Builder 构建器对 SubscribeReq 的属性进行设置，对于集合类型，通过 addAllXXX() 方法可以将集合对象设置到对应的属性中。

编码时通过调用 SubscribeReqProto.SubscribeReq 实例的 toByteArray 即可将 SubscribeReq 编码为 byte 数组，使用非常方便。

解码时通过 SubscribeReqProto.SubscribeReq 的静态方法 parseFrom 将二进制 byte 数组解码为原始的对象。

由于 Protobuf 支持复杂 POJO 对象编解码，所以代码都是通过工具自动生成，相比于传统的 POJO 对象的赋值操作，其使用略微复杂一些，但是习惯之后也不会带来额外的工作量，主要差异还是编程习惯的不同。

Protobuf 的编解码接口非常简单和实用，但是功能和性能却非常强大，这也是它流行的一个重要原因。

下个小节我们将执行 TestSubscribeReqProto，看它的功能是否正常。

8.1.3 运行 Protobuf 例程

我们运行上一小节编写的 TestSubscribeReqProto 程序，看经过编解码后的对象是否和编码之前的初始对象等价，代码执行结果如图 8-7 所示。

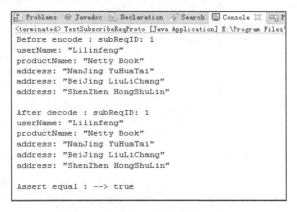

图 8-7　Protobuf 编解码运行结果

运行结果表明，经过 Protobuf 编解码后，生成的 SubscribeReqProto.SubscribeReq 与编码前原始的 SubscribeReqProto.SubscribeReq 等价。

至此，我们已经学会了如何搭建 Protobuf 的开发和运行环境，并初步掌握了 Protobuf 的编解码接口的使用方法，而且通过实际 demo 的开发和运行巩固了所学的知识。从下一节开始，我们将学习使用 Netty 的 Protobuf 编解码框架。

8.2　Netty 的 Protobuf 服务端开发

我们仍旧以商品订购例程作为 demo 进行学习，看看如何开发出一个 Protobuf 版本的图书订购程序。

8.2.1 Protobuf 版本的图书订购服务端开发

对 SubReqServer 进行升级，代码如下。

代码清单 8-2 Protobuf 版本图书订购代码 SubReqServer

```
20. public class SubReqServer {
21.     public void bind(int port) throws Exception {
22.         // 配置服务端的NIO线程组
23.         EventLoopGroup bossGroup = new NioEventLoopGroup();
24.         EventLoopGroup workerGroup = new NioEventLoopGroup();
25.         try {
26.             ServerBootstrap b = new ServerBootstrap();
27.             b.group(bossGroup, workerGroup)
28.                 .channel(NioServerSocketChannel.class)
29.                 .option(ChannelOption.SO_BACKLOG, 100)
30.                 .handler(new LoggingHandler(LogLevel.INFO))
31.                 .childHandler(new ChannelInitializer<SocketChannel>() {
32.                     @Override
33.                     public void initChannel(SocketChannel ch) {
34.                         ch.pipeline().addLast(
35.                             new ProtobufVarint32FrameDecoder());
36.                         ch.pipeline().addLast(
37.                             new ProtobufDecoder(
38.                                 SubscribeReqProto.SubscribeReq
39.                                     .getDefaultInstance()));
40. ch.pipeline().addLast(
41.                             new ProtobufVarint32LengthFieldPrepender());
42.                         ch.pipeline().addLast(new ProtobufEncoder());
43.                         ch.pipeline().addLast(new SubReqServerHandler());
44.                     }
45.                 });
46.
47.             // 绑定端口，同步等待成功
48.             ChannelFuture f = b.bind(port).sync();
49.
50.             // 等待服务端监听端口关闭
51.             f.channel().closeFuture().sync();
52.         } finally {
53.             // 优雅退出，释放线程池资源
54.             bossGroup.shutdownGracefully();
55.             workerGroup.shutdownGracefully();
```

```
56.        }
57.    }
58.
59.    public static void main(String[] args) throws Exception {
60.        int port = 8080;
61.        if (args != null && args.length > 0) {
62.            try {
63.                port = Integer.valueOf(args[0]);
64.            } catch (NumberFormatException e) {
65.                // 采用默认值
66.            }
67.        }
68.        new SubReqServer().bind(port);
69.    }
70. }
```

第 34 行首先向 ChannelPipeline 添加 ProtobufVarint32FrameDecoder，它主要用于半包处理，随后继续添加 ProtobufDecoder 解码器，它的参数是 com.google.protobuf.MessageLite，实际上就是要告诉 ProtobufDecoder 需要解码的目标类是什么，否则仅仅从字节数组中是无法判断出要解码的目标类型信息的。

下面我们继续看 SubReqServerHandler 的实现。

代码清单 8-3 Protobuf 版本图书订购代码 SubReqServerHandler

```
11. @Sharable
12. public class SubReqServerHandler extends ChannelHandlerAdapter {
13.
14.     @Override
15.     public void channelRead(ChannelHandlerContext ctx, Object msg)
16.             throws Exception {
17.         SubscribeReqProto.SubscribeReq req = (SubscribeReqProto.SubscribeReq) msg;
18.         if ("Lilinfeng".equalsIgnoreCase(req.getUserName())) {
19.             System.out.println("Service accept client subscribe req : ["
20.                 + req.toString() + "]");
21.             ctx.writeAndFlush(resp(req.getSubReqID()));
22.         }
23.     }
24.
25.     private SubscribeRespProto.SubscribeResp resp(int subReqID) {
26.         SubscribeRespProto.SubscribeResp.Builder builder =
```

```
SubscribeRespProto.SubscribeResp
27.             .newBuilder();
28.        builder.setSubReqID(subReqID);
29.        builder.setRespCode(0);
30.        builder.setDesc("Netty book order succeed, 3 days later, sent to the designated address");
31.        return builder.build();
32.    }
33.
34.    @Override
35.    public void exceptionCaught(ChannelHandlerContext ctx, Throwable cause) {
36.        cause.printStackTrace();
37.        ctx.close();// 发生异常，关闭链路
38.    }
39. }
```

由于 ProtobufDecoder 已经对消息进行了自动解码，因此接收到的订购请求消息可以直接使用。对用户名进行校验，校验通过后构造应答消息返回给客户端，由于使用了 ProtobufEncoder，所以不需要对 SubscribeRespProto.SubscribeResp 进行手工编码。

下个小节我们继续看客户端的代码实现。

8.2.2 Protobuf 版本的图书订购客户端开发

对订购请求消息使用 Protobuf 进行消息编解码。

代码清单 8-4 Protobuf 版本图书订购代码 SubReqClient

```
20. public class SubReqClient {
21.
22.     public void connect(int port, String host) throws Exception {
23.         // 配置客户端NIO线程组
24.         EventLoopGroup group = new NioEventLoopGroup();
25.         try {
26.             Bootstrap b = new Bootstrap();
27.             b.group(group).channel(NioSocketChannel.class)
28.                 .option(ChannelOption.TCP_NODELAY, true)
29.                 .handler(new ChannelInitializer<SocketChannel>() {
30.                     @Override
31.                     public void initChannel(SocketChannel ch)
```

```
32.              throws Exception {
33.                  ch.pipeline().addLast(
34.                      new ProtobufVarint32FrameDecoder());
35.                  ch.pipeline().addLast(
36.                      new ProtobufDecoder(
37.                          SubscribeRespProto.SubscribeResp
38.                              .getDefaultInstance()));
39.                  ch.pipeline().addLast(
40.                      new ProtobufVarint32LengthFieldPrepender());
41.                  ch.pipeline().addLast(new ProtobufEncoder());
42.                  ch.pipeline().addLast(new SubReqClientHandler());
43.              }
44.          });
45.
46.          // 发起异步连接操作
47.          ChannelFuture f = b.connect(host, port).sync();
48.
49.          // 等待客户端链路关闭
50.          f.channel().closeFuture().sync();
51.      } finally {
52.          // 优雅退出，释放NIO线程组
53.          group.shutdownGracefully();
54.      }
55.  }
56.
57.  /**
58.   * @param args
59.   * @throws Exception
60.   */
61.  public static void main(String[] args) throws Exception {
62.      int port = 8080;
63.      if (args != null && args.length > 0) {
64.          try {
65.              port = Integer.valueOf(args[0]);
66.          } catch (NumberFormatException e) {
67.              // 采用默认值
68.          }
69.      }
70.      new SubReqClient().connect(port, "127.0.0.1");
71.  }
72. }
```

需要指出的是客户端需要解码的对象是订购响应，所以第 37～38 行使用 SubscribeResp Proto.SubscribeResp 的实例做入参。

代码清单 8-5　Protobuf 版本图书订购代码　SubReqClientHandler

```
13.    public class SubReqClientHandler extends ChannelHandlerAdapter {
14.
15.        /**
16.         * Creates a client-side handler.
17.         */
18.        public SubReqClientHandler() {
19.        }
20.
21.        @Override
22.        public void channelActive(ChannelHandlerContext ctx) {
23.            for (int i = 0; i < 10; i++) {
24.                ctx.write(subReq(i));
25.            }
26.            ctx.flush();
27.        }
28.
29.        private SubscribeReqProto.SubscribeReq subReq(int i) {
30.            SubscribeReqProto.SubscribeReq.Builder builder = SubscribeReqProto.SubscribeReq
31.                    .newBuilder();
32.            builder.setSubReqID(i);
33.            builder.setUserName("Lilinfeng");
34.            builder.setProductName("Netty Book For Protobuf");
35.            List<String> address = new ArrayList<>();
36.            address.add("NanJing YuHuaTai");
37.            address.add("BeiJing LiuLiChang");
38.            address.add("ShenZhen HongShuLin");
39.            builder.addAllAddress(address);
40.            return builder.build();
41.        }
42.
43.        @Override
44.        public void channelRead(ChannelHandlerContext ctx, Object msg)
45.                throws Exception {
46.            System.out.println("Receive server response : [" + msg + "]");
47.        }
48.
49.        @Override
```

```
50.     public void channelReadComplete(ChannelHandlerContext ctx) throws Exception {
51.         ctx.flush();
52.     }
53.
54.     @Override
55.     public void exceptionCaught(ChannelHandlerContext ctx, Throwable cause) {
56.         cause.printStackTrace();
57.         ctx.close();
58.     }
59. }
```

客户端接收到服务端的应答消息之后会直接打印，按照设计，应该打印 10 次。下面我们就测试下 Protobuf 的服务端和客户端，看它是否能正常运行。

8.2.3　Protobuf 版本的图书订购程序功能测试

分别运行服务端和客户端，运行结果如下。

服务端运行结果如下。

```
Service accept client subscribe req : [subReqID: 0
userName: "Lilinfeng"
productName: "Netty Book For Protobuf"
address: "NanJing YuHuaTai"
address: "BeiJing LiuLiChang"
address: "ShenZhen HongShuLin"
]
Service accept client subscribe req : [subReqID: 1
userName: "Lilinfeng"
productName: "Netty Book For Protobuf"
address: "NanJing YuHuaTai"
address: "BeiJing LiuLiChang"
address: "ShenZhen HongShuLin"
]
Service accept client subscribe req : [subReqID: 2
userName: "Lilinfeng"
productName: "Netty Book For Protobuf"
address: "NanJing YuHuaTai"
address: "BeiJing LiuLiChang"
address: "ShenZhen HongShuLin"
```

```
]
//中间省略部分代码……
]
Service accept client subscribe req : [subReqID: 9
userName: "Lilinfeng"
productName: "Netty Book For Protobuf"
address: "NanJing YuHuaTai"
address: "BeiJing LiuLiChang"
address: "ShenZhen HongShuLin"
]
```

客户端运行结果如下。

```
Receive server response : [subReqID: 0
respCode: 0
desc: "Netty book order succeed, 3 days later, sent to the designated address"
]
Receive server response : [subReqID: 1
respCode: 0
desc: "Netty book order succeed, 3 days later, sent to the designated address"
]
Receive server response : [subReqID: 2
//中间省略部分代码……
respCode: 0
desc: "Netty book order succeed, 3 days later, sent to the designated address"
]
Receive server response : [subReqID: 9
respCode: 0
desc: "Netty book order succeed, 3 days later, sent to the designated address"
]
```

运行结果表明，我们基于 Netty Protobuf 编解码框架开发的图书订购程序可以正常工作。利用 Netty 提供的 Protobuf 编解码能力，我们在不需要了解 Protobuf 实现和使用细节的情况下就能轻松支持 Protobuf 编解码，可以方便地实现跨语言的远程服务调用和与周边的异构系统进行通信对接。

8.3 Protobuf 的使用注意事项

ProtobufDecoder 仅仅负责解码，它不支持读半包。因此，在 ProtobufDecoder 前面，一定要有能够处理读半包的解码器，有以下三种方式可以选择。

1）使用 Netty 提供的 ProtobufVarint32FrameDecoder，它可以处理半包消息；

2）继承 Netty 提供的通用半包解码器 LengthFieldBasedFrameDecoder；

3）继承 ByteToMessageDecoder 类，自己处理半包消息。

如果你只使用 ProtobufDecoder 解码器而忽略对半包消息的处理，程序是不能正常工作的。以前面的图书订购为例对服务端代码进行修改，注释掉 ProtobufVarint32Frame Decoder，代码修改如图 8-8 所示。

```
b.group(bossGroup, workerGroup)
    .channel(NioServerSocketChannel.class)
    .option(ChannelOption.SO_BACKLOG, 100)
    .handler(new LoggingHandler(LogLevel.INFO))
    .childHandler(new ChannelInitializer<SocketChannel>() {
    @Override
    public void initChannel(SocketChannel ch) {
        // ch.pipeline().addLast(
        // new ProtobufVarint32FrameDecoder());
        ch.pipeline().addLast(
            new ProtobufDecoder(
                SubscribeReqProto.SubscribeReq
                    .getDefaultInstance()));
        ch.pipeline().addLast(
            new ProtobufVarint32LengthFieldPrepender());
        ch.pipeline().addLast(new ProtobufEncoder());
        ch.pipeline().addLast(new SubReqServerHandler());
```

图 8-8　注释掉 ProtobufVarint32FrameDecoder

运行程序，结果如图 8-9 所示，运行出错。

```
io.netty.handler.codec.DecoderException: com.google.protobuf.InvalidProtocolBufferException: Protocol message tag had invalid wire type.
    at io.netty.handler.codec.MessageToMessageDecoder.channelRead(MessageToMessageDecoder.java:99)
    at io.netty.channel.ChannelHandlerInvokerUtil.invokeChannelReadNow(ChannelHandlerInvokerUtil.java:74)
    at io.netty.channel.DefaultChannelHandlerInvoker.invokeChannelRead(DefaultChannelHandlerInvoker.java:138)
    at io.netty.channel.DefaultChannelHandlerContext.fireChannelRead(DefaultChannelHandlerContext.java:320)
    at io.netty.channel.DefaultChannelPipeline.fireChannelRead(DefaultChannelPipeline.java:846)
    at io.netty.channel.nio.AbstractNioByteChannel$NioByteUnsafe.read(AbstractNioByteChannel.java:127)
    at io.netty.channel.nio.NioEventLoop.processSelectedKey(NioEventLoop.java:485)
    at io.netty.channel.nio.NioEventLoop.processSelectedKeysOptimized(NioEventLoop.java:452)
    at io.netty.channel.nio.NioEventLoop.run(NioEventLoop.java:346)
    at io.netty.util.concurrent.SingleThreadEventExecutor$5.run(SingleThreadEventExecutor.java:794)
    at java.lang.Thread.run(Thread.java:744)
Caused by: com.google.protobuf.InvalidProtocolBufferException: Protocol message tag had invalid wire type.
    at com.google.protobuf.InvalidProtocolBufferException.invalidWireType(InvalidProtocolBufferException.java:99)
    at com.google.protobuf.UnknownFieldSet$Builder.mergeFieldFrom(UnknownFieldSet.java:498)
    at com.google.protobuf.GeneratedMessage.parseUnknownField(GeneratedMessage.java:193)
    at com.phei.netty.codec.protobuf.SubscribeReqProto$SubscribeReq.<init>(SubscribeReqProto.java:119)
    at com.phei.netty.codec.protobuf.SubscribeReqProto$SubscribeReq.<init>(SubscribeReqProto.java:102)
    at com.phei.netty.codec.protobuf.SubscribeReqProto$SubscribeReq$1.parsePartialFrom(SubscribeReqProto.java:181)
    at com.phei.netty.codec.protobuf.SubscribeReqProto$SubscribeReq$1.parsePartialFrom(SubscribeReqProto.java:1)
    at com.google.protobuf.AbstractParser.parsePartialFrom(AbstractParser.java:141)
    at com.google.protobuf.AbstractParser.parseFrom(AbstractParser.java:176)
    at com.google.protobuf.AbstractParser.parseFrom(AbstractParser.java:182)
    at com.google.protobuf.AbstractParser.parseFrom(AbstractParser.java:49)
    at io.netty.handler.codec.protobuf.ProtobufDecoder.decode(ProtobufDecoder.java:114)
    at io.netty.handler.codec.protobuf.ProtobufDecoder.decode(ProtobufDecoder.java:1)
    at io.netty.handler.codec.MessageToMessageDecoder.channelRead(MessageToMessageDecoder.java:89)
    ... 10 more
```

图 8-9　注释掉 ProtobufVarint32FrameDecoder 运行出错

8.4 总结

本章首先介绍了 Protobuf 的入门知识，通过一个简单的样例代码开发让读者熟悉了如何使用 Protobuf 对 POJO 对象进行编解码；在掌握了 Protobuf 的基础知识后，讲解如何使用 Netty 的 Protobuf 编解码框架进行客户端和服务端的开发；最后，对 Protobuf 解码器的使用陷阱进行了说明，并给出了正确的使用建议。

在下一章中，我们继续学习另一种序列化技术——Jboss 的 Marshalling 序列化框架，它是 Jboss 内部使用的序列化框架，Netty 提供了 Marshalling 编码和解码器，方便用户在 Netty 中使用 Marshalling。第 9 章适用于对 Marshalling 框架感兴趣的读者，如果你想直接学习后面的知识，也可以跳过第 9 章。

第 9 章

JBoss Marshalling 编解码

JBoss Marshalling 是一个 Java 对象序列化包，对 JDK 默认的序列化框架进行了优化，但又保持跟 java.io.Serializable 接口的兼容，同时增加了一些可调的参数和附加的特性，这些参数和特性可通过工厂类进行配置。

本章主要内容包括：

◎ Marshalling 开发环境准备

◎ Netty 的 Marshalling 服务端开发

◎ Netty 的 Marshalling 客户端开发

◎ 运行 Marshalling 例程

9.1 Marshalling 开发环境准备

首先下载相关的 Marshalling 类库，由于我们只是用到了它的序列化类库，因此，只需要下载 jboss-marshalling-1.3.0 和 jboss-marshalling-serial-1.3.0 类库即可，下载网址如下：

https://www.jboss.org/jbossmarshalling/downloads

在页面上选择 JBoss Marshalling API 和 JBoss Marshalling Serial Protocol 进行下载，如

图 9-1 所示。

Name	Version	Size	Release date	License	Download
JBoss Marshalling API	1.3.0.CR9	226 kB	2011-04-27	LGPL	jboss-marshalling-1.3.0.CR9.jar Downloads: 321
JBoss Marshalling API Sources	1.3.0.CR9	171 kB	2011-04-27	LGPL	jboss-marshalling-1.3.0.CR9-sources.jar Downloads: 103
JBoss Marshalling River Protocol	1.3.0.CR9	79 kB	2011-04-27	LGPL	jboss-marshalling-river-1.3.0.CR9.jar Downloads: 84
JBoss Marshalling River Protocol Sources	1.3.0.CR9	45 kB	2011-04-27	LGPL	jboss-marshalling-river-1.3.0.CR9-sources.jar Downloads: 53
JBoss Marshalling Serial Protocol	1.3.0.CR9	68 kB	2011-04-27	LGPL	jboss-marshalling-serial-1.3.0.CR9.jar Downloads: 50

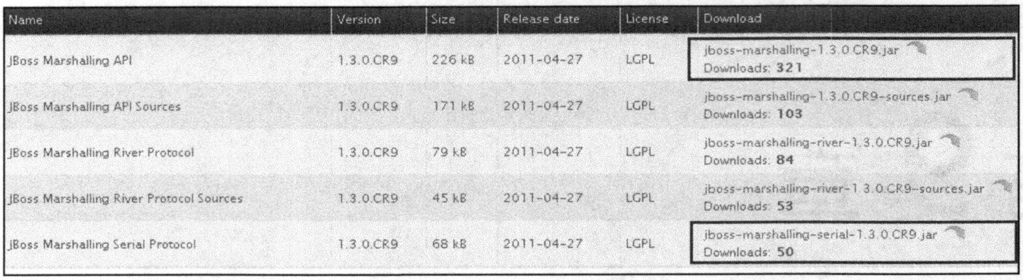

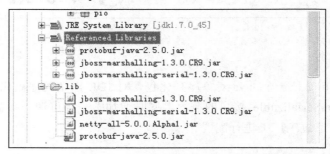

图 9-1　Marshalling 类库下载

将下载的类库 build 到 classpath 中，如图 9-2 所示。

图 9-2　将 Marshalling 添加到引用类库中

Marshalling 开发环境搭建完成之后，我们开始学习 Marshalling 服务端的开发。

9.2　Netty 的 Marshalling 服务端开发

首先定义 POJO 对象，由于 JBoss 的 Marshalling 完全兼容 JDK 序列化，因此我们继续使用商品订购例程中定义的 SubscribeReq 和 SubscribeResp 对象。通过 JBoss 提供的序列化 API，对 SubscribeReq 和 SubscribeResp 进行编解码。由于 Netty 对 JBoss 的编解码类库进行了封装，下面通过图书订购的实例看看如何使用 Netty 的 Marshalling 编解码类对消息进行序列化和反序列化。

服务端开发示例

首先看下服务端启动类的开发，代码如下。

代码清单 9-1 Marshalling 版本图书订购代码 SubReqServer

```
18. public class SubReqServer {
19.     public void bind(int port) throws Exception {
20.         // 配置服务端的NIO线程组
21.         EventLoopGroup bossGroup = new NioEventLoopGroup();
22.         EventLoopGroup workerGroup = new NioEventLoopGroup();
23.         try {
24.             ServerBootstrap b = new ServerBootstrap();
25.             b.group(bossGroup, workerGroup)
26.              .channel(NioServerSocketChannel.class)
27.              .option(ChannelOption.SO_BACKLOG, 100)
28.              .handler(new LoggingHandler(LogLevel.INFO))
29.              .childHandler(new ChannelInitializer<SocketChannel>() {
30.                  @Override
31.                  public void initChannel(SocketChannel ch) {
32.                      ch.pipeline().addLast(
33.                          MarshallingCodeCFactory
34.                              .buildMarshallingDecoder());
35.                      ch.pipeline().addLast(
36.                          MarshallingCodeCFactory
37.                              .buildMarshallingEncoder());
38.                      ch.pipeline().addLast(new SubReqServerHandler());
39.                  }
40.              });
41.
42.             // 绑定端口，同步等待成功
43.             ChannelFuture f = b.bind(port).sync();
44.
45.             // 等待服务端监听端口关闭
46.             f.channel().closeFuture().sync();
47.         } finally {
48.             // 优雅退出，释放线程池资源
49.             bossGroup.shutdownGracefully();
50.             workerGroup.shutdownGracefully();
51.         }
52.     }
53.
54.     public static void main(String[] args) throws Exception {
55.         int port = 8080;
56.         if (args != null && args.length > 0) {
57.             try {
```

```
58.            port = Integer.valueOf(args[0]);
59.        } catch (NumberFormatException e) {
60.            // 采用默认值
61.        }
62.     }
63.     new SubReqServer().bind(port);
64.    }
65. }
```

第 32～34 行通过 MarshallingCodeCFactory 工厂类创建了 MarshallingDecoder 解码器，并将其加入到 ChannelPipeline 中；第 35～37 行通过工厂类创建 MarshallingEncoder 编码器，并添加到 ChannelPipeline 中。

下面继续看 MarshallingCodeCFactory 是如何实现的，代码如下。

代码清单 9-2　Marshalling 版本图书订购代码 MarshallingCodeCFactory

```
18. public final class MarshallingCodeCFactory {
19.
20.    /**
21.     * 创建JBoss Marshalling 解码器MarshallingDecoder
22.     *
23.     * @return
24.     */
25.    public static MarshallingDecoder buildMarshallingDecoder() {
26.        final MarshallerFactory marshallerFactory = Marshalling
27.             .getProvidedMarshallerFactory("serial");
28.        final MarshallingConfiguration configuration = new MarshallingConfiguration();
29.        configuration.setVersion(5);
30.        UnmarshallerProvider provider = new DefaultUnmarshallerProvider(
31.            marshallerFactory, configuration);
32.        MarshallingDecoder decoder = new MarshallingDecoder(provider, 1024);
33.        return decoder;
34.    }
35.
36.    /**
37.     * 创建JBoss Marshalling 编码器MarshallingEncoder
38.     *
39.     * @return
40.     */
```

```
41.    public static MarshallingEncoder buildMarshallingEncoder() {
42.        final MarshallerFactory marshallerFactory = Marshalling
43.            .getProvidedMarshallerFactory("serial");
44.        final MarshallingConfiguration configuration = new MarshallingConfiguration();
45.        configuration.setVersion(5);
46.        MarshallerProvider provider = new DefaultMarshallerProvider(
47.            marshallerFactory, configuration);
48.        MarshallingEncoder encoder = new MarshallingEncoder(provider);
49.        return encoder;
50.    }
51. }
```

第 26～27 行首先通过 Marshalling 工具类的 getProvidedMarshallerFactory 静态方法获取 MarshallerFactory 实例，参数"serial"表示创建的是 Java 序列化工厂对象，它由 jboss-marshalling-serial-1.3.0.CR9.jar 提供。

第 28 行创建了 MarshallingConfiguration 对象，将它的版本号设置为 5，然后根据 MarshallerFactory 和 MarshallingConfiguration 创建 UnmarshallerProvider 实例，最后通过构造函数创建 Netty 的 MarshallingDecoder 对象，它有两个参数，分别是 UnmarshallerProvider 和单个消息序列化后的最大长度。

第 42～45 行同样是构造 MarshallerFactory 和 MarshallingConfiguration，第 46 行创建 MarshallerProvider 对象，它用于创建 Netty 提供的 MarshallingEncoder 实例，MarshallingEncoder 用于将实现序列化接口的 POJO 对象序列化为二进制数组。

由于 SubReqServerHandler 的实现与第 8 章例程中的 SubReqServerHandler 实现完全相同，因此这里不再给出源码。

9.3 Netty 的 Marshalling 客户端开发

首先由客户端发送订购请求消息，为了测试 TCP 粘包/拆包是否能被正确处理，采取连续发送 10 条请求消息的策略。在客户端的 ChannelPipeline 中添加 MarshallingEncoder 编码器对 POJO 对象进行编码。接收服务端应答消息的时候需要对经过 Marshalling 序列化后的码流进行解码，因此也需要添加 MarshallingDecoder，下面我们看下客户端代码的具体实现。

客户端开发示例

首先看客户端启动类。

代码清单 9-3　Marshalling 版本图书订购代码 SubReqClient

```
16. public class SubReqClient {
17.
18.     public void connect(int port, String host) throws Exception {
19.         // 配置客户端NIO线程组
20.         EventLoopGroup group = new NioEventLoopGroup();
21.         try {
22.             Bootstrap b = new Bootstrap();
23.             b.group(group).channel(NioSocketChannel.class)
24.                 .option(ChannelOption.TCP_NODELAY, true)
25.                 .handler(new ChannelInitializer<SocketChannel>() {
26.                     @Override
27.                     public void initChannel(SocketChannel ch)
28.                         throws Exception {
29.                         ch.pipeline().addLast(
30.                             MarshallingCodeCFactory
31.                                 .buildMarshallingDecoder());
32.                         ch.pipeline().addLast(
33.                             MarshallingCodeCFactory
34.                                 .buildMarshallingEncoder());
35.                         ch.pipeline().addLast(new SubReqClientHandler());
36.                     }
37.                 });
38.
39.             // 发起异步连接操作
40.             ChannelFuture f = b.connect(host, port).sync();
41.
42.             // 等待客户端链路关闭
43.             f.channel().closeFuture().sync();
44.         } finally {
45.             // 优雅退出，释放NIO线程组
46.             group.shutdownGracefully();
47.         }
48.     }
49.
50.     /**
51.      * @param args
```

```
52.      * @throws Exception
53.      */
54.     public static void main(String[] args) throws Exception {
55.         int port = 8080;
56.         if (args != null && args.length > 0) {
57.             try {
58.                 port = Integer.valueOf(args[0]);
59.             } catch (NumberFormatException e) {
60.                 // 采用默认值
61.             }
62.         }
63.         new SubReqClient().connect(port, "127.0.0.1");
64.     }
65. }
```

第 29～34 行分别创建 Marshalling 编码器和解码器，并将其添加到 ChannelPipeline 中。

SubReqClientHandler 相比于第 8 章的例程，仅仅修改了订购应答消息的产品名称，具体修改如图 9-3 所示。

```
private SubscribeReq subReq(int i) {
    SubscribeReq req = new SubscribeReq();
    req.setAddress("NanJing YuHuaTai");
    req.setPhoneNumber("138xxxxxxxx");
    req.setProductName("Netty Book For Marshalling");
    req.setSubReqID(i);
    req.setUserName("Lilinfeng");
    return req;
}
```

图 9-3　构造订购成功应答消息

9.4　运行 Marshalling 客户端和服务端例程

分别运行图书订购服务端和客户端例程，运行结果如下。

服务端运行结果如下。

```
Service accept client subscrib req : [SubscribeReq [subReqID=0, userName=
Lilinfeng, productName=Netty Book For Marshalling, phoneNumber=138xxxxxxxx,
address=NanJing YuHuaTai]]
Service accept client subscrib req : [SubscribeReq [subReqID=1, userName=
```

```
Lilinfeng, productName=Netty Book For Marshalling, phoneNumber=138xxxxxxxxx,
address=NanJing YuHuaTai]]
    Service accept client subscrib req : [SubscribeReq [subReqID=2, userName=
Lilinfeng, productName=Netty Book For Marshalling, phoneNumber=138xxxxxxxxx,
address=NanJing YuHuaTai]]
    //此处省略部分代码……
    Service accept client subscrib req : [SubscribeReq [subReqID=9, userName=
Lilinfeng, productName=Netty Book For Marshalling, phoneNumber=138xxxxxxxxx,
address=NanJing YuHuaTai]]
```

服务端共收到了 10 条客户端请求消息，subReqID 为从 0 到 9，与客户端发送的订购请求消息完全一致，服务端运行结果正确。

客户端运行结果如下。

```
    Receive server response : [SubscribeResp [subReqID=0, respCode=0, desc=Netty
book order succeed, 3 days later, sent to the designated address]]
    Receive server response : [SubscribeResp [subReqID=1, respCode=0, desc=Netty
book order succeed, 3 days later, sent to the designated address]]
    Receive server response : [SubscribeResp [subReqID=2, respCode=0, desc=Netty
book order succeed, 3 days later, sent to the designated address]]
    //此处省略部分代码……
    Receive server response : [SubscribeResp [subReqID=9, respCode=0, desc=Netty
book order succeed, 3 days later, sent to the designated address]]
```

客户端成功收到了服务端返回的 10 条应答消息，subReqID 为从 0 到 9，与服务端发送的应答消息完全一致，测试表明客户端运行结果正确。

由于我们模拟了 TCP 的粘包/拆包场景，但是程序的运行结果仍然正确，说明 Netty 的 Marshalling 编解码器支持半包和粘包的处理，对于开发者而言，只需要正确地将 Marshalling 编码器和解码器加入到 ChannelPipeline 中，就能实现对 Marshalling 序列化的支持。

利用 Netty 的 Marshalling 编解码器，可以轻松地开发出与 JBoss 内部模块进行远程通信的程序，而且支持异步非阻塞，这无疑降低了基于 Netty 开发的应用程序与 JBoss 内部模块对接的难度。

9.5　总结

本章介绍了如何使用 Netty 的 Marshalling 编码器和解码器对 POJO 对象进行序列化。

通过使用 Netty 的 Marshalling 编解码器，我们可以轻松地开发出支持 JBoss Marshalling 序列化的客户端和服务端程序，方便地对接 JBoss 的内部模块，同时也有利于对已有使用 Jboss Marshalling 框架做通信协议的模块的桥接和重用。

　　从下一章开始，我们将学习如何使用 Netty 的 HTTP 协议栈进行 HTTP 服务端和客户端的开发。由于 HTTP 协议目前仍然是各个行业主流的系统间通信协议，因此，Netty 的 HTTP 协议栈的应用空间非常广泛。

高级篇

Netty 多协议开发和应用

第 10 章　HTTP 协议开发应用

第 11 章　WebSocket 协议开发

第 12 章　私有协议栈开发

第 13 章　服务端创建

第 14 章　客户端创建

第 10 章

HTTP 协议开发应用

HTTP（超文本传输协议）协议是建立在 TCP 传输协议之上的应用层协议，它的发展是万维网协会和 Internet 工作小组 IETF 合作的结果。HTTP 是一个属于应用层的面向对象的协议，由于其简捷、快速的方式，适用于分布式超媒体信息系统。它于 1990 年提出，经过多年的使用和发展，得到了不断的完善和扩展。

由于 HTTP 协议是目前 Web 开发的主流协议，基于 HTTP 的应用非常广泛，因此，掌握 HTTP 的开发非常重要。本章将重点介绍如何基于 Netty 的 HTTP 协议栈进行 HTTP 服务端和客户端开发。由于 Netty 的 HTTP 协议栈是基于 Netty 的 NIO 通信框架开发的，因此，Netty 的 HTTP 协议也是异步非阻塞的。

本章主要内容包括：

◎ HTTP 协议介绍

◎ Netty HTTP 服务端入门开发

◎ HTTP + XML 应用开发

◎ HTTP 附件处理

10.1 HTTP 协议介绍

HTTP 是一个属于应用层的面向对象的协议，由于其简捷、快速的方式，适用于分布式超媒体信息系统。

HTTP 协议的主要特点如下。

- ◎ 支持 Client/Server 模式；
- ◎ 简单——客户向服务器请求服务时，只需指定服务 URL，携带必要的请求参数或者消息体；
- ◎ 灵活——HTTP 允许传输任意类型的数据对象，传输的内容类型由 HTTP 消息头中的 Content-Type 加以标记；
- ◎ 无状态——HTTP 协议是无状态协议，无状态是指协议对于事务处理没有记忆能力。缺少状态意味着如果后续处理需要之前的信息，则它必须重传，这样可能导致每次连接传送的数据量增大。另一方面，在服务器不需要先前信息时它的应答就较快，负载较轻。

10.1.1 HTTP 协议的 URL

HTTP URL（URL 是一种特殊类型的 URI，包含了用于查找某个资源的足够的信息）的格式如下。

```
http://host[":"port][abs_path]
```

其中，http 表示要通过 HTTP 协议来定位网络资源；host 表示合法的 Internet 主机域名或者 IP 地址；port 指定一个端口号，为空则使用默认端口 80；abs_path 指定请求资源的 URI，如果 URL 中没有给出 abs_path，那么当它作为请求 URI 时，必须以"/"的形式给出，通常这点工作浏览器会自动帮我们完成。

10.1.2 HTTP 请求消息（HttpRequest）

HTTP 请求由三部分组成，具体如下。

- ◎ HTTP 请求行；
- ◎ HTTP 消息头；
- ◎ HTTP 请求正文。

请求行以一个方法符开头，以空格分开，后面跟着请求的 URI 和协议的版本，格式为：Method　Request-URI　HTTP-Version CRLF。

其中 Method 表示请求方法，Request-URI 是一个统一资源标识符，HTTP-Version 表示请求的 HTTP 协议版本，CRLF 表示回车和换行（除了作为结尾的 CRLF 外，不允许出现单独的 CR 或 LF 字符）。

请求方法有多种，各方法的作用如下。

- ◎ GET：请求获取 Request-URI 所标识的资源；
- ◎ POST：在 Request-URI 所标识的资源后附加新的提交数据；
- ◎ HEAD：请求获取由 Request-URI 所标识的资源的响应消息报头；
- ◎ PUT：请求服务器存储一个资源，并用 Request-URI 作为其标识；
- ◎ DELETE：请求服务器删除 Request-URI 所标识的资源；
- ◎ TRACE：请求服务器回送收到的请求信息，主要用于测试或诊断；
- ◎ CONNECT：保留将来使用；
- ◎ OPTIONS：请求查询服务器的性能，或者查询与资源相关的选项和需求。

GET 方法：以在浏览器的地址栏中输入网址的方式访问网页时，浏览器采用 GET 方法向服务器获取资源。例如，我们直接在浏览器中输入 http://localhost:8080/netty-5.0.0，如图 10-1 所示。

图 10-1　通过浏览器访问 Netty HTTP 服务端

通过服务端抓包,打印 HTTP 请求消息头,内容如下。

```
GET /netty5.0 HTTP/1.1
Host: localhost:8080
Connection: keep-alive
User-Agent: Mozilla/5.0 (Windows NT 5.1) AppleWebKit/537.1 (KHTML, like Gecko) Chrome/21.0.1180.89 Safari/537.1
Accept: text/html,application/xhtml+xml,application/xml;q=0.9,*/*;q=0.8
Accept-Encoding: gzip,deflate,sdch
Accept-Language: zh-CN,zh;q=0.8
Accept-Charset: GBK,utf-8;q=0.7,*;q=0.3
Content-Length: 0
```

我们可以看到第一行请求行使用的是 GET 方法。

POST 方法要求被请求服务器接受附在请求后面的数据,常用于提交表单。GET 一般用于获取/查询资源信息,而 POST 一般用于更新资源信息。GET 和 POST 的主要区别如下。

(1) 根据 HTTP 规范,GET 用于信息获取,而且应该是安全的和幂等的;POST 则表示可能改变服务器上的资源的请求。

(2) GET 提交,请求的数据会附在 URL 之后,就是把数据放置在请求行(request line)中,以"?"分隔 URL 和传输数据,多个参数用"&"连接;而 POST 提交会把提交的数据放置在 HTTP 消息的包体中,数据不会在地址栏中显示出来。

(3) 传输数据的大小不同。特定浏览器和服务器对 URL 长度有限制,例如 IE 对 URL 长度的限制是 2083 字节(2KB+35B),因此 GET 携带的参数的长度会受到浏览器的限制;而 POST 由于不是通过 URL 传值,理论上数据长度不会受限。

(4) 安全性。POST 的安全性要比 GET 的安全性高。比如通过 GET 提交数据,用户名和密码将明文出现在 URL 上。因为 1) 登录页面有可能被浏览器缓存;2) 其他人查看浏览器的历史记录,那么别人就可以拿到你的账号和密码了。除此之外,使用 GET 提交数据还可能会造成 Cross-site request forgery 攻击。POST 提交的内容由于在消息体中传输,因此不存在上述安全问题。

请求报头允许客户端向服务器端传递请求的附加信息以及客户端自身的信息。常用的请求报头如表 10-1 所示。

HTTP 请求消息体是可选的,比较常用的 HTTP+XML 协议就是通过 HTTP 请求和响应消息体来承载 XML 信息的。

表 10-1　HTTP 的部分请求消息头列表

名称（KEY）	作　用
Accept	用于指定客户端接受哪些类型的信息。 例如：Accept:image/gif，表示客户端希望接受 GIF 图像格式的资源
Accept-Charset	用于指定客户端接受的字符集。 例如：Accept-Charset:iso-8859-1,gb2312，如果在请求消息中没有设置这个域，默认是任何字符集都可以接受
Accept-Encoding	类似于 Accept，但是它用于指定可接受的内容编码。 例如：Accept-Encoding:gzip.deflate，如果请求消息中没有设置这个域，则服务器假定客户端对各种内容编码都可以接受
Accept-Language	类似于 Accept，但是它用于指定一种自然语言。 例如：Accept-Language:zh-cn，如果请求消息中没有设置这个报头域，则服务器假定客户端对各种语言都可以接受
Authorization	主要用于证明客户端有权查看某个资源。当浏览器访问一个页面时，如果收到服务器的响应代码为 401（未授权），可以发送一个包含 Authorization 请求报头域的请求，要求服务器对其进行认证
Host	发送请求时，该报头域是必需的，用于指定被请求资源的 Internet 主机和端口号，它通常是从 HTTP URL 中提取出来的
User-Agent	允许客户端将它的操作系统、浏览器和其他属性告诉服务器
Content-Length	请求消息体的长度
Content-Type	表示后面的文档属于什么 MIME 类型。Servlet 默认为 text/plain，但通常需要显式地指定为 text/html。由于经常要设置 Content-Type，因此 HttpServletResponse 提供了一个专用的方法 setContentType
Connection	连接类型

10.1.3　HTTP 响应消息（HttpResponse）

处理完 HTTP 客户端的请求之后，HTTP 服务端返回响应消息给客户端，HTTP 响应也是由三个部分组成，分别是：状态行、消息报头、响应正文。

状态行的格式为：HTTP-Version Status-Code Reason-Phrase CRLF，其中 HTTP-Version 表示服务器 HTTP 协议的版本，Status-Code 表示服务器返回的响应状态代码。

状态代码由三位数字组成，第一个数字定义了响应的类别，它有 5 种可能的取值。

(1) 1xx：指示信息。表示请求已接收，继续处理；

(2) 2xx：成功。表示请求已被成功接收、理解、接受；

(3) 3xx：重定向。要完成请求必须进行更进一步的操作；

(4) 4xx：客户端错误。请求有语法错误或请求无法实现；

(5) 5xx：服务器端错误。服务器未能处理请求。

常见的状态代码、状态描述如表 10-2 所示。

表 10-2　HTTP 响应状态代码和描述信息

状态码	状态描述
200	OK：客户端请求成功
400	Bad Request：客户端请求有语法错误，不能被服务器所理解
401	Unauthorized：请求未经授权，这个状态代码必须和 WWW-Authenticate 报头域一起使用
403	Forbidden：服务器收到请求，但是拒绝提供服务
404	Not Found：请求资源不存在
500	Internal Server Error：服务器发生不可预期的错误
503	Server Unavailable：服务器当前不能处理客户端的请求，一段时间后可能恢复正常

响应报头允许服务器传递不能放在状态行中的附加响应信息，以及关于服务器的信息和对 Request-URI 所标识的资源进行下一步访问的信息。常用的响应报头如表 10-3 所示。

表 10-3　常用的响应报头

名称（KEY）	作　　用
Location	用于重定向接收者到一个新的位置，Location 响应报头域常用于更换域名的时候
Server	包含了服务器用来处理请求的软件信息，与 User-Agent 请求报头域是相对应的
WWW-Authenticate	必须被包含在 401（未授权的）响应消息中，客户端收到 401 响应消息，并发送 Authorization 报头域请求服务器对其进行验证时，服务端响应报头就包含该报头域

10.2　Netty HTTP 服务端入门开发

从本节开始我们学习如何使用 Netty 的 HTTP 协议栈开发 HTTP 服务端和客户端应用

程序。由于 Netty 天生是异步事件驱动的架构，因此基于 NIO TCP 协议栈开发的 HTTP 协议栈也是异步非阻塞的。

Netty 的 HTTP 协议栈无论在性能还是可靠性上，都表现优异，非常适合在非 Web 容器的场景下应用，相比于传统的 Tomcat、Jetty 等 Web 容器，它更加轻量和小巧，灵活性和定制性也更好。

10.2.1 HTTP 服务端例程场景描述

我们以文件服务器为例学习 Netty 的 HTTP 服务端入门开发，例程场景如下：文件服务器使用 HTTP 协议对外提供服务，当客户端通过浏览器访问文件服务器时，对访问路径进行检查，检查失败时返回 HTTP 403 错误，该页无法访问；如果校验通过，以链接的方式打开当前文件目录，每个目录或者文件都是个超链接，可以递归访问。

如果是目录，可以继续递归访问它下面的子目录或者文件，如果是文件且可读，则可以在浏览器端直接打开，或者通过【目标另存为】下载该文件。

介绍完了样例程序的开发场景，下面我们一起看看如何开发一个基于 Netty 的 HTTP 程序。

10.2.2 HTTP 服务端开发

首先看下 HTTP 文件服务器的启动类是如何实现的。

代码清单 10-1　HTTP 文件服务器 启动类 HttpFileServer

```
19. public class HttpFileServer {
20.
21.     private static final String DEFAULT_URL = "/src/com/phei/netty/";
22.
23.     public void run(final int port, final String url) throws Exception {
24.         EventLoopGroup bossGroup = new NioEventLoopGroup();
25.         EventLoopGroup workerGroup = new NioEventLoopGroup();
26.         try {
27.             ServerBootstrap b = new ServerBootstrap();
28.             b.group(bossGroup, workerGroup)
29.               .channel(NioServerSocketChannel.class)
30.               .childHandler(new ChannelInitializer<SocketChannel>() {
```

```
31.            @Override
32.            protected void initChannel(SocketChannel ch)
33.                throws Exception {
34.                ch.pipeline().addLast("http-decoder",
35.                    new HttpRequestDecoder());
36.                ch.pipeline().addLast("http-aggregator",
37.                    new HttpObjectAggregator(65536));
38.                ch.pipeline().addLast("http-encoder",
39.                    new HttpResponseEncoder());
40.                ch.pipeline().addLast("http-chunked",
41.                    new ChunkedWriteHandler());
42.                ch.pipeline().addLast("fileServerHandler",
43.                    new HttpFileServerHandler(url));
44.            }
45.        });
46.        ChannelFuture future = b.bind("192.168.1.102", port).sync();
47.        System.out.println("HTTP 文件目录服务器启动，网址是 : " + "http://192.168.1.102:"
48.            + port + url);
49.        future.channel().closeFuture().sync();
50.    } finally {
51.        bossGroup.shutdownGracefully();
52.        workerGroup.shutdownGracefully();
53.    }
54. }
55.
56. public static void main(String[] args) throws Exception {
57.    int port = 8080;
58.    if (args.length > 0) {
59.        try {
60.            port = Integer.parseInt(args[0]);
61.        } catch (NumberFormatException e) {
62.            e.printStackTrace();
63.        }
64.    }
65.    String url = DEFAULT_URL;
66.    if (args.length > 1)
67.        url = args[1];
68.    new HttpFileServer().run(port, url);
69.  }
70. }
```

首先我们看 main 函数，它有两个参数：第一个是端口，第二个是 HTTP 服务端的 URL 路径。如果启动的时候没有配置，则使用默认值，默认端口是 8080，默认的 URL 路径是"/src/com/phei/netty/"。

重点关注第 34~43 行，首先向 ChannelPipeline 中添加 HTTP 请求消息解码器，随后，又添加了 HttpObjectAggregator 解码器，它的作用是将多个消息转换为单一的 FullHttpRequest 或者 FullHttpResponse，原因是 HTTP 解码器在每个 HTTP 消息中会生成多个消息对象。

（1）HttpRequest / HttpResponse；

（2）HttpContent；

（3）LastHttpContent。

第 38~39 行新增 HTTP 响应编码器，对 HTTP 响应消息进行编码；第 40~41 行新增 Chunked handler，它的主要作用是支持异步发送大的码流（例如大的文件传输），但不占用过多的内存，防止发生 Java 内存溢出错误。

最后添加 HttpFileServerHandler，用于文件服务器的业务逻辑处理。下面我们具体看看它是如何实现的。

代码清单 10-2　HTTP 文件服务器 处理类 HttpFileServerHandler

```
47. public class HttpFileServerHandler extends
48.     SimpleChannelInboundHandler<FullHttpRequest> {
55.     @Override
56.     public void messageReceived(ChannelHandlerContext ctx,
57.         FullHttpRequest request) throws Exception {
58.     if (!request.getDecoderResult().isSuccess()) {
59.         sendError(ctx, BAD_REQUEST);
60.         return;
61.     }
62.     if (request.getMethod() != GET) {
63.         sendError(ctx, METHOD_NOT_ALLOWED);
64.         return;
65.     }
66.     final String uri = request.getUri();
67.     final String path = sanitizeUri(uri);
68.     if (path == null) {
69.         sendError(ctx, FORBIDDEN);
```

```
70.            return;
71.        }
72.        File file = new File(path);
73.        if (file.isHidden() || !file.exists()) {
74.            sendError(ctx, NOT_FOUND);
75.            return;
76.        }
77.        此处代码省略......
85.        if (!file.isFile()) {
86.            sendError(ctx, FORBIDDEN);
87.            return;
88.        }
89.        RandomAccessFile randomAccessFile = null;
90.        try {
91.            randomAccessFile = new RandomAccessFile(file, "r");// 以只读的
方式打开文件
92.        } catch (FileNotFoundException fnfe) {
93.            sendError(ctx, NOT_FOUND);
94.            return;
95.        }
96.        long fileLength = randomAccessFile.length();
97.        HttpResponse response = new DefaultHttpResponse(HTTP_1_1, OK);
98.        setContentLength(response, fileLength);
99.        setContentTypeHeader(response, file);
100.        if (isKeepAlive(request)) {
101.            response.headers().set(CONNECTION,    HttpHeaders.Values.
KEEP_ALIVE);
102.        }
103.        ctx.write(response);
104.        ChannelFuture sendFileFuture;
105.        sendFileFuture=ctx.write(new ChunkedFile(randomAccessFile, 0,
106.            fileLength, 8192), ctx.newProgressivePromise());
107.        sendFileFuture.addListener(new    ChannelProgressiveFutureListener() {
142.        此处代码省略......
143.        private String sanitizeUri(String uri) {
144.        try {
145.            uri = URLDecoder.decode(uri, "UTF-8");
146.        } catch (UnsupportedEncodingException e) {
147.            try {
148.                uri = URLDecoder.decode(uri, "ISO-8859-1");
```

```
149.            } catch (UnsupportedEncodingException e1) {
150.                throw new Error();
151.            }
152.        }
153.        此处代码省略......
159.        uri = uri.replace('/', File.separatorChar);
160.        if (uri.contains(File.separator + '.')
161.            || uri.contains('.'+File.separator)||uri.startsWith(".")
162.            || uri.endsWith(".")||INSECURE_URI.matcher(uri).matches()){
163.            return null;
164.        }
165.        return System.getProperty("user.dir") + File.separator + uri;
166.    }
167.
168.    private static final Pattern ALLOWED_FILE_NAME = Pattern
169.            .compile("[A-Za-z0-9][-_A-Za-z0-9\\.]*");
170.
171.    private static void sendListing(ChannelHandlerContext ctx, File dir) {
172.        FullHttpResponse response = new DefaultFullHttpResponse(HTTP_1_1, OK);
173.        response.headers().set(CONTENT_TYPE,"text/html;charset=UTF-8");
174.        StringBuilder buf = new StringBuilder();
            此处代码省略......
185.        buf.append("<li>链接: <a href=\"../\">..</a></li>\r\n");
186.        for (File f : dir.listFiles()) {
            此处代码省略......
194.            buf.append("<li>链接: <a href=\"");
            此处代码省略......
199.        }
200.        buf.append("</ul></body></html>\r\n");
201.        ByteBuf buffer = Unpooled.copiedBuffer(buf, CharsetUtil.UTF_8);
202.        response.content().writeBytes(buffer);
203.        buffer.release();
204.        ctx.writeAndFlush(response).addListener(ChannelFutureListener.CLOSE);
205.    }
206.    此处代码省略......
227. }
```

首先从消息接入方法看起，第58~61行首先对HTTP请求消息的解码结果进行判断，如果解码失败，直接构造HTTP 400错误返回。第62~65行对请求行中的方法进行判断，如果不是从浏览器或者表单设置为 GET 发起的请求（例如 POST），则构造 HTTP 405 错误返回。

第 67 行对请求 URL 进行包装，然后对 sanitizeUri 方法展开分析。跳到第 145 行，首先使用 JDK 的 java.net.URLDecoder 对 URL 进行解码，使用 UTF-8 字符集，解码成功之后对 URI 进行合法性判断，如果 URI 与允许访问的 URI 一致或者是其子目录（文件），则校验通过，否则返回空。第 159 行将硬编码的文件路径分隔符替换为本地操作系统的文件路径分隔符。第 161～164 行对新的 URI 做二次合法性校验，如果校验失败则直接返回空。最后对文件进行拼接，使用当前运行程序所在的工程目录 + URI 构造绝对路径返回。

第 68～71 行，如果构造的 URI 不合法，则返回 HTTP 403 错误。第 72 行使用新组装的 URI 路径构造 File 对象。第 73～76 行，如果文件不存在或者是系统隐藏文件，则构造 HTTP 404 异常返回。如果文件是目录，则发送目录的链接给客户端浏览器。下面我们重点分析返回文件链接响应给客户端的代码。

第 172 行首先创建成功的 HTTP 响应消息，随后设置消息头的类型为 "text/html; charset=UTF-8"。第 174 行用于构造响应消息体，由于需要将响应结果显示在浏览器上，所以采用了 HTML 的格式。由于大家对 HTML 的语法已经非常熟悉，这里不再详细介绍。我们挑重点的代码进行分析，第 185 行打印了一个 ".." 的链接。第 186～199 行用于展示根目录下的所有文件和文件夹，同时使用超链接来标识。第 201 行分配对应消息的缓冲对象，第 202 行将缓冲区中的响应消息存放到 HTTP 应答消息中，然后释放缓冲区，最后调用 writeAndFlush 将响应消息发送到缓冲区并刷新到 SocketChannel 中。

如果用户在浏览器上点击超链接直接打开或者下载文件，代码会执行第 85 行，对超链接的文件进行合法性判断，如果不是合法文件，则返回 HTTP 403 错误。校验通过后，第 85～95 行使用随机文件读写类以只读的方式打开文件，如果文件打开失败，则返回 HTTP 404 错误。

第 96 行获取文件的长度，构造成功的 HTTP 应答消息，然后在消息头中设置 content length 和 content type，判断是否是 Keep-Alive，如果是，则在应答消息头中设置 Connection 为 Keep-Alive。第 103 行发送响应消息。第 105～106 行通过 Netty 的 ChunkedFile 对象直接将文件写入到发送缓冲区中。最后为 sendFileFuture 增加 GenericFutureListener，如果发送完成，打印 "Transfer complete."。

如果使用 chunked 编码，最后需要发送一个编码结束的空消息体，将 LastHttpContent 的 EMPTY_LAST_CONTENT 发送到缓冲区中，标识所有的消息体已经发送完成，同时调用 flush 方法将之前在发送缓冲区的消息刷新到 SocketChannel 中发送给对方。

如果是非 Keep-Alive 的,最后一包消息发送完成之后,服务端要主动关闭连接。

服务端的代码已经介绍完毕,下面让我们看看运行结果。

10.2.3 Netty HTTP 文件服务器例程运行结果

启动 HTTP 文件服务器,通过浏览器进行访问,运行结果如下。

首先,启动文件服务器,运行结果如图 10-1 所示。

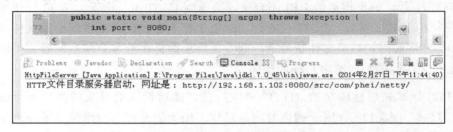

图 10-1 HTTP 文件服务器启动结果

我们首先进行异常场景的测试,输入错误的 URL 网址:

http://192.168.1.102:8080/abcde/get?123

运行结果如图 10-2 所示。

图 10-2 输入错误的网址,返回 403 错误

结果分析：由于输入的 URL 路径不是个合法的文件或者目录，所以程序会执行第 69 行，返回 HTTP 403 错误。

我们继续测试正常场景，在浏览器中输入正确的网址：

http://192.168.1.102:8080/src/com/phei/netty/

浏览器显示结果如图 10-3 所示。

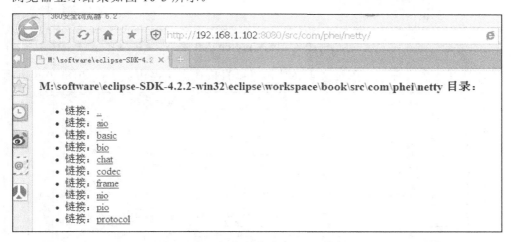

图 10-3　Netty 文件服务器目录展示

单击 codec 文件链接，显示结果如图 10-4 所示。

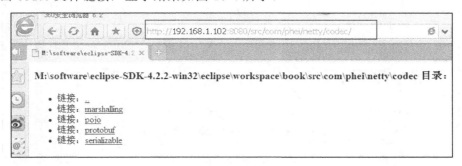

图 10-4　单击目录链接，进入下一级目录

查看浏览器的网址，已经进入下一级目录，继续单击 protobuf 目录，显示如图 10-5 所示。

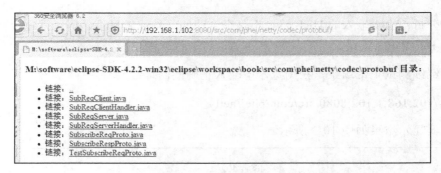

图 10-5 进入 protobuf 目录

随便打开一个文件,由于是文本文件,可以直接在浏览器中显示,如图 10-6 所示。

```
/*
 * Copyright 2013-2018 Lilinfeng.
 * 
 * Licensed under the Apache License, Version 2.0 (the "License");
 * you may not use this file except in compliance with the License.
 * You may obtain a copy of the License at
 * 
 *      http://www.apache.org/licenses/LICENSE-2.0
 * 
 * Unless required by applicable law or agreed to in writing, software
 * distributed under the License is distributed on an "AS IS" BASIS,
 * WITHOUT WARRANTIES OR CONDITIONS OF ANY KIND, either express or implied.
 * See the License for the specific language governing permissions and
 * limitations under the License.
 */
package com.phei.netty.codec.protobuf;

import java.util.ArrayList;
import java.util.List;

import com.google.protobuf.InvalidProtocolBufferException;

/**
 * @author Administrator
 * @date 2014年2月23日
 * @version 1.0
 */
public class TestSubscribeReqProto {

    private static byte[] encode(SubscribeReqProto.SubscribeReq req) {
        return req.toByteArray();
    }
```

图 10-6 打开 TestSubscribeReqProto.java 文件

如果内容显示有乱码,说明浏览器使用的编码方式和源代码中的不一致,直接在打开的网页中单击右键,在弹出菜单中选择编码方式为 UTF-8 即可,如图 10-7 所示。

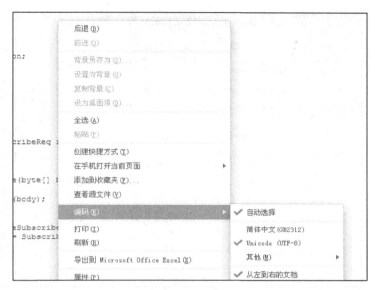

图 10-7　设置浏览器的编码方式为 UTF-8 解决中文乱码问题

我们也可以通过右键【目标另存为】从文件服务器下载文件，如图 10-8 所示。

图 10-8　从文件服务器下载源文件

下载完成后，打开下载的文件，看内容是否正确，如图 10-9 所示。

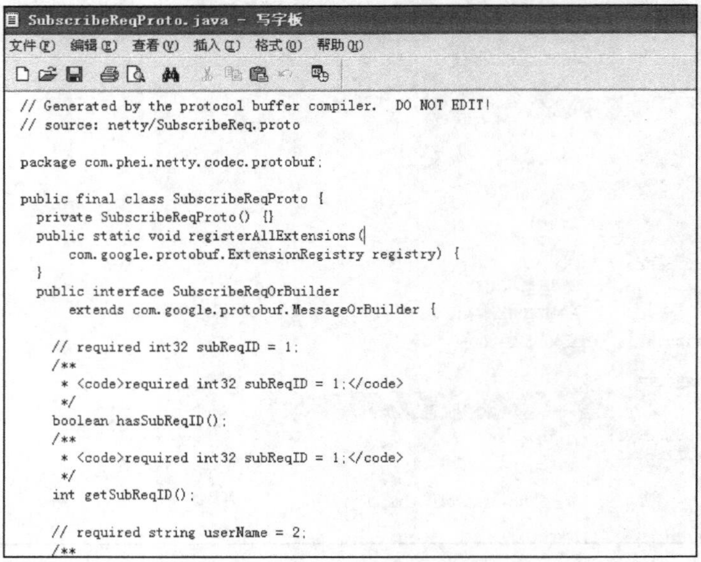


```java
// Generated by the protocol buffer compiler.  DO NOT EDIT!
// source: netty/SubscribeReq.proto

package com.phei.netty.codec.protobuf;

public final class SubscribeReqProto {
  private SubscribeReqProto() {}
  public static void registerAllExtensions(
      com.google.protobuf.ExtensionRegistry registry) {
  }
  public interface SubscribeReqOrBuilder
      extends com.google.protobuf.MessageOrBuilder {

    // required int32 subReqID = 1;
    /**
     * <code>required int32 subReqID = 1;</code>
     */
    boolean hasSubReqID();
    /**
     * <code>required int32 subReqID = 1;</code>
     */
    int getSubReqID();

    // required string userName = 2;
    /**
```

图 10-9　查看从文件服务器下载的源文件

对比服务器上的源文件，内容完全一致，说明 HTTP 文件服务器文件下载功能正常。

至此，作为入门级的 Netty HTTP 协议栈的应用——HTTP 文件服务器已经开发完毕，相信通过本节的例程学习，大家已经初步掌握了基于 Netty 的 HTTP 服务端应用开发。

下一节，我们将学习目前最流行的 HTTP+XML 开发。HTTP+XML 应用非常广泛，一旦我们掌握了如何在 Netty 中实现通用的 HTTP+XML 协议栈，后续相关的应用层开发和维护将会变得非常简单。

10.3　Netty HTTP+XML 协议栈开发

由于 HTTP 协议的通用性，很多异构系统间的通信交互采用 HTTP 协议，通过 HTTP 协议承载业务数据进行消息交互，例如非常流行的 HTTP+XML 或者 RESTful+JSON。

在 Java 领域，最常用的 HTTP 协议栈就是基于 Servlet 规范的 Tomcat 等 Web 容器，由于谷歌等业界大佬的强力推荐，Jetty 等轻量级的 Web 容器也得到了广泛的应用。但是，很多基于 HTTP 的应用都是后台应用，HTTP 仅仅是承载数据交换的一个通道，是一个载体而不是 Web 容器，在这种场景下，一般不需要类似 Tomcat 这样的重量型 Web 容器。

在网络安全日益严峻的今天，重量级的 Web 容器由于功能繁杂，会存在很多安全漏洞，典型的如 Tomcat。如果你的客户是安全敏感型的，这意味着你需要为 Web 容器做很多安全加固工作去修补这些漏洞，然而你并没有使用到这些功能，这会带来开发和维护成本的增加。在这种场景下，一个更加轻量级的 HTTP 协议栈是个更好的选择。

本章节将介绍如何利用 Netty 提供的基础 HTTP 协议栈功能，扩展开发 HTTP+XML 协议栈。

10.3.1 开发场景介绍

作为一个示例程序，我们先模拟一个简单的用户订购系统。客户端填写订单，通过 HTTP 客户端向服务端发送订购请求，请求消息放在 HTTP 消息体中，以 XML 承载，即采用 HTTP+XML 的方式进行通信。HTTP 服务端接收到订购请求后，对订单请求进行修改，然后通过 HTTP+XML 的方式返回应答消息。双方采用 HTTP1.1 协议，连接类型为 CLOSE 方式，即双方交互完成，由 HTTP 服务端主动关闭链路，随后客户端也关闭链路并退出。

订购请求消息定义如表 10-3 所示。

表 10-3 订购请求消息定义（Order）

字段名称	类　　型	备　　注
订购数量	Int64	订购的商品数量
客户信息	Customer	客户信息，负责 POJO 对象
账单地址	Address	账单的地址
寄送方式	Shipping	枚举类型如下： 普通邮递 宅急送 国际邮递 国内快递 国际快递
送货地址	Address	送货地址
总价	float	商品总价

客户信息定义如表 10-4 所示。

表10-4 客户信息定义（Customer）

字段名称	类型	备注
客户ID	Int64	客户ID，长整型
姓	String	客户姓氏，字符串
名	String	客户名字，字符串
全名	List<String>	客户全称，字符列表

地址信息定义如表10-5所示。

表10-5 地址信息定义（Address）

字段名称	类型	备注
街道1	String	
街道2	String	
城市	String	
省份	String	
邮政编码	String	
国家	String	

邮递方式定义如表10-6所示。

表10-6 邮递方式定义（Shipping）

字段名称	类型	备注
普通邮递	枚举类型	
宅急送	枚举类型	
国际邮递	枚举类型	
国内快递	枚举类型	
国际快递	枚举类型	

数据定义完成之后，接着看订购请求消息的XML Schema定义。

```
<xs:schema xmlns:xs="http://www.w3.org/2001/XMLSchema" xmlns:tns="http://
phei.com/netty/protocol/http/xml/pojo"        elementFormDefault="qualified"
targetNamespace="http://phei.com/netty/protocol/http/xml/pojo">
    <xs:element type="tns:order" name="order"/>
    <xs:complexType name="address">
      <xs:sequence>
        <xs:element type="xs:string" name="street1" minOccurs="0"/>
```

```xml
        <xs:element type="xs:string" name="street2" minOccurs="0"/>
        <xs:element type="xs:string" name="city" minOccurs="0"/>
        <xs:element type="xs:string" name="state" minOccurs="0"/>
        <xs:element type="xs:string" name="postCode" minOccurs="0"/>
        <xs:element type="xs:string" name="country" minOccurs="0"/>
      </xs:sequence>
    </xs:complexType>
    <xs:complexType name="order">
      <xs:sequence>
        <xs:element name="customer" minOccurs="0">
          <xs:complexType>
            <xs:sequence>
              <xs:element type="xs:string" name="firstName" minOccurs="0"/>
              <xs:element type="xs:string" name="lastName" minOccurs="0"/>
              <xs:element type="xs:string" name="middleName" minOccurs="0" maxOccurs="unbounded"/>
            </xs:sequence>
            <xs:attribute type="xs:long" use="required" name="customerNumber"/>
          </xs:complexType>
        </xs:element>
        <xs:element type="tns:address" name="billTo" minOccurs="0"/>
        <xs:element name="shipping" minOccurs="0">
          <xs:simpleType>
            <xs:restriction base="xs:string">
              <xs:enumeration value="STANDARD_MAIL"/>
              <xs:enumeration value="PRIORITY_MAIL"/>
              <xs:enumeration value="INTERNATIONAL_MAIL"/>
              <xs:enumeration value="DOMESTIC_EXPRESS"/>
              <xs:enumeration value="INTERNATIONAL_EXPRESS"/>
            </xs:restriction>
          </xs:simpleType>
        </xs:element>
        <xs:element type="tns:address" name="shipTo" minOccurs="0"/>
      </xs:sequence>
      <xs:attribute type="xs:long" use="required" name="orderNumber"/>
      <xs:attribute type="xs:float" name="total"/>
    </xs:complexType></xs:schema>
```

熟悉 XML 和 Schema 的读者理解上面的 Schema 定义没有什么困难，如果你对 XML 的相关知识还不太了解，建议先找一本 XML 开发入门方面的书籍进行学习，或者登录 W3C 的网站学习 XML 的相关知识。

开发背景介绍完毕，下面我们进入设计环节，看看如何设计和开发 HTTP+XML 协议栈。

10.3.2 HTTP+XML 协议栈设计

通过商品订购的流程图看下订购的关键步骤和主要技术点，找出当前 Netty HTTP 协议栈的功能不足之后，通过扩展的方式完成 HTTP+XML 协议栈的开发。如图 10-10 所示。

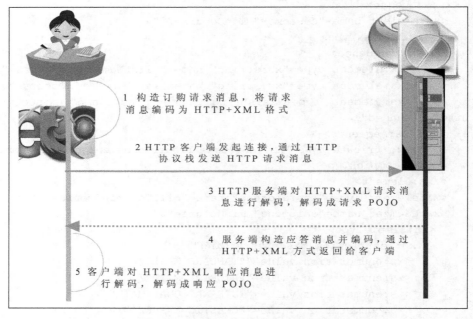

图 10-10　HTTP+XML 订购流程图

首先对订购流程图进行分析，先看步骤 1，构造订购请求消息并将其编码为 HTTP+XML 形式。Netty 的 HTTP 协议栈提供了构造 HTTP 请求消息的相关接口，但是无法将普通的 POJO 对象转换为 HTTP+XML 的 HTTP 请求消息，需要自定义 HTTP+XML 格式的请求消息编码器。

再看步骤 2，利用 Netty 的 HTTP 协议栈，可以支持 HTTP 链路的建立和请求消息的发送，所以不需要额外开发，直接重用 Netty 的能力即可。

步骤 3，HTTP 服务端需要将 HTTP+XML 格式的订购请求消息解码为订购请求 POJO 对象，同时获取 HTTP 请求消息头信息。利用 Netty 的 HTTP 协议栈服务端，可以完成 HTTP

请求消息的解码，但是，如果消息体为 XML 格式，Netty 无法支持将其解码为 POJO 对象，需要在 Netty 协议栈的基础上扩展实现。

步骤 4，服务端对订购请求消息处理完成后，重新将其封装成 XML，通过 HTTP 应答消息体携带给客户端，Netty 的 HTTP 协议栈不支持直接将 POJO 对象的应答消息以 XML 方式发送，需要定制。

步骤 5，HTTP 客户端需要将 HTTP+XML 格式的应答消息解码为订购 POJO 对象，同时能够获取应答消息的 HTTP 头信息，Netty 的协议栈不支持自动的消息解码。

通过分析，我们可以了解到哪些能力是 Netty 支持的，哪些需要扩展开发实现。下面给出设计思路。

（1）需要一套通用、高性能的 XML 序列化框架，它能够灵活地实现 POJO-XML 的互相转换，最好能够通过工具自动生成绑定关系，或者通过 XML 的方式配置双方的映射关系；

（2）作为通用的 HTTP+XML 协议栈，XML-POJO 对象的映射关系应该非常灵活，支持命名空间和自定义标签；

（3）提供 HTTP+XML 请求消息编码器，供 HTTP 客户端发送请求消息自动编码使用；

（4）提供 HTTP+XML 请求消息解码器，供 HTTP 服务端对请求消息自动解码使用；

（5）提供 HTTP+XML 响应消息编码器，供 HTTP 服务端发送响应消息自动编码使用；

（6）提供 HTTP+XML 响应消息解码器，供 HTTP 客户端对应答消息进行自动解码使用；

（7）协议栈使用者不需要关心 HTTP+XML 的编解码，对上层业务零侵入，业务只需要对上层的业务 POJO 对象进行编排。

下个小节我们将讲述 XML 框架的选型和开发，它是 HTTP+XML 协议栈的关键技术点。

10.3.3　高效的 XML 绑定框架 JiBx

JiBX 是一款非常优秀的 XML（Extensible Markup Language）数据绑定框架。它提供灵活的绑定映射文件，实现数据对象与 XML 文件之间的转换，并不需要修改既有的 Java 类。另外，它的转换效率是目前很多其他开源项目都无法比拟的。

1. JiBx 入门

XML 已经成为目前程序开发配置的重要组成部分了，可以用来操作 XML 文件的开源项目也已经成熟，比如说流行的 Digester、XStream、Castor、JDOM、dom4j 和 Xalan 等，当然也少不了专门为 Java 语言设计的 XML 数据绑定框架 JiBX。它的主要优点包括：转换效率高、配置绑定文件简单、不需要操作 xpath 文件、不需要写属性的 get/set 方法、对象属性名与 XML 文件 element 名可以不同，等等。

使用 JiBX 绑定 XML 文档与 Java 对象需要分两步走：第一步是绑定 XML 文件，也就是映射 XML 文件与 Java 对象之间的对应关系；第二步是在运行时，实现 XML 文件与 Java 实例之间的互相转换。这时，它已经与绑定文件无关了，可以说是完全脱耦了。

在运行程序之前，需要先配置绑定文件并进行绑定，在绑定过程中它将会动态地修改程序中相应的 class 文件，主要是生成对应对象实例的方法和添加被绑定标记的属性 JiBX_bindingList 等。它使用的技术是 BCEL（Byte Code Engineering Library），BCEL 是 Apache Software Foundation 的 Jakarta 项目的一部分，也是目前 Java classworking 最广泛使用的一种框架，它可以让你深入 JVM 汇编语言进行类操作。在 JiBX 运行时，它使用了目前比较流行的一个技术 XPP（Xml Pull Parsing），这也正是 JiBX 如此高效的原因。

JiBx 有两个比较重要的概念：Unmarshal（数据分解）和 Marshal（数据编排）。从字面意思也很容易理解，Unmarshal 是将 XML 文件转换成 Java 对象，而 Marshal 则是将 Java 对象编排成规范的 XML 文件。JiBX 在 Unmarshal/Marshal 上如此高效，这要归功于使用了 XPP 技术，而不是使用基于树型（tree-based）方式，将整个文档写入内存，然后进行操作的 DOM（Document Object Model），也不是使用基于事件流（event stream）的 SAX（Simple API for Xml）。XPP 使用的是不断增加的数据流处理方式，同时允许在解析 XML 文件时中断。

介绍完了 JiBx 的基础概念，下面我们就结合订购例程，来学习下如何使用 JiBx 进行 XML 的开发。

2. POJO 对象定义

通过 JiBx 提供的工具 jar 包，可以根据 Schema 自动生成 POJO 对象，也可以根据普通的 POJO 对象生成 JiBx 绑定文件和 Schema 定义 XSD。

考虑到大多数人的编码习惯，我们采用先定义 POJO 对象，再生成 XML 和对象的绑定文件的方式。JiBx 对 POJO 对象没有特殊要求，只要符合 Java Bean 的规则即可，下面我们以订购例程为例，看下主要类的定义。

代码清单10-3　HTTP+XML POJO 类定义 Order

```
2.  public class Order {
3.      private long orderNumber;
4.      private Customer customer;
5.
6.      /** Billing address information. */
7.      private Address billTo;
8.
9.      private Shipping shipping;
10.
11.     /**
12.      * Shipping address information. If missing, the billing address is also
13.      * used as the shipping address.
14.      */
15.     private Address shipTo;
16.
17.     private Float total;
……//定义 set 和 get 方法
53. }
```

代码清单10-4　HTTP+XML POJO 类定义 Customer

```
2.  import java.util.List;
3.  public class Customer {
4.      private long customerNumber;
5.      /** Personal name. */
6.      private String firstName;
7.
8.      /** Family name. */
9.      private String lastName;
10.     /** Middle name(s), if any. */
11.     private List<String> middleNames;
……//定义 set 和 get 方法
35.     }
```

代码清单10-5　HTTP+XML POJO 类定义 Address

```
2.  public class Address {
3.      /** First line of street information (required). */
4.      private String street1;
5.      /** Second line of street information (optional). */
6.      private String street2;
7.      private String city;
8.
```

```
 9.     /**
10.      * State abbreviation (required for the U.S. and Canada, optional
11.      * otherwise).
12.      */
13.     private String state;
14.
15.     /** Postal code(required for the U.S.and Canada,optional otherwise).*/
16.     private String postCode;
17.     /** Country name (optional, U.S. assumed if not supplied). */
18.     private String country;
……//定义set和get方法
54.     }
```

代码清单10-6　HTTP+XML POJO 类定义 Shipping

```
1. package com.phei.netty.protocol.http.xml.pojo;
2.
3. public enum Shipping {
4.     STANDARD_MAIL, PRIORITY_MAIL, INTERNATIONAL_MAIL, DOMESTIC_EXPRESS,
INTERNATIONAL_EXPRESS
5. }
```

POJO 对象定义完成之后，通过 Ant 脚本来生成 XML 和 POJO 对象的绑定关系文件，同时也附加生成 XML 的 Schema 定义文件。下个小节开始介绍 Ant 脚本的编写。

3. 通过 Ant 脚步生成 XML 和对象的绑定关系

首先确认你使用的 Eclipse 安装了 Ant，通过【Window】主菜单选择【Preferences】，在弹出的窗口中可以查看是否支持 Ant。如图 10-11 所示。

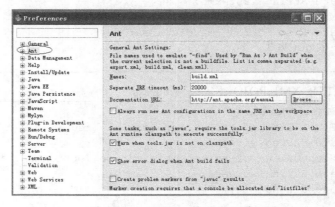

图 10-11　Eclipse 的 Ant 配置

第 10 章 HTTP 协议开发应用

目前主流的 Eclipse 版本默认都支持 Ant，如果你使用其他的开发工具，可以安装对应的 Ant 插件。

由于本书的重点并非讲解 Ant 的使用，所以，如果你对 Ant 感兴趣或者作为初学者学习 Ant，可以选择相关的 Ant 教材或者直接通过 Ant 的官方文档进行学习。

JiBx 的绑定和编译，重点关注两个 task。如图 10-12 所示。

```
<!-- generate default binding and schema -->
<target name="bindgen">
  <echo message="Running BindGen tool"/>
  <java classpathref="classpath" fork= "true" failonerror="true"
classname="org.jibx.binding.generator.BindGen">
    <arg value="-s"/>
    <arg value="${basedir}/src/com/phei/netty/protocol/http/xml/pojo"/>
    <arg value="com.phei.netty.protocol.http.xml.pojo.Order"/>
  </java>
</target>
```

图 10-12　使用 BindGen 命令生成绑定文件

通过 JiBx 的 org.jibx.binding.generator.BindGen 工具类可以将指定的 POJO 对象 Order 类生成绑定文件和 Schema 定义文件，执行结果如图 10-13 所示。

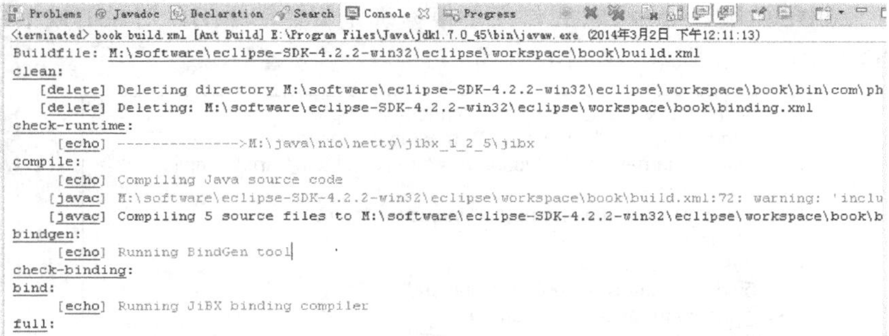

图 10-13　执行 Ant 脚本，生成 XML 绑定关系

执行成功后，在当前的工程目录下生成 binding.xml 和 pojo.xsd 文件，结果如图 10-14 所示。

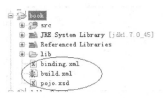

图 10-14　生成的 XML 绑定文件

打开绑定文件，我们可以发现绑定文件实际就是 XML 的元素和 POJO 对象字段之间的映射关系，生成之后的文件可以手工调整，例如修改 XML 中的元素名称、必选和可选标识等。如果你熟悉绑定规则，绑定文件也可以自己手工配置开发。XML 和 Order 对象的映射关系如图 10-15 所示。

```
<mapping abstract="true" type-name="ns1:order"
class="com.phei.netty.protocol.http.xml.pojo.Order">
  <value style="attribute" name="orderNumber" field="orderNumber"/>
  <structure field="customer" usage="optional" name="customer">
    <value style="attribute" name="customerNumber" field="customerNumber"/>
    <value style="element" name="firstName" field="firstName" usage="optional"/>
    <value style="element" name="lastName" field="lastName" usage="optional"/>
    <collection field="middleNames" usage="optional" create-type="java.util.ArrayList">
      <value name="middleName" type="java.lang.String"/>
    </collection>
  </structure>
......此处省略
</mapping>
```

图 10-15　XML 和 Order 对象的映射关系

再看一下 JiBx 的编译命令，它的作用是根据绑定文件和 POJO 对象的映射关系和规则动态修改 POJO 类，Ant 脚本如图 10-16 所示。

```
<target name="bind" depends="check-binding">
<echo message="Running JiBX binding compiler"/>
<taskdef name="bind" classname="org.jibx.binding.ant.CompileTask">
  <classpath>
    <fileset dir="${jibx-home}/lib" includes="*.jar"/>
  </classpath>
</taskdef>
<bind binding="${basedir}/binding.xml">
  <classpath refid="classpath"/>
</bind>
</target>
```

图 10-16　编译绑定脚本和动态修改 Class 的 Ant 脚本

编译结果如图 10-17 所示。

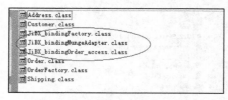

图 10-17　动态修改 Class 文件

到此为止，JiBx 相关的准备工作已经完成，下个小节通过一个简单的测试程序来学习 JiBx 类库的使用。

4．JiBx 的类库使用

JiBx 的类库使用非常简单，下面直接看代码。

代码清单 10-7　HTTP+XML POJO 测试类定义　TestOrder

```
18. public class TestOrder {
19.     private IBindingFactory factory = null;
20.     private StringWriter writer = null;
21.     private StringReader reader = null;
22.     private final static String CHARSET_NAME = "UTF-8";
23.     private String encode2Xml(Order order) throws JiBXException, IOException {
24.         factory = BindingDirectory.getFactory(Order.class);
25.         writer = new StringWriter();
26.         IMarshallingContext mctx = factory.createMarshallingContext();
27.         mctx.setIndent(2);
28.         mctx.marshalDocument(order, CHARSET_NAME, null, writer);
29.         String xmlStr = writer.toString();
30.         writer.close();
31.         System.out.println(xmlStr.toString());
32.         return xmlStr;
33.     }
34.
35.     private Order decode2Order(String xmlBody) throws JiBXException {
36.         reader = new StringReader(xmlBody);
37.         IUnmarshallingContext uctx = factory.createUnmarshallingContext();
38.         Order order = (Order) uctx.unmarshalDocument(reader);
39.         return order;
40.     }
41.
42.     public static void main(String[] args) throws JiBXException, IOException {
43.         TestOrder test = new TestOrder();
44.         Order order = OrderFactory.create(123);
45.         String body = test.encode2Xml(order);
46.         Order order2 = test.decode2Order(body);
47.         System.out.println(order2);
```

```
48.    }
49. }
```

首先看第 24 行，根据 Order 的 Class 实例构造 IBindingFactory 对象。第 25 行创建新的 StringWriter 对象，通过 IBindingFactory 构造 Marshalling 上下文，最后通过 marshalDocument 将 Order 序列化为 StringWriter，通过 StringWriter 的 toString()方法可以返回 String 类型的 XML 对象。

解码与编码类似，不同的是它使用 StringReader 来读取 String 类型的 XML 对象，然后通过 unmarshalDocument 方法将其反序列化为 Order 对象。

执行结果如下。

```
<?xml version="1.0" encoding="UTF-8"?>
<order xmlns="http://phei.com/netty/protocol/http/xml/pojo" orderNumber="123" total="9999.999">
  <customer customerNumber="123">
    <firstName>李</firstName>
    <lastName>林峰</lastName>
  </customer>
  <billTo>
    <street1>龙眠大道</street1>
    <city>南京市</city>
    <state>江苏省</state>
    <postCode>123321</postCode>
    <country>中国</country>
  </billTo>
  <shipping>INTERNATIONAL_MAIL</shipping>
  <shipTo>
    <street1>龙眠大道</street1>
    <city>南京市</city>
    <state>江苏省</state>
    <postCode>123321</postCode>
    <country>中国</country>
  </shipTo>
</order>
Order [orderNumber=123, customer=Customer [customerNumber=123, firstName=李, lastName=林峰, middleNames=null], billTo=Address [street1=龙眠大道, street2=null, city=南京市, state=江苏省, postCode=123321, country=中国], shipping=INTERNATIONAL_MAIL, shipTo=Address [street1=龙眠大道, street2=null, city=南京市, state=江苏省, postCode=123321, country=中国], total=9999.999]
```

通过上面的执行结果可以发现，XML 序列化和反序列化后的结果与预期一致，我们开发的 JiBx 应用可以正常工作。

XML 绑定框架选型和开发完成之后，下面继续 Netty HTTP+XM 编解码框架的开发。

10.3.4　HTTP+XML 编解码框架开发

本节会有 6 个部分来讲解如何基于 Netty 开发 HTTP+XML 协议栈，在 Netty 提供的 HTTP 基础协议栈上进行扩展和封装，以实现对上层业务的零侵入。下面我们一起学习如何进行开发。

1．HTTP+XML 请求消息编码类

对于上层业务侧，构造订购请求消息后，以 HTTP+XML 协议将消息发送给服务端，如果要实现对业务零侵入或者尽可能少的侵入，协议层和应用层应该解耦。

考虑到 HTTP+XML 协议栈需要一定的定制扩展能力，例如通过 HTTP 消息头携带业务自定义字段，所以，应该允许业务利用 Netty 的 HTTP 协议栈接口自行构造私有的 HTTP 消息头。

HTTP+XML 的协议编码仍然采用 ChannelPipeline 中增加对应的编码 handler 类实现。

下面我们来一起看下 HTTP+XML 请求消息编码类的源码实现。

代码清单 10-8　HTTP+XML HTTP 请求消息编码类

```
11. public class HttpXmlRequestEncoder extends
12.     AbstractHttpXmlEncoder<HttpXmlRequest> {
13.
14.     @Override
15.     protected void encode(ChannelHandlerContext ctx, HttpXmlRequest msg,
16.         List<Object> out) throws Exception {
17.         ByteBuf body = encode0(ctx, msg.getBody());
18.         FullHttpRequest request = msg.getRequest();
19.         if (request == null) {
20.             request = new DefaultFullHttpRequest(HttpVersion.HTTP_1_1,
21.                 HttpMethod.GET, "/do", body);
22.         HttpHeaders headers = request.headers();
23.         headers.set(HttpHeaders.Names.HOST,
InetAddress.getLocalHost()
```

```
24.                .getHostAddress());
25.            headers.set(HttpHeaders.Names.CONNECTION,
HttpHeaders.Values.CLOSE);
26.            headers.set(HttpHeaders.Names.ACCEPT_ENCODING,
27.                HttpHeaders.Values.GZIP.toString() + ','
28.                + HttpHeaders.Values.DEFLATE.toString());
29.            headers.set(HttpHeaders.Names.ACCEPT_CHARSET,
30.                "ISO-8859-1,utf-8;q=0.7,*;q=0.7");
31.            headers.set(HttpHeaders.Names.ACCEPT_LANGUAGE, "zh");
32.            headers.set(HttpHeaders.Names.USER_AGENT,
33.                "Netty xml Http Client side");
34.            headers.set(HttpHeaders.Names.ACCEPT,
"text/html,application/xhtml+xml,application/xml;q=0.9,*/*;q=0.8");
35.        }
36.        HttpHeaders.setContentLength(request, body.readableBytes());
37.        out.add(request);
38.    }
39. }
```

第 17 行首先调用父类的 encode0,将业务需要发送的 POJO 对象 Order 实例通过 JiBx 序列化为 XML 字符串,随后将它封装成 Netty 的 ByteBuf。第 18 行对消息头进行判断,如果业务侧自定义和定制了消息头,则使用业务侧设置的 HTTP 消息头,如果业务侧没有设置,则构造新的 HTTP 消息头。

第 20~35 行用来构造和设置默认的 HTTP 消息头,由于通常情况下 HTTP 通信双方更关注消息体本身,所以这里采用了硬编码的方式,如果要产品化,可以做成 XML 配置文件,允许业务自定义配置,以提升定制的灵活性。

第 36 行很重要,由于请求消息消息体不为空,也没有使用 Chunk 方式,所以在 HTTP 消息头中设置消息体的长度 Content-Length,完成消息体的 XML 序列化后将重新构造的 HTTP 请求消息加入到 out 中,由后续 Netty 的 HTTP 请求编码器继续对 HTTP 请求消息进行编码。

下面我们来看父类 AbstractHttpXmlEncoder 的实现。

代码清单 10-9　HTTP+XML HTTP 请求消息编码基类　AbstractHttpXmlEncoder

```
14. public abstract class AbstractHttpXmlEncoder<T> extends
15.     MessageToMessageEncoder<T> {
16.     IBindingFactory factory = null;
17.     StringWriter writer = null;
18.     final static String CHARSET_NAME = "UTF-8";
```

```
19.     final static Charset UTF_8 = Charset.forName(CHARSET_NAME);
20.
21.     protected ByteBuf encode0(ChannelHandlerContext ctx, Object body)
22.         throws Exception {
23.         factory = BindingDirectory.getFactory(body.getClass());
24.         writer = new StringWriter();
25.         IMarshallingContext mctx = factory.createMarshallingContext();
26.         mctx.setIndent(2);
27.         mctx.marshalDocument(body, CHARSET_NAME, null, writer);
28.         String xmlStr = writer.toString();
29.         writer.close();
30.         writer = null;
31.         ByteBuf encodeBuf = Unpooled.copiedBuffer(xmlStr, UTF_8);
32.         return encodeBuf;
33.     }
34.
35.     @Override
36.     public void exceptionCaught(ChannelHandlerContext ctx, Throwable cause)
37.         throws Exception {
38.         // 释放资源
39.         if (writer != null) {
40.             writer.close();
41.             writer = null;
42.         }
43.     }
44. }
```

首先看下第 23～30 行，代码很熟悉，在 JiBx 章节已经介绍了 XML 序列化和反序列化的相关类库使用，在此将业务的 Order 实例序列化为 XML 字符串。第 31 行将 XML 字符串包装成 Netty 的 ByteBuf 并返回，实现了 HTTP 请求消息的 XML 编码。

下面继续看下 HttpXmlRequest 是如何实现的。

代码清单 10-10　HTTP+XML 请求消息 HttpXmlRequest

```
9.  public class HttpXmlRequest {
10.     private FullHttpRequest request;
11.     private Object body;
12.
13.     public HttpXmlRequest(FullHttpRequest request, Object body) {
14.         this.request = request;
```

```
15.         this.body = body;
16.     }
17.
18.     /**
19.      * @return the request
20.      */
21.     public final FullHttpRequest getRequest() {
22.         return request;
23.     }
24.
25.     /**
26.      * @param request
27.      *              the request to set
28.      */
29.     public final void setRequest(FullHttpRequest request) {
30.         this.request = request;
31.     }
32.
33.     /**
34.      * @return the object
35.      */
36.     public final Object getBody() {
37.         return body;
38.     }
39.
40.     /**
41.      * @param object
42.      *              the object to set
43.      */
44.     public final void setBody(Object body) {
45.         this.body = body;
46.     }
47. }
```

它包含两个成员变量 FullHttpRequest 和编码对象 Object，用于实现和协议栈之间的解耦。

2. HTTP+XML 请求消息解码类

HTTP 服务端接收到 HTTP+XML 请求消息后，需要从 HTTP 消息体中获取请求码流，通过 JiBx 框架对它进行反序列化，得到请求 POJO 对象，然后对结果进行封装，回调到业务 handler 对象，业务得到的就是解码后的 POJO 对象和 HTTP 消息头。

下面就看下具体实现。

代码清单 10-11　HTTP+XML HTTP 请求消息解码类　HttpXmlRequestDecoder

```
20. public class HttpXmlRequestDecoder extends
21.     AbstractHttpXmlDecoder<FullHttpRequest> {
22.
23.     public HttpXmlRequestDecoder(Class<?> clazz) {
24.       this(clazz, false);
25.     }
26.
27.     public HttpXmlRequestDecoder(Class<?> clazz, boolean isPrint) {
28.       super(clazz, isPrint);
29.     }
30.
31.     @Override
32.     protected void decode(ChannelHandlerContext arg0, FullHttpRequest arg1,
33.         List<Object> arg2) throws Exception {
34.       if (!arg1.getDecoderResult().isSuccess()) {
35.         sendError(arg0, BAD_REQUEST);
36.         return;
37.       }
38.       HttpXmlRequest request = new HttpXmlRequest(arg1, decode0(arg0,
39.           arg1.content()));
40.       arg2.add(request);
41.     }
42.
43.     private static void sendError(ChannelHandlerContext ctx,
44.         HttpResponseStatus status) {
45.       FullHttpResponse response = new DefaultFullHttpResponse(HTTP_1_1,
46.         status, Unpooled.copiedBuffer("Failure: " + status.toString()
47.           + "\r\n", CharsetUtil.UTF_8));
48.       response.headers().set(CONTENT_TYPE, "text/plain; charset=UTF-8");
49.       ctx.writeAndFlush(response).addListener(ChannelFutureListener.CLOSE);
50.     }
51. }
```

HttpXmlRequestDecoder 有两个参数，分别为需要解码的对象的类型信息和是否打印

HTTP 消息体码流的码流开关,码流开关默认关闭。第 34～37 行首先对 HTTP 请求消息本身的解码结果进行判断,如果已经解码失败,再对消息体进行二次解码已经没有意义。

第 43～50 行,如果 HTTP 消息本身解码失败,则构造处理结果异常的 HTTP 应答消息返回给客户端。作为演示程序,本例程没有考虑 XML 消息解码失败后的异常封装和处理,在商用项目中需要统一的异常处理机制,提升协议栈的健壮性和可靠性。

第 38 行通过 HttpXmlRequest 和反序列化后的 Order 对象构造 HttpXmlRequest 实例,最后将它添加到解码结果 List 列表中。

继续看下它的父类 AbstractHttpXmlDecoder 的实现。

代码清单 10-12　HTTP+XML HTTP 请求消息解码类　AbstractHttpXmlDecoder

```
18.  public abstract class AbstractHttpXmlDecoder<T> extends
19.      MessageToMessageDecoder<T> {
20.     private IBindingFactory factory;
21.     private StringReader reader;
22.     private Class<?> clazz;
23.     private boolean isPrint;
24.     private final static String CHARSET_NAME = "UTF-8";
25.     private final static Charset UTF_8 = Charset.forName(CHARSET_NAME);
26.
27.     protected AbstractHttpXmlDecoder(Class<?> clazz) {
28.       this(clazz, false);
29.     }
30.
31.     protected AbstractHttpXmlDecoder(Class<?> clazz, boolean isPrint) {
32.       this.clazz = clazz;
33.       this.isPrint = isPrint;
34.     }
35.
36.     protected Object decode0(ChannelHandlerContext arg0, ByteBuf body)
37.         throws Exception {
38.       factory = BindingDirectory.getFactory(clazz);
39.       String content = body.toString(UTF_8);
40.       if (isPrint)
41.         System.out.println("The body is : " + content);
42.       reader = new StringReader(content);
43.       IUnmarshallingContext uctx = factory.createUnmarshallingContext();
44.       Object result = uctx.unmarshalDocument(reader);
45.       reader.close();
```

```
46.         reader = null;
47.         return result;
48.     }
49.     @Override
50.     public void exceptionCaught(ChannelHandlerContext ctx, Throwable cause)
51.         throws Exception {
52.         // 释放资源
53.         if (reader != null) {
54.             reader.close();
55.             reader = null;
56.         }
57.     }
58. }
```

第 38~47 行从 HTTP 的消息体中获取请求码流，然后通过 JiBx 类库将 XML 转换成 POJO 对象。最后，根据码流开关决定是否打印消息体码流。增加码流开关往往是为了方便问题定位，在实际项目中，需要打印到日志中。

第 53~55 行，如果解码发生异常，要判断 StringReader 是否已经关闭，如果没有关闭，则关闭输入流并通知 JVM 对其进行垃圾回收。

3. HTTP+XML 响应消息编码类

对于响应消息，用户可能并不关心 HTTP 消息头之类的，它将业务处理后的 POJO 对象丢给 HTTP+XML 协议栈，由基础协议栈进行后续的处理。为了降低业务的定制开发难度，我们首先封装一个全新的 HTTP XML 应答对象，它的实现如下。

代码清单 10-13　HTTP+XML HTTP XML 应答消息 HttpXmlResponse

```
9.  public class HttpXmlResponse {
10.     private FullHttpResponse httpResponse;
11.     private Object result;
12.
13.     public HttpXmlResponse(FullHttpResponse httpResponse,Object result){
14.         this.httpResponse = httpResponse;
15.         this.result = result;
16.     }
17.
18.     /**
19.      * @return the httpResponse
```

```
20.     */
21.     public final FullHttpResponse getHttpResponse() {
22.       return httpResponse;
23.     }
24.
25.     /**
26.      * @param httpResponse
27.      *            the httpResponse to set
28.      */
29.     public final void setHttpResponse(FullHttpResponse httpResponse) {
30.       this.httpResponse = httpResponse;
31.     }
32.
33.     /**
34.      * @return the body
35.      */
36.     public final Object getResult() {
37.       return result;
38.     }
39.
40.     /**
41.      * @param body
42.      *            the body to set
43.      */
44.     public final void setResult(Object result) {
45.       this.result = result;
46.     }
47. }
```

它包含两个成员变量：FullHttpResponse 和 Object，Object 就是业务需要发送的应答 POJO 对象。

下面继续看应答消息的 XML 编码类实现。

代码清单 10-14　HTTP+XML 应答消息编码类 HttpXmlResponseEncoder

```
17. public class HttpXmlResponseEncoder extends
18.     AbstractHttpXmlEncoder<HttpXmlResponse> {
19.
20.     /*
21.      * (non-Javadoc)
22.      *
```

```
23.     * @see
24.     * io.netty.handler.codec.MessageToMessageEncoder#encode(io.netty.channel
25.     * .ChannelHandlerContext, java.lang.Object, java.util.List)
26.     */
27.    protected void encode(ChannelHandlerContext ctx,HttpXmlResponse msg,
28.        List<Object> out) throws Exception {
29.     ByteBuf body = encode0(ctx, msg.getResult());
30.     FullHttpResponse response = msg.getHttpResponse();
31.     if (response == null) {
32.         response = new DefaultFullHttpResponse(HTTP_1_1, OK, body);
33.     } else {
34.         response = new DefaultFullHttpResponse(msg.getHttpResponse()
35.             .getProtocolVersion(), msg.getHttpResponse().getStatus(),
36.             body);
37.     }
38.     response.headers().set(CONTENT_TYPE, "text/xml");
39.     setContentLength(response, body.readableBytes());
40.     out.add(response);
41.    }
42. }
```

它的实现非常简单,第 30 行对应答消息进行判断,如果业务侧已经构造了 HTTP 应答消息,则利用业务已有应答消息重新复制一个新的 HTTP 应答消息。无法重用业务侧自定义 HTTP 应答消息的主要原因,是 Netty 的 DefaultFullHttpResponse 没有提供动态设置消息体 content 的接口,只能在第一次构造的时候设置内容。由于这个局限,导致我们的实现有点麻烦。

作为示例程序并没有提供更多的 API 供业务侧灵活设置 HTTP 应答消息头,在实际商用时,可以基于本书提供的基础协议栈进行扩展。

第 38 行设置消息体内容格式为"text/xml",然后在消息头中设置消息体的长度。第 40 行把编码后的 DefaultFullHttpResponse 对象添加到编码结果列表中,由后续 Netty 的 HTTP 编码类进行二次编码。

4. HTTP+XML 应答消息解码

客户端接收到 HTTP+XML 应答消息后,对消息进行解码,获取 HttpXmlResponse 对象,源码如下。

代码清单 10-15　HTTP+XML 应答消息解码类 HttpXmlResponseDecoder

```
11. public class HttpXmlResponseDecoder extends
12.     AbstractHttpXmlDecoder<DefaultFullHttpResponse> {
13.
14.     public HttpXmlResponseDecoder(Class<?> clazz) {
15.         this(clazz, false);
16.     }
17.
18.     public HttpXmlResponseDecoder(Class<?> clazz, boolean isPrintlog) {
19.         super(clazz, isPrintlog);
20.     }
21.
22.     @Override
23.     protected void decode(ChannelHandlerContext ctx,
24.         DefaultFullHttpResponse msg, List<Object> out) throws Exception {
25.     HttpXmlResponse resHttpXmlResponse = new HttpXmlResponse(msg, decode0(
26.         ctx, msg.content()));
27.     out.add(resHttpXmlResponse);
28.     }
29. }
```

第 25 行通过 DefaultFullHttpResponse 和 HTTP 应答消息反序列化后的 POJO 对象构造 HttpXmlResponse，并将其添加到解码结果列表中。

5. HTTP+XML 客户端开发

客户端的功能如下。

（1）发起 HTTP 连接请求；

（2）构造订购请求消息，将其编码成 XML，通过 HTTP 协议发送给服务端；

（3）接收 HTTP 服务端的应答消息，将 XML 应答消息反序列化为订购消息 POJO 对象；

（4）关闭 HTTP 连接。

基于它的功能定位，我们首先开始主程序的开发。

代码清单 10-16　HTTP+XML 客户端启动类 HttpXmlClient

```
23. public class HttpXmlClient {
24.
25.     public void connect(int port) throws Exception {
```

```
26.        // 配置客户端NIO线程组
27.        EventLoopGroup group = new NioEventLoopGroup();
28.        try {
29.            Bootstrap b = new Bootstrap();
30.            b.group(group).channel(NioSocketChannel.class)
31.                .option(ChannelOption.TCP_NODELAY, true)
32.                .handler(new ChannelInitializer<SocketChannel>() {
33.                    @Override
34.                    public void initChannel(SocketChannel ch)
35.                        throws Exception {
36.                        ch.pipeline().addLast("http-decoder",
37.                            new HttpResponseDecoder());
38.                        ch.pipeline().addLast("http-aggregator",
39.                            new HttpObjectAggregator(65536));
40.                        // XML 解码器
41.                        ch.pipeline().addLast(
42.                            "xml-decoder",
43.                            new HttpXmlResponseDecoder(Order.class,
44.                                true));
45.                        ch.pipeline().addLast("http-encoder",
46.                            new HttpRequestEncoder());
47.                        ch.pipeline().addLast("xml-encoder",
48.                            new HttpXmlRequestEncoder());
49.                        ch.pipeline().addLast("xmlClientHandler",
50.                            new HttpXmlClientHandle());
51.                    }
52.                });
53.
54.            // 发起异步连接操作
55.            ChannelFuture f=b.connect(new InetSocketAddress(port)).sync();
56.
57.            // 等待客户端链路关闭
58.            f.channel().closeFuture().sync();
59.        } finally {
60.            // 优雅退出，释放NIO线程组
61.            group.shutdownGracefully();
62.        }
63.    }
64.
65.    /**
66.     * @param args
```

```
67.     * @throws Exception
68.     */
69.    public static void main(String[] args) throws Exception {
70.        int port = 8080;
71.        if (args != null && args.length > 0) {
72.            try {
73.                port = Integer.valueOf(args[0]);
74.            } catch (NumberFormatException e) {
75.                // 采用默认值
76.            }
77.        }
78.        new HttpXmlClient().connect(port);
79.    }
80. }
```

在 ChannelPipeline 中新增了 HttpResponseDecoder，它负责将二进制码流解码成为 HTTP 的应答消息；随后第 38 行新增了 HttpObjectAggregator，它负责将 1 个 HTTP 请求消息的多个部分合并成一条完整的 HTTP 消息；第 41～44 行将前面开发的 XML 解码器 HttpXmlResponseDecoder 添加到 ChannelPipeline 中，它有两个参数，分别是解码对象的类型信息和码流开关，这样就实现了 HTTP+XML 应答消息的自动解码。

第 45～46 行将 HttpRequestEncoder 编码器添加到 ChannelPipeline 中时，需要注意顺序，编码的时候是按照从尾到头的顺序调度执行的，它后面放的是我们自定义开发的 HTTP+XML 请求消息编码器 HttpXmlRequestEncoder。

最后是业务的逻辑编排类 HttpXmlClientHandle，我们继续分析它的实现。

代码清单 10-17　HTTP+XML 客户端业务逻辑编排类 HttpXmlClientHandle

```
7. public class HttpXmlClientHandle extends
8.     SimpleChannelInboundHandler<HttpXmlResponse> {
9.
10.    @Override
11.    public void channelActive(ChannelHandlerContext ctx) {
12.     HttpXmlRequest request = new HttpXmlRequest(null,
13.         OrderFactory.create(123));
14.     ctx.writeAndFlush(request);
15.    }
16.
17.    @Override
18.    public void exceptionCaught(ChannelHandlerContext ctx, Throwable
```

```
cause) {
19.         cause.printStackTrace();
20.         ctx.close();
21.     }
22.
23.     @Override
24.     protected void messageReceived(ChannelHandlerContext ctx,
25.         HttpXmlResponse msg) throws Exception {
26.         System.out.println("The client receive response of http header is : "
27.             + msg.getHttpResponse().headers().names());
28.         System.out.println("The client receive response of http body is : "
29.             + msg.getResult());
30.     }
31. }
```

客户端的实现非常简单,第 12 行构造 HttpXmlRequest 对象,调用 ChannelHandlerContext 的 writeAndFlush 发送 HttpXmlRequest。

第 24～30 行用于接收服务端的应答消息,从接口看,它接收到的已经是自动解码后的 HttpXmlResponse 对象了;第 28～29 行将应答 POJO 消息打印出来,可以与服务端发送的原始对象进行比对,两者的内容将完全一致。

最后,看下订购对象工厂类的实现。

代码清单 10-18 HTTP+XML 订购对象工厂类 OrderFactory

```
2.  public class OrderFactory {
3.      public static Order create(long orderID) {
4.          Order order = new Order();
5.          order.setOrderNumber(orderID);
6.          order.setTotal(9999.999f);
7.          Address address = new Address();
8.          address.setCity("南京市");
9.          address.setCountry("中国");
10.         address.setPostCode("123321");
11.         address.setState("江苏省");
12.         address.setStreet1("龙眠大道");
13.         order.setBillTo(address);
14.         Customer customer = new Customer();
15.         customer.setCustomerNumber(orderID);
16.         customer.setFirstName("李");
17.         customer.setLastName("林峰");
```

```
18.        order.setCustomer(customer);
19.        order.setShipping(Shipping.INTERNATIONAL_MAIL);
20.        order.setShipTo(address);
21.        return order;
22.    }
23. }
```

6. HTTP+XML 服务端开发

HTTP 服务端的功能如下。

（1）接收 HTTP 客户端的连接；

（2）接收 HTTP 客户端的 XML 请求消息，并将其解码为 POJO 对象；

（3）对 POJO 对象进行业务处理，构造应答消息返回；

（4）通过 HTTP+XML 的格式返回应答消息；

（5）主动关闭 HTTP 连接。

下面我们首先看下服务端监听主程序的实现。

代码清单 10-19　HTTP+XML 服务端主程序 HttpXmlServer

```
22. public class HttpXmlServer {
23.     public void run(final int port) throws Exception {
24.         EventLoopGroup bossGroup = new NioEventLoopGroup();
25.         EventLoopGroup workerGroup = new NioEventLoopGroup();
26.         try {
27.             ServerBootstrap b = new ServerBootstrap();
28.             b.group(bossGroup, workerGroup)
29.                 .channel(NioServerSocketChannel.class)
30.                 .childHandler(new ChannelInitializer<SocketChannel>() {
31.                     @Override
32.                     protected void initChannel(SocketChannel ch)
33.                         throws Exception {
34.                         ch.pipeline().addLast("http-decoder",
35.                             new HttpRequestDecoder());
36.                         ch.pipeline().addLast("http-aggregator",
37.                             new HttpObjectAggregator(65536));
38.                         ch.pipeline()
39.                             .addLast(
```

```
40.                    "xml-decoder",
41.                    new HttpXmlRequestDecoder(
42.                        Order.class, true));
43.              ch.pipeline().addLast("http-encoder",
44.                    new HttpResponseEncoder());
45.              ch.pipeline().addLast("xml-encoder",
46.                    new HttpXmlResponseEncoder());
47.              ch.pipeline().addLast("xmlServerHandler",
48.                    new HttpXmlServerHandler());
49.            }
50.          });
51.      ChannelFuture future = b.bind(new InetSocketAddress(port)).sync();
52.      System.out.println("HTTP 订购服务器启动，网址是 : " + "http://localhost:"
53.            + port);
54.      future.channel().closeFuture().sync();
55.    } finally {
56.      bossGroup.shutdownGracefully();
57.      workerGroup.shutdownGracefully();
58.    }
59.  }
60.
61.  public static void main(String[] args) throws Exception {
62.    int port = 8080;
63.    if (args.length > 0) {
64.      try {
65.        port = Integer.parseInt(args[0]);
66.      } catch (NumberFormatException e) {
67.        e.printStackTrace();
68.      }
69.    }
70.    new HttpXmlServer().run(port);
71.  }
72. }
```

HTTP 服务端的启动与之前一样，在此不再详述，我们具体看下编解码 Handler 是如何设置的。

第 34～37 行用于绑定 HTTP 请求消息解码器；第 38～42 行将我们自定义的 HttpXmlRequestDecoder 添加到 HTTP 解码器之后；第 45～47 行添加自定义的 HttpXmlResponseEncoder

编码器用于响应消息的编码。

下面我们继续看 HttpXmlServerHandler 的实现。

代码清单 10-20　HTTP+XML 服务端处理类 HttpXmlServerHandler

```
30.  public class HttpXmlServerHandler extends
31.      SimpleChannelInboundHandler<HttpXmlRequest> {
32.
33.      @Override
34.      public void messageReceived(final ChannelHandlerContext ctx,
35.          HttpXmlRequest xmlRequest) throws Exception {
36.      HttpRequest request = xmlRequest.getRequest();
37.      Order order = (Order) xmlRequest.getBody();
38.      System.out.println("Http server receive request : " + order);
39.      dobusiness(order);
40.      ChannelFuture future = ctx.writeAndFlush(new HttpXmlResponse(null,
41.          order));
42.      if (!isKeepAlive(request)) {
43.          future.addListener(new  GenericFutureListener<Future<? super Void>>() {
44.              public void operationComplete(Future future) throws Exception {
45.                  ctx.close();
46.              }
47.          });
48.      }
49.  }
50.
51.  private void dobusiness(Order order) {
52.      order.getCustomer().setFirstName("狄");
53.      order.getCustomer().setLastName("仁杰");
54.      List<String> midNames = new ArrayList<String>();
55.      midNames.add("李元芳");
56.      order.getCustomer().setMiddleNames(midNames);
57.      Address address = order.getBillTo();
58.      address.setCity("洛阳");
59.      address.setCountry("大唐");
60.      address.setState("河南道");
61.      address.setPostCode("123456");
62.      order.setBillTo(address);
63.      order.setShipTo(address);
64.  }
```

```
65.
66.     @Override
67.     public void exceptionCaught(ChannelHandlerContext ctx, Throwable cause)
68.         throws Exception {
69.      cause.printStackTrace();
70.      if (ctx.channel().isActive()) {
71.          sendError(ctx, INTERNAL_SERVER_ERROR);
72.      }
73.     }
74.
75.     private static void sendError(ChannelHandlerContext ctx,
76.         HttpResponseStatus status) {
77.      FullHttpResponse response = new DefaultFullHttpResponse(HTTP_1_1,
78.         status, Unpooled.copiedBuffer("失败: " + status.toString()
79.             + "\r\n", CharsetUtil.UTF_8));
80.      response.headers().set(CONTENT_TYPE,"text/plain; charset=UTF-8");
81.      ctx.writeAndFlush(response).addListener(ChannelFutureListener.CLOSE);
82.     }
83. }
```

通过 messageReceived 的方法入参 HttpXmlRequest，可以看出服务端业务处理类接收到的已经是解码后的业务消息了。第 37 行用于获取请求消息对象；随后将它打印出来，可以与客户端发送的原始消息进行对比；第 39 行对订购请求消息进行业务逻辑编排；第 40～47 行用于发送应答消息，并且在发送成功之后主动关闭 HTTP 连接。

第 70～71 行，在发生异常并且链路没有关闭的情况下，构造内部异常消息发送给客户端，发送完成之后关闭 HTTP 链路。

到此，HTTP+XML 的协议栈开发工作全部完成，下个小节我们看下运行结果。

10.3.5　HTTP+XML 协议栈测试

本小节对前面几节开发的 HTTP+XML 协议栈进行测试。

首先对工程进行编译，然后执行 JiBx 的 Ant 脚本，对涉及的 POJO 对象进行二次编译，执行完成之后，首先运行 HTTP 服务端，然后再运行客户端，执行结果如下。

1. 服务端

服务端接收到的请求消息码流打印如下。

```
The body is : <?xml version="1.0" encoding="UTF-8"?>
<order xmlns="http://phei.com/netty/protocol/http/xml/pojo" orderNumber="123" total="9999.999">
    <customer customerNumber="123">
      <firstName>李</firstName>
      <lastName>林峰</lastName>
    </customer>
    <billTo>
      <street1>龙眠大道</street1>
      <city>南京市</city>
      <state>江苏省</state>
      <postCode>123321</postCode>
      <country>中国</country>
    </billTo>
    <shipping>INTERNATIONAL_MAIL</shipping>
    <shipTo>
      <street1>龙眠大道</street1>
      <city>南京市</city>
      <state>江苏省</state>
      <postCode>123321</postCode>
      <country>中国</country>
    </shipTo>
</order>
```

服务端解码后的业务对象如下。

```
Http server receive request : Order [orderNumber=123, customer=Customer [customerNumber=123, firstName=李, lastName=林峰, middleNames=null], billTo=Address [street1=龙眠大道, street2=null, city=南京市, state=江苏省, postCode=123321, country=中国], shipping=INTERNATIONAL_MAIL, shipTo=Address [street1=龙眠大道, street2=null, city=南京市, state=江苏省, postCode=123321, country=中国], total=9999.999]
```

2. 客户端

客户端接收到的响应消息体码流如下。

```
    The body is : <?xml version="1.0" encoding="UTF-8"?>
    <order xmlns="http://phei.com/netty/protocol/http/xml/pojo" orderNumber=
"123" total="9999.999">
      <customer customerNumber="123">
        <firstName>狄</firstName>
        <lastName>仁杰</lastName>
        <middleName>李元芳</middleName>
      </customer>
      <billTo>
        <street1>龙眠大道</street1>
        <city>洛阳</city>
        <state>河南道</state>
        <postCode>123456</postCode>
        <country>大唐</country>
      </billTo>
      <shipping>INTERNATIONAL_MAIL</shipping>
      <shipTo>
        <street1>龙眠大道</street1>
        <city>洛阳</city>
        <state>河南道</state>
        <postCode>123456</postCode>
        <country>大唐</country>
      </shipTo>
    </order>
```

解码后的响应消息如下。

```
    The client receive response of http body is : Order [orderNumber=123,
customer=Customer [customerNumber=123, firstName=狄, lastName=仁杰,
middleNames=[李元芳]], billTo=Address [street1=龙眠大道, street2=null, city=洛
阳, state=河南道, postCode=123456, country=大唐], shipping=INTERNATIONAL_MAIL,
shipTo=Address [street1=龙眠大道, street2=null, city=洛阳, state=河南道,
postCode=123456, country=大唐], total=9999.999]
```

测试结果表明，HTTP+XML 协议栈功能正常，达到了设计预期。

10.3.6　小结

需要指出的是，尽管本章节开发的 HTTP+XML 协议栈是个高性能、通用的协议栈，但是，作为例程我们忽略了一些异常场景的处理、可扩展性的 API 和一些配置能力。所以，

如果你打算在商用项目中使用本章节开发的 HTTP+XML 协议栈,仍需要做一些产品化的完善工作。

10.4　总结

本章节重点介绍了 HTTP 协议以及如何使用 Netty 的 HTTP 协议栈开发基于 HTTP 的应用程序,最后通过 HTTP+XML 协议栈的开发向读者展示了如何基于 Netty 提供的 HTTP 协议栈做二次定制开发。

本章节的 HTTP+XML 协议栈在实际项目中非常有用,如果读者打算以它为基础进行商业应用,需要补齐一些产品化的能力,例如配置能力、容错能力、更丰富的 API。

第 11 章

WebSocket 协议开发

一直以来，网络在很大程度上都是围绕着 HTTP 的请求/响应模式而构建的。客户端加载一个网页，然后直到用户单击下一页之前，什么都不会发生。在 2005 年左右，AJAX 开始让网络变得更加动态了。但所有的 HTTP 通信仍然是由客户端控制的，这就需要用户进行互动或定期轮询，以便从服务器加载新数据。

长期以来存在着各种技术让服务器得知有新数据可用时，立即将数据发送到客户端。这些技术种类繁多，例如"推送"或 Comet。最常用的一种黑客手段是对服务器发起连接创建假象，被称为长轮询。利用长轮询，客户端可以打开指向服务器的 HTTP 连接，而服务器会一直保持连接打开，直到发送响应。服务器只要实际拥有新数据，就会发送响应（其他技术包括 Flash、XHR multipart 请求和所谓的 HTML Files）。长轮询和其他技术都非常好用，在 Gmail 聊天等应用中会经常使用它们。

但是，这些解决方案都存在一个共同的问题：由于 HTTP 协议的开销，导致它们不适用于低延迟应用。

为了解决这些问题，WebSocket 将网络套接字引入到了客户端和服务端，浏览器和服务器之间可以通过套接字建立持久的连接，双方随时都可以互发数据给对方，而不是之前由客户端控制的一请求一应答模式。

本章主要内容包括：

◎ HTTP 协议的弊端

◎ WebSocket 入门

◎ Netty WebSocket 协议开发

11.1　HTTP 协议的弊端

将 HTTP 协议的主要弊端总结如下。

（1）HTTP 协议为半双工协议。半双工协议指数据可以在客户端和服务端两个方向上传输，但是不能同时传输。它意味着在同一时刻，只有一个方向上的数据传送；

（2）HTTP 消息冗长而繁琐。HTTP 消息包含消息头、消息体、换行符等，通常情况下采用文本方式传输，相比于其他的二进制通信协议，冗长而繁琐；

（3）针对服务器推送的黑客攻击。例如长时间轮询。

现在，很多网站为了实现消息推送，所用的技术都是轮询。轮询是在特定的的时间间隔（如每 1 秒），由浏览器对服务器发出 HTTP request，然后由服务器返回最新的数据给客户端浏览器。这种传统的模式具有很明显的缺点，即浏览器需要不断地向服务器发出请求，然而 HTTP request 的 Header 是非常冗长的，里面包含的可用数据比例可能非常低，这会占用很多的带宽和服务器资源。

比较新的一种轮询技术是 Comet，使用了 AJAX。这种技术虽然可达到双向通信，但依然需要发出请求，而且在 Comet 中，普遍采用了长连接，这也会大量消耗服务器带宽和资源。

为了解决 HTTP 协议效率低下的问题，HTML5 定义了 WebSocket 协议，能更好地节省服务器资源和带宽并达到实时通信，下个小节让我们一起来学习 WebSocket 的入门知识。

11.2　WebSocket 入门

WebSocket 是 HTML5 开始提供的一种浏览器与服务器间进行全双工通信的网络技

术，WebSocket 通信协议于 2011 年被 IETF 定为标准 RFC6455，WebSocket API 被 W3C 定为标准。

在 WebSocket API 中，浏览器和服务器只需要做一个握手的动作，然后，浏览器和服务器之间就形成了一条快速通道，两者就可以直接互相传送数据了。WebSocket 基于 TCP 双向全双工进行消息传递，在同一时刻，既可以发送消息，也可以接收消息，相比 HTTP 的半双工协议，性能得到很大提升。

下面总结一下 WebSocket 的特点。

◎ 单一的 TCP 连接，采用全双工模式通信；
◎ 对代理、防火墙和路由器透明；
◎ 无头部信息、Cookie 和身份验证；
◎ 无安全开销；
◎ 通过"ping/pong"帧保持链路激活；
◎ 服务器可以主动传递消息给客户端，不再需要客户端轮询。

11.2.1　WebSocket 背景

WebSocket 设计出来的目的就是要取代轮询和 Comet 技术，使客户端浏览器具备像 C/S 架构下桌面系统一样的实时通信能力。浏览器通过 JavaScript 向服务器发出建立 WebSocket 连接的请求，连接建立以后，客户端和服务器端可以通过 TCP 连接直接交换数据。因为 WebSocket 连接本质上就是一个 TCP 连接，所以在数据传输的稳定性和数据传输量的大小方面，和轮询以及 Comet 技术相比，具有很大的性能优势。Websocket.org 网站对传统的轮询方式和 WebSocket 调用方式作了一个详细的测试和比较，将一个简单的 Web 应用分别通过轮询方式和 WebSocket 方式来实现，在这里引用一下测试结果，如图 11-1 所示。

通过对比图可以清楚地看出，在流量和负载增大的情况下，WebSocket 方案相比传统的 AJAX 轮询方案有极大的性能优势。这也是我们认为 WebSocket 是未来实时 Web 应用的首选方案的原因。

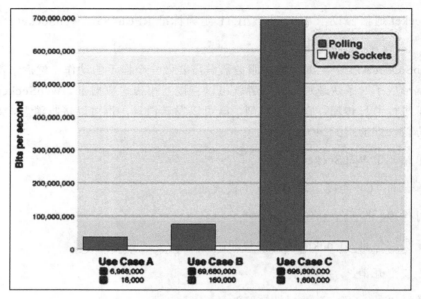

图 11-1　轮询和 WebSocket 网络负载对比图

11.2.2　WebSocket 连接建立

客户端和服务端连接建立的示意图如图 11-2 所示。

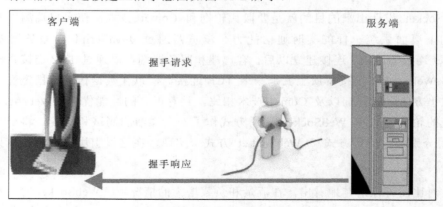

图 11-2　客户端和服务端握手连接

建立 WebSocket 连接时，需要通过客户端或者浏览器发出握手请求，请求消息示例如图 11-3 所示。

```
GET /chat HTTP/1.1
Host: server.example.com
Upgrade: websocket
Connection: Upgrade
Sec-WebSocket-Key: dGhlIHNhbXBsZSBub25jZQ==
Origin: http://example.com
Sec-WebSocket-Protocol: chat, superchat
Sec-WebSocket-Version: 13
```

图 11-3　WebSocket 客户端握手请求消息

为了建立一个 WebSocket 连接，客户端浏览器首先要向服务器发起一个 HTTP 请求，这个请求和通常的 HTTP 请求不同，包含了一些附加头信息，其中附加头信息 "Upgrade: WebSocket" 表明这是一个申请协议升级的 HTTP 请求。服务器端解析这些附加的头信息，然后生成应答信息返回给客户端，客户端和服务器端的 WebSocket 连接就建立起来了，双方可以通过这个连接通道自由地传递信息，并且这个连接会持续存在直到客户端或者服务器端的某一方主动关闭连接。

服务端返回给客户端的应答消息如图 11-4 所示。

```
HTTP/1.1 101 Switching Protocols
Upgrade: websocket
Connection: Upgrade
Sec-WebSocket-Accept: s3pPLMBiTxaQ9kYGzzhZRbK+xOo=
Sec-WebSocket-Protocol: chat
```

图 11-4　WebSocket 服务端返回的握手应答消息

请求消息中的 "Sec-WebSocket-Key" 是随机的，服务器端会用这些数据来构造出一个 SHA-1 的信息摘要，把 "Sec-WebSocket-Key" 加上一个魔幻字符串 "258EAFA5-E914-47DA-95CA-C5AB0DC85B11"。使用 SHA-1 加密，然后进行 BASE-64 编码，将结果做为 "Sec-WebSocket-Accept" 头的值，返回给客户端。

11.2.3　WebSocket 生命周期

握手成功之后，服务端和客户端就可以通过 "messages" 的方式进行通信了，一个消息由一个或者多个帧组成，WebSocket 的消息并不一定对应一个特定网络层的帧，它可以被分割成多个帧或者被合并。

帧都有自己对应的类型，属于同一个消息的多个帧具有相同类型的数据。从广义上讲，数据类型可以是文本数据（UTF-8[RFC3629]文字）、二进制数据和控制帧（协议级信令，如信号）。

WebSocket 连接生命周期示意图如图 11-5 所示。

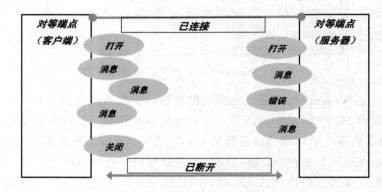

图 11-5　WebSocket 生命周期示意图

11.2.4　WebSocket 连接关闭

为关闭 WebSocket 连接，客户端和服务端需要通过一个安全的方法关闭底层 TCP 连接以及 TLS 会话。如果合适，丢弃任何可能已经接收的字节，必要时（比如受到攻击）可以通过任何可用的手段关闭连接。

底层的 TCP 连接，在正常情况下，应该首先由服务器关闭。在异常情况下（例如在一个合理的时间周期后没有接收到服务器的 TCP Close），客户端可以发起 TCP Close。因此，当服务器被指示关闭 WebSocket 连接时，它应该立即发起一个 TCP Close 操作；客户端应该等待服务器的 TCP Close。

WebSocket 的握手关闭消息带有一个状态码和一个可选的关闭原因，它必须按照协议要求发送一个 Close 控制帧，当对端接收到关闭控制帧指令时，需要主动关闭 WebSocket 连接。

通过本节的描述，相信读者对 WebSocket 的基础知识有了一定的了解，大家如果对 WebSocket 规范感兴趣，可以访问 WebSocket 的官网去了解更多的相关知识。

下个小节我们将一起学习如何使用 Netty 开发 WebSocket 服务端。

11.3 Netty WebSocket 协议开发

Netty 基于 HTTP 协议栈开发了 WebSocket 协议栈，利用 Netty 的 WebSocket 协议栈可以非常方便地开发出 WebSocket 客户端和服务端。本节通过一个 Netty 服务端实例的开发，向读者讲解如何使用 Netty 进行 WebSocket 开发。

11.3.1 WebSocket 服务端功能介绍

WebSocket 服务端的功能如下：支持 WebSocket 的浏览器通过 WebSocket 协议发送请求消息给服务端，服务端对请求消息进行判断，如果是合法的 WebSocket 请求，则获取请求消息体（文本），并在后面追加字符串"欢迎使用 Netty WebSocket 服务，现在时刻：系统时间"。

客户端 HTML 通过内嵌的 JS 脚本创建 WebSocket 连接，如果握手成功，在文本框中打印"打开 WebSocket 服务正常，浏览器支持 WebSocket!"。客户端界面如图 11-6 所示。

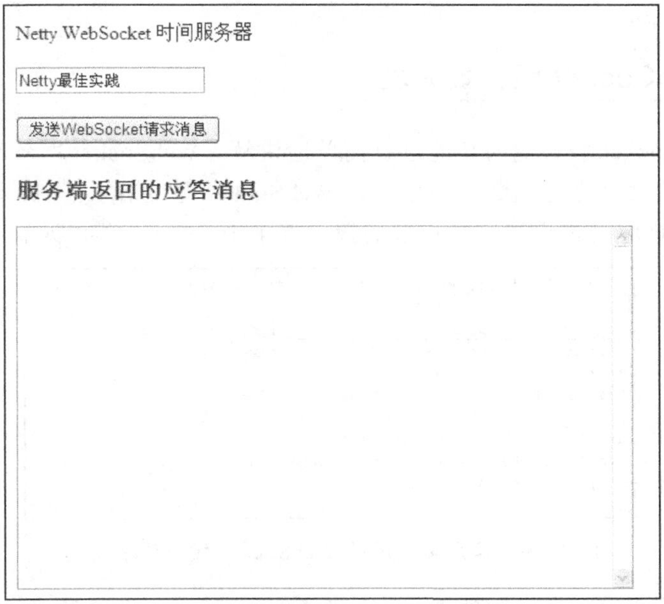

图 11-6 WebSocket 客户端 HTML

当前支持 WebSocket 的浏览器如图 11-7 所示。从图中可以看出，目前主流的浏览器都已经支持 WebSocket，但是在运行本例程之前你仍然需要确认自己使用的浏览器版本是否已经支持 WebSocket，否则会提示"抱歉，您的浏览器不支持 WebSocket 协议!"。

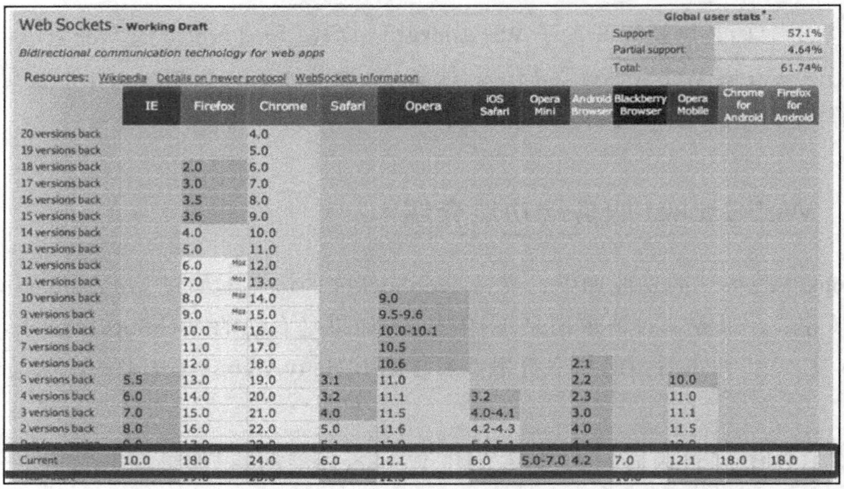

图 11-7　支持 WebSocket 的浏览器及其版本

11.3.2　WebSocket 服务端开发

首先对 WebSocket 服务端的功能进行简单地讲解。WebSocket 服务端接收到请求消息之后，先对消息的类型进行判断，如果不是 WebSocket 握手请求消息，则返回 HTTP 400 BAD REQUEST 响应给客户端。客户端的握手请求消息如图 11-8 所示。

图 11-8　客户端发送的 WebSocket 握手请求消息

服务端对握手请求消息进行处理，构造握手响应返回，双方的 Socket 连接正式建立，服务端返回的握手应答消息如图 11-9 所示。

```
DefaultFullHttpResponse(decodeResult: success)
HTTP/1.1 101 Switching Protocols
Upgrade: websocket
Connection: Upgrade
Sec-WebSocket-Accept: XPORwKeQ8rU8v+fcTLFEw+AcI1Y=
```

图 11-9 WebSocket 握手应答消息

连接建立成功后，到被关闭之前，双方都可以主动向对方发送消息，这点跟 HTTP 的一请求一应答模式存在很大的差别。相比于 HTTP，它的网络利用率更高，可以通过全双工的方式进行消息发送和接收。

下面一起来看下服务端代码的具体实现，首先看服务启动类。

代码清单 11-1 WebSocket 服务端启动类 WebSocketServer

```
1.  public class WebSocketServer {
2.      public void run(int port) throws Exception {
3.      EventLoopGroup bossGroup = new NioEventLoopGroup();
4.      EventLoopGroup workerGroup = new NioEventLoopGroup();
5.      try {
6.          ServerBootstrap b = new ServerBootstrap();
7.          b.group(bossGroup, workerGroup)
8.              .channel(NioServerSocketChannel.class)
9.              .childHandler(new ChannelInitializer<SocketChannel>() {
10.
11.                 @Override
12.                 protected void initChannel(SocketChannel ch)
13.                     throws Exception {
14.                     ChannelPipeline pipeline = ch.pipeline();
15.                     pipeline.addLast("http-codec",
16.                         new HttpServerCodec());
17.                     pipeline.addLast("aggregator",
18.                         new HttpObjectAggregator(65536));
19.                     ch.pipeline().addLast("http-chunked",
20.                         new ChunkedWriteHandler());
21.                     pipeline.addLast("handler",
22.                         new WebSocketServerHandler());
23.                 }
24.             });
25.
26.         Channel ch = b.bind(port).sync().channel();
```

```
27.         System.out.println("Web socket server started at port " + port
28.             + '.');
29.         System.out
30.             .println("Open your browser and navigate to http://localhost:"
31.                 + port + '/');
32.
33.         ch.closeFuture().sync();
34.     } finally {
35.         bossGroup.shutdownGracefully();
36.         workerGroup.shutdownGracefully();
37.     }
38. }
39.
40. public static void main(String[] args) throws Exception {
41.     int port = 8080;
42.     if (args.length > 0) {
43.         try {
44.          port = Integer.parseInt(args[0]);
45.         } catch (NumberFormatException e) {
46.          e.printStackTrace();
47.         }
48.     }
49.     new WebSocketServer().run(port);
50.   }
51. }
```

第 15～16 行首先添加 HttpServerCodec，将请求和应答消息编码或者解码为 HTTP 消息；第 17～18 行增加 HttpObjectAggregator，它的目的是将 HTTP 消息的多个部分组合成一条完整的 HTTP 消息；第 19～20 行添加 ChunkedWriteHandler，来向客户端发送 HTML5 文件，它主要用于支持浏览器和服务端进行 WebSocket 通信；最后 21～22 行增加 WebSocket 服务端 Handler。

看了 WebSocket 的服务启动类，很多读者会心存疑惑：怎么 WebSocket 服务端的代码跟 HTTP 协议的非常类似呢？没有看到在 ChannelPipeline 中增加 WebSocket 的 Handler，那如何处理 WebSocket 消息？这个疑问很好，下面就一起来从 WebSocketServerHandler 的实现中寻找答案。

代码清单 11-2　WebSocket 服务端处理类 WebSocketServerHandler

```
1.  public class WebSocketServerHandler extends SimpleChannel
InboundHandler<Object> {
2.      private static final Logger logger = Logger
3.          .getLogger(WebSocketServerHandler.class.getName());
4. 
5.      private WebSocketServerHandshaker handshaker;
6. 
7.      @Override
8.      public void messageReceived(ChannelHandlerContext ctx, Object msg)
9.          throws Exception {
10.         // 传统的 HTTP 接入
11.         if (msg instanceof FullHttpRequest) {
12.             handleHttpRequest(ctx, (FullHttpRequest) msg);
13.         }
14.         // WebSocket 接入
15.         else if (msg instanceof WebSocketFrame) {
16.             handleWebSocketFrame(ctx, (WebSocketFrame) msg);
17.         }
18.     }
19. 
20.     @Override
21.     public void channelReadComplete(ChannelHandlerContext ctx) throws
Exception {
22.         ctx.flush();
23.     }
24. 
25.     private void handleHttpRequest(ChannelHandlerContext ctx,
26.         FullHttpRequest req) throws Exception {
27. 
28.         // 如果 HTTP 解码失败，返回 HTTP 异常
29.         if (!req.getDecoderResult().isSuccess()
30.             || (!"websocket".equals(req.headers().get("Upgrade")))) {
31.             sendHttpResponse(ctx, req, new DefaultFullHttpResponse(HTTP_1_1,
32.                 BAD_REQUEST));
33.             return;
34.         }
35. 
36.         // 构造握手响应返回，本机测试
37.         WebSocketServerHandshakerFactory wsFactory = new WebSocketServer
HandshakerFactory(
```

```
38.            "ws://localhost:8080/websocket", null, false);
39.        handshaker = wsFactory.newHandshaker(req);
40.        if (handshaker == null) {
41.            WebSocketServerHandshakerFactory
42.                .sendUnsupportedWebSocketVersionResponse(ctx.channel());
43.        } else {
44.            handshaker.handshake(ctx.channel(), req);
45.        }
46.    }
47.
48.    private void handleWebSocketFrame(ChannelHandlerContext ctx,
49.        WebSocketFrame frame) {
50.
51.        // 判断是否是关闭链路的指令
52.        if (frame instanceof CloseWebSocketFrame) {
53.            handshaker.close(ctx.channel(),
54.                (CloseWebSocketFrame) frame.retain());
55.            return;
56.        }
57.        // 判断是否是Ping消息
58.        if (frame instanceof PingWebSocketFrame) {
59.            ctx.channel().write(
60.                new PongWebSocketFrame(frame.content().retain()));
61.            return;
62.        }
63.        // 本例程仅支持文本消息，不支持二进制消息
64.        if (!(frame instanceof TextWebSocketFrame)) {
65.            throw new UnsupportedOperationException(String.format(
66.                "%s frame types not supported", frame.getClass().getName()));
67.        }
68.
69.        // 返回应答消息
70.        String request = ((TextWebSocketFrame) frame).text();
71.        if (logger.isLoggable(Level.FINE)) {
72.            logger.fine(String.format("%s received %s", ctx.channel(), request));
73.        }
74.        ctx.channel().write(
75.            new TextWebSocketFrame(request
76.                + " ，欢迎使用Netty WebSocket服务，现在时刻："
77.                + new java.util.Date().toString()));
```

```
78.     }
79.
80.     private static void sendHttpResponse(ChannelHandlerContext ctx,
81.         FullHttpRequest req, FullHttpResponse res) {
82.         // 返回应答给客户端
83.         if (res.getStatus().code() != 200) {
84.             ByteBuf buf = Unpooled.copiedBuffer(res.getStatus().toString(),
85.                 CharsetUtil.UTF_8);
86.             res.content().writeBytes(buf);
87.             buf.release();
88.             setContentLength(res, res.content().readableBytes());
89.         }
90.
91.         // 如果是非 Keep-Alive，关闭连接
92.         ChannelFuture f = ctx.channel().writeAndFlush(res);
93.         if (!isKeepAlive(req) || res.getStatus().code() != 200) {
94.             f.addListener(ChannelFutureListener.CLOSE);
95.         }
96.     }
97.
98.     @Override
99.     public void exceptionCaught(ChannelHandlerContext ctx, Throwable cause)
100.            throws Exception {
101.         cause.printStackTrace();
102.         ctx.close();
103.     }
104. }
```

首先从第 11 行看起，第一次握手请求消息由 HTTP 协议承载，所以它是一个 HTTP 消息，执行 handleHttpRequest 方法来处理 WebSocket 握手请求。第 29～34 行首先对握手请求消息进行判断，如果消息头中没有包含 Upgrade 字段或者它的值不是 websocket，则返回 HTTP 400 响应。

握手请求简单校验通过之后，开始构造握手工厂，创建握手处理类 WebSocketServerHandshaker，通过它构造握手响应消息返回给客户端，同时将 WebSocket 相关的编码和解码类动态添加到 ChannelPipeline 中，用于 WebSocket 消息的编解码，代码如图 11-10 所示。

添加 WebSocket Encoder 和 WebSocket Decoder 之后，服务端就可以自动对 WebSocket 消息进行编解码了，后面的业务 handler 可以直接对 WebSocket 对象进行操作。下面继续

分析链路建立成功之后的操作：客户端通过文本框提交请求消息给服务端，WebSocket ServerHandler 接收到的是已经解码后的 WebSocketFrame 消息。第 48～96 行对 WebSocket 请求消息进行处理，首先需要对控制帧进行判断，如果是关闭链路的控制消息，就调用 WebSocketServerHandshaker 的 close 方法关闭 WebSocket 连接；如果是维持链路的 Ping 消息，则构造 Pong 消息返回。由于本例程的 WebSocket 通信双方使用的都是文本消息，所以对请求消息的类型进行判断，不是文本的抛出异常。

```
if (ctx == null) {
    // this means the user use a HttpServerCodec
    ctx = p.context(HttpServerCodec.class);
    if (ctx == null) {
        promise.setFailure(
            new IllegalStateException("No HttpDecoder and no "
                + "HttpServerCodec in the pipeline"));
        return promise;
    }
    p.addBefore(ctx.name(), "wsdecoder", newWebsocketDecoder());
    p.addBefore(ctx.name(), "wsencoder", newWebSocketEncoder());
    encoderName = ctx.name();
} else {
    p.replace(ctx.name(), "wsdecoder", newWebsocketDecoder());

    encoderName = p.context(HttpResponseEncoder.class).name();
    p.addBefore(encoderName, "wsencoder", newWebSocketEncoder());
}
```

图 11-10　WebSocket 握手应答时动态增加编解码 handler

最后，从 TextWebSocketFrame 中获取请求消息字符串，对它处理后通过构造新的 TextWebSocketFrame 消息返回给客户端，由于握手应答时动态增加了 TextWebSocketFrame 的编码类，所以，可以直接发送 TextWebSocketFrame 对象。

客户端浏览器接收到服务端的应答消息后，将其内容取出展示到浏览器页面中。我们简单分析客户端 WebSocketServer.html 的源码。

```
<html>
<head>
<meta charset="UTF-8">
Netty WebSocket 时间服务器
</head>
<br>
<body>
<br>
<script type="text/javascript">
var socket;
if (!window.WebSocket)
{
    window.WebSocket = window.MozWebSocket;
```

```
        }
    if (window.WebSocket) {
        socket = new WebSocket("ws://localhost:8080/websocket");
        socket.onmessage = function(event) {
            var ta = document.getElementById('responseText');
            ta.value="";
            ta.value = event.data
        };
        socket.onopen = function(event) {
            var ta = document.getElementById('responseText');
            ta.value = "打开WebSocket服务正常,浏览器支持WebSocket!";
        };
        socket.onclose = function(event) {
            var ta = document.getElementById('responseText');
            ta.value = "";
            ta.value = "WebSocket 关闭!";
        };
    }
    else
        {
        alert("抱歉,您的浏览器不支持WebSocket协议!");
        }

    function send(message) {
        if (!window.WebSocket) { return; }
        if (socket.readyState == WebSocket.OPEN) {
            socket.send(message);
        }
        else
            {
            alert("WebSocket 连接没有建立成功!");
            }
    }
</script>
<form onsubmit="return false;">
<input type="text" name="message" value="Netty 最佳实践"/>
<br><br>
<input type="button" value=" 发 送 WebSocket 请 求 消 息 " onclick="send
(this.form.message.value)"/>
<hr color="blue"/>
<h3>服务端返回的应答消息</h3>
```

```
<textarea id="responseText" style="width:500px;height:300px;"></textarea>
    </form>
  </body>
</html>
```

由于通过 JS 开发 WebSocket 的接口比较简单，所以此处不再赘述，对此感兴趣的同学可以通过 JSR356、相关的技术书籍或者网站进行学习。

好，基于 Netty 的 WebSocket 服务端已经开发完毕，下个小节我们一起来测试下本节开发的程序，看它的各项功能是否能够达到设计预期。

11.3.3 运行 WebSocket 服务端

启动 WebSocket 服务，采用支持 WebSocket 的浏览器访问 WebSocketServer.html，显示结果如图 11-11 所示。

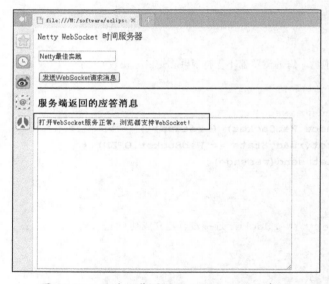

图 11-11　通过浏览器访问 WebSocket 服务端

文本框中显示"打开 WebSocket 服务正常，浏览器支持 WebSocket!"，说明 WebSocket 链路建立成功，然后单击【发送 WebSocket 请求消息】按钮，显示结果如图 11-12 所示。

当使用不支持 WebSocket 的老版本 IE 浏览器打开 WebSocketServer.html 时，运行结果如图 11-13 所示。

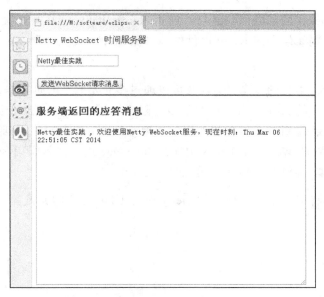

图 11-12　打印 WebSocket 服务端返回的结果

图 11-13　不支持 WebSocket 的浏览器运行效果图

11.4　总结

本章首先介绍了 HTTP 协议的弊端和产生 WebSocket 的一些技术背景，随后对 WebSocket 的优势和基础入门知识进行了介绍，包括 WebSocket 的握手请求和响应、连接的建立和关闭、WebSocket 的生命周期等。

学习了 WebSocket 的基础知识之后，通过 Netty WebSocket 时间服务器的开发，读者朋友可以更好地掌握如何利用 Netty 提供的 WebSocket 协议栈进行 WebSocket 应用程序的开发。

由于 WebSocket 本身的复杂性，以及可以通过多种形式（例如文本方式、二进制方式）承载消息，所以，它的 API 和用法也非常多，限于篇幅，本书无法对这些场景一一枚举。希望本章的例程能够起到抛砖引玉的作用，想要掌握更多的用法，需要读者结合 Netty 的测试用例和示例，以及 WebSocket 相关类库的 Java Doc 进行深入学习和实践，相信通过不断地实践很快就能掌握更多的功能和用法。

下一章，我们将继续学习如何利用 Netty 进行 UDP 协议的开发。

第 12 章

私有协议栈开发

广义上区分,通信协议可以分为公有协议和私有协议。由于私有协议的灵活性,它往往会在某个公司或者组织内部使用,按需定制,也因为如此,升级起来会非常方便,灵活性好。

绝大多数的私有协议传输层都基于 TCP/IP,所以利用 Netty 的 NIO TCP 协议栈可以非常方便地进行私有协议的定制和开发。本章节通过一个私有协议的设计和开发,让读者能够熟悉和掌握这方面的知识。

本章主要内容包括:

◎ 私有协议介绍

◎ 基于 Netty 的私有协议栈设计

◎ 私有协议栈开发

12.1 私有协议介绍

私有协议本质上是厂商内部发展和采用的标准,除非授权,其他厂商一般无权使用该协议。私有协议也称非标准协议,就是未经国际或国家标准化组织采纳或批准,由某个企业自己制订，协议实现细节不愿公开，只在企业自己生产的设备之间使用的协议。私有协议具有封闭性、垄断性、排他性等特点。如果网上大量存在私有（非标准）协议，现行网络或用户一旦使用了它，后进入的厂家设备就必须跟着使用这种非标准协议，才能够互连互通，否则根本不可能进入现行网络。这样，使用非标准协议的厂家就实现了垄断市场的愿望。

尽管私有协议具有垄断性的特征，但并非所有的私有协议设计者的初衷就是为了垄断。由于现代软件系统的复杂性，一个大型软件系统往往会被人为地拆分成多个模块，另外随着移动互联网的兴起，网站的规模也越来越大，业务的功能越来越多，为了能够支撑业务的发展，往往需要集群和分布式部署，这样，各个模块之间就要进行跨节点通信。

在传统的 Java 应用中，通常使用以下 4 种方式进行跨节点通信。

（1）通过 RMI 进行远程服务调用；

（2）通过 Java 的 Socket+Java 序列化的方式进行跨节点调用；

（3）利用一些开源的 RPC 框架进行远程服务调用，例如 Facebook 的 Thrift、Apache 的 Avro 等；

（4）利用标准的公有协议进行跨节点服务调用，例如 HTTP+XML、RESTful+JSON 或者 WebService。

跨节点的远程服务调用，除了链路层的物理连接外，还需要对请求和响应消息进行编解码。在请求和应答消息本身以外，也需要携带一些其他控制和管理类指令，例如链路建立的握手请求和响应消息、链路检测的心跳消息等。当这些功能组合到一起之后，就会形成私有协议。

事实上，私有协议并没有标准的定义，只要是能够用于跨进程、跨主机数据交换的非标准协议，都可以称为私有协议。通常情况下，正规的私有协议都有具体的协议规范文档，类似于《XXXX 协议 VXX 规范》，但是在实际的项目中，内部使用的私有协议往往是口头约定的规范，由于并不需要对外呈现或者被外部调用，所以一般不会单独写相关的内部私有协议规范文档。

本章使用 Netty 提供的异步 TCP 协议栈开发一个私有协议栈，该协议栈被命名为 Netty 协议栈。从下个小节开始，我们将详细介绍 Netty 协议的设计和开发。

12.2　Netty 协议栈功能设计

Netty 协议栈用于内部各模块之间的通信,它基于 TCP/IP 协议栈,是一个类 HTTP 协议的应用层协议栈,相比于传统的标准协议栈,它更加轻巧、灵活和实用。

12.2.1　网络拓扑图

如图 12-1 所示,在分布式组网环境下,每个 Netty 节点(Netty 进程)之间建立长连接,使用 Netty 协议进行通信。Netty 节点并没有服务端和客户端的区分,谁首先发起连接,谁就作为客户端,另一方自然就成为服务端。一个 Netty 节点既可以作为客户端连接另外的 Netty 节点,也可以作为 Netty 服务端被其他 Netty 节点连接,这完全取决于使用者的业务场景和具体的业务组网。

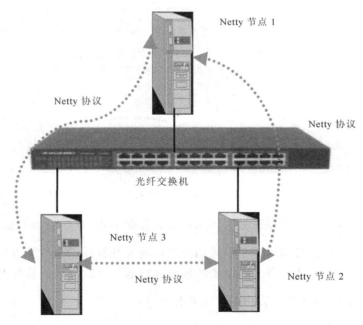

图 12-1　Netty 协议网络拓扑示意图

12.2.2 协议栈功能描述

Netty 协议栈承载了业务内部各模块之间的消息交互和服务调用，它的主要功能如下。

（1）基于 Netty 的 NIO 通信框架，提供高性能的异步通信能力；

（2）提供消息的编解码框架，可以实现 POJO 的序列化和反序列化；

（3）提供基于 IP 地址的白名单接入认证机制；

（4）链路的有效性校验机制；

（5）链路的断连重连机制。

12.2.3 通信模型

Netty 协议栈通信模型如图 12-2 所示。

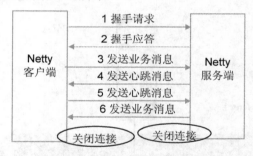

图 12-2　Netty 协议栈通信交互图

具体步骤如下。

（1）Netty 协议栈客户端发送握手请求消息，携带节点 ID 等有效身份认证信息；

（2）Netty 协议栈服务端对握手请求消息进行合法性校验，包括节点 ID 有效性校验、节点重复登录校验和 IP 地址合法性校验，校验通过后，返回登录成功的握手应答消息；

（3）链路建立成功之后，客户端发送业务消息；

（4）链路成功之后，服务端发送心跳消息；

（5）链路建立成功之后，客户端发送心跳消息；

（6）链路建立成功之后，服务端发送业务消息；

（7）服务端退出时，服务端关闭连接，客户端感知对方关闭连接后，被动关闭客户端连接。

备注：需要指出的是，Netty 协议通信双方链路建立成功之后，双方可以进行全双工通信，无论客户端还是服务端，都可以主动发送请求消息给对方，通信方式可以是 TWO WAY 或者 ONE WAY。双方之间的心跳采用 Ping-Pong 机制，当链路处于空闲状态时，客户端主动发送 Ping 消息给服务端，服务端接收到 Ping 消息后发送应答消息 Pong 给客户端，如果客户端连续发送 N 条 Ping 消息都没有接收到服务端返回的 Pong 消息，说明链路已经挂死或者对方处于异常状态，客户端主动关闭连接，间隔周期 T 后发起重连操作，直到重连成功。

12.2.4 消息定义

Netty 协议栈消息定义包含两部分：

◎ 消息头；

◎ 消息体。

其具体定义分别如表 12-1 和表 12-2 所示。

表 12-1　Netty 消息定义表（NettyMessage）

名　称	类　型	长　度	描　述
header	Header	变长	消息头定义
body	Object	变长	对于请求消息，它是方法的参数（作为示例，只支持携带一个参数）；对于响应消息，它是返回值

表 12-2　Netty 协议消息头定义（Header）

名　称	类　型	长　度	描　述
crcCode	整型 int	32	Netty 消息的校验码，它由三部分组成： 1）0xABEF：固定值，表明该消息是 Netty 协议消息，2 个字节； 2）主版本号：1～255，1 个字节； 3）次版本号：1～255，1 个字节。 crcCode = 0xABEF + 主版本号 + 次版本号

续表

名称	类型	长度	描述
length	整型 int	32	消息长度,整个消息,包括消息头和消息体
sessionID	长整型 long	64	集群节点内全局唯一,由会话 ID 生成器生成
type	Byte	8	0:业务请求消息 1:业务响应消息 2:业务 ONE WAY 消息(既是请求又是响应消息) 3:握手请求消息 4:握手应答消息 5:心跳请求消息 6:心跳应答消息
priority	Byte	8	消息优先级:0~255
attachment	Map<String, Object>	变长	可选字段,用于扩展消息头

12.2.5 Netty 协议支持的字段类型

Netty 协议支持的数据类型如表 12-3 所示。

表 12-3 Netty 协议支持的数据类型

字段类型	备注说明
boolean	包括它的包装类型 Integer
byte	包括它的包装类型 Byte
int	对应于 C/C++ 的 int32
char	包括它的包装类型 Character
short	对应 C/C++ 的 int16
long	对应 C/C++ 的 int64
float	包括它的包装类型 Float
double	包括它的包装类型 Double
string	对应 C/C++ 的 String
list	支持各种 List 的实现
array	支持各种数组的实现
map	支持 Map 的嵌套和泛型
set	支持 Set 的嵌套和泛型

12.2.6 Netty 协议的编解码规范

1. Netty 协议的编码

Netty 协议 NettyMessage 的编码规范如下。

（1）crcCode：java.nio.ByteBuffer.putInt(int value)，如果采用其他缓冲区实现，必须与其等价；

（2）length：java.nio.ByteBuffer.putInt(int value)，如果采用其他缓冲区实现，必须与其等价；

（3）sessionID：java.nio.ByteBuffer.putLong(long value)，如果采用其他缓冲区实现，必须与其等价；

（4）type: java.nio.ByteBuffer.put(byte b)，如果采用其他缓冲区实现，必须与其等价；

（5）priority：java.nio.ByteBuffer.put(byte b)，如果采用其他缓冲区实现，必须与其等价；

（6）attachment：它的编码规则为——如果 attachment 长度为 0，表示没有可选附件，则将长度编码设为 0，java.nio.ByteBuffer.putInt(0)；如果大于 0，说明有附件需要编码，具体的编码规则如下。

- ◎ 首先对附件的个数进行编码，java.nio.ByteBuffer.putInt(attachment.size());
- ◎ 然后对 Key 进行编码，先编码长度，再将它转换成 byte 数组之后编码内容，具体代码如下。

```
String key = null;
byte[] value = null;
for (Map.Entry<String, Object> param : attachment.entrySet()) {
        key = param.getKey();
        buffer.writeString(key);
        value = marshaller.writeObject(param.getValue());
        buffer.writeBinary(value);
    }
        key = null;
        value = null;
```

需要说明的是，将 String 字符串写入 ByteBuffer 和通过 Jboss Marshalling 将 Object 序列化为 byte 数组，此处没有详细展开介绍，后续代码开发章节会给出具体实现。

（7）body 的编码：通过 JBoss Marshalling 将其序列化为 byte 数组，然后调用 java.nio.ByteBuffer.put(byte [] src)将其写入 ByteBuffer 缓冲区中。

由于整个消息的长度必须等全部字段都编码完成之后才能确认，所以最后需要更新消息头中的 length 字段，将其重新写入 ByteBuffer 中。

2. Netty 协议的解码

相对于 NettyMessage 的编码，仍旧以 java.nio.ByteBuffer 为例，给出 Netty 协议的解码规范。

（1）crcCode：通过 java.nio.ByteBuffer.getInt()获取校验码字段，其他缓冲区需要与其等价；

（2）length：通过 java.nio.ByteBuffer.getInt()获取 Netty 消息的长度，其他缓冲区需要与其等价；

（3）sessionID：通过 java.nio.ByteBuffer.getLong()获取会话 ID，其他缓冲区需要与其等价；

（4）type：通过 java.nio.ByteBuffer.get()获取消息类型，其他缓冲区需要与其等价；

（5）priority：通过 java.nio.ByteBuffer.get()获取消息优先级，其他缓冲区需要与其等价；

（6）attachment：它的解码规则为——首先创建一个新的 attachment 对象，调用 java.nio.ByteBuffer.getInt()获取附件的长度，如果为 0，说明附件为空，解码结束，继续解消息体；如果非空，则根据长度通过 for 循环进行解码。

```
String key = null;
Object value = null;
for (int i = 0; i < size; i++) {
    key = buffer.readString();
    value = unmarshaller.readObject(buffer.readBinary());
    this.attachment.put(key, value);
}
    key = null;
    value = null;
```

后续的代码开发章节会给出附件解码的具体实现，此处不再详细展开，仅仅给出解码的规则。

（7）body：通过 JBoss 的 marshaller 对其进行解码。

12.2.7　链路的建立

Netty 协议栈支持服务端和客户端，对于使用 Netty 协议栈的应用程序而言，不需要刻意区分到底是客户端还是服务端，在分布式组网环境中，一个节点可能既是服务端也是客户端，这个依据具体的用户场景而定。

Netty 协议栈对客户端的说明如下：如果 A 节点需要调用 B 节点的服务，但是 A 和 B 之间还没有建立物理链路，则由调用方主动发起连接，此时，调用方为客户端，被调用方为服务端。

考虑到安全，链路建立需要通过基于 IP 地址或者号段的黑白名单安全认证机制，作为样例，本协议使用基于 IP 地址的安全认证，如果有多个 IP，通过逗号进行分割。在实际商用项目中，安全认证机制会更加严格，例如通过密钥对用户名和密码进行安全认证。

客户端与服务端链路建立成功之后，由客户端发送握手请求消息，握手请求消息的定义如下。

（1）消息头的 type 字段值为 3；

（2）可选附件为个数为 0；

（3）消息体为空；

（4）握手消息的长度为 22 个字节。

服务端接收到客户端的握手请求消息之后，如果 IP 校验通过，返回握手成功应答消息给客户端，应用层链路建立成功。握手应答消息定义如下。

（1）消息头的 type 字段值为 4；

（2）可选附件个数为 0；

（3）消息体为 byte 类型的结果，"0" 表示认证成功；"-1" 表示认证失败。

链路建立成功之后，客户端和服务端就可以互相发送业务消息了。

12.2.8 链路的关闭

由于采用长连接通信,在正常的业务运行期间,双方通过心跳和业务消息维持链路,任何一方都不需要主动关闭连接。

但是,在以下情况下,客户端和服务端需要关闭连接。

(1) 当对方宕机或者重启时,会主动关闭链路,另一方读取到操作系统的通知信号,得知对方 REST 链路,需要关闭连接,释放自身的句柄等资源。由于采用 TCP 全双工通信,通信双方都需要关闭连接,释放资源;

(2) 消息读写过程中,发生了 I/O 异常,需要主动关闭连接;

(3) 心跳消息读写过程中发生了 I/O 异常,需要主动关闭连接;

(4) 心跳超时,需要主动关闭连接;

(5) 发生编码异常等不可恢复错误时,需要主动关闭连接。

12.2.9 可靠性设计

Netty 协议栈可能会运行在非常恶劣的网络环境中,网络超时、闪断、对方进程僵死或者处理缓慢等情况都有可能发生。为了保证在这些极端异常场景下 Netty 协议栈仍能够正常工作或者自动恢复,需要对它的可靠性进行统一规划和设计。

1. 心跳机制

在凌晨等业务低谷期时段,如果发生网络闪断、连接被 Hang 住等网络问题时,由于没有业务消息,应用进程很难发现。到了白天业务高峰期时,会发生大量的网络通信失败,严重的会导致一段时间进程内无法处理业务消息。为了解决这个问题,在网络空闲时采用心跳机制来检测链路的互通性,一旦发现网络故障,立即关闭链路,主动重连。

具体的设计思路如下。

(1) 当网络处于空闲状态持续时间达到 T(连续周期 T 没有读写消息)时,客户端主动发送 Ping 心跳消息给服务端。

(2) 如果在下一个周期 T 到来时客户端没有收到对方发送的 Pong 心跳应答消息或者

读取到服务端发送的其他业务消息,则心跳失败计数器加 1。

(3)每当客户端接收到服务的业务消息或者 Pong 应答消息时,将心跳失败计数器清零;连续 N 次没有接收到服务端的 Pong 消息或者业务消息,则关闭链路,间隔 INTERVAL 时间后发起重连操作。

(4)服务端网络空闲状态持续时间达到 T 后,服务端将心跳失败计数器加 1;只要接收到客户端发送的 Ping 消息或者其他业务消息,计数器清零。

(5)服务端连续 N 次没有接收到客户端的 Ping 消息或者其他业务消息,则关闭链路,释放资源,等待客户端重连。

通过 Ping-Pong 双向心跳机制,可以保证无论通信哪一方出现网络故障,都能被及时地检测出来。为了防止由于对方短时间内繁忙没有及时返回应答造成的误判,只有连续 N 次心跳检测都失败才认定链路已经损害,需要关闭链路并重建链路。

当读或者写心跳消息发生 I/O 异常的时候,说明链路已经中断,此时需要立即关闭链路,如果是客户端,需要重新发起连接。如果是服务端,需要清空缓存的半包信息,等待客户端重连。

2. 重连机制

如果链路中断,等待 INTERVAL 时间后,由客户端发起重连操作,如果重连失败,间隔周期 INTERVAL 后再次发起重连,直到重连成功。

为了保证服务端能够有充足的时间释放句柄资源,在首次断连时客户端需要等待 INTERVAL 时间之后再发起重连,而不是失败后就立即重连。

为了保证句柄资源能够及时释放,无论什么场景下的重连失败,客户端都必须保证自身的资源被及时释放,包括但不限于 SocketChannel、Socket 等。

重连失败后,需要打印异常堆栈信息,方便后续的问题定位。

3. 重复登录保护

当客户端握手成功之后,在链路处于正常状态下,不允许客户端重复登录,以防止客户端在异常状态下反复重连导致句柄资源被耗尽。

服务端接收到客户端的握手请求消息之后,首先对 IP 地址进行合法性检验,如果校验成功,在缓存的地址表中查看客户端是否已经登录,如果已经登录,则拒绝重复登录,返回错误码-1,同时关闭 TCP 链路,并在服务端的日志中打印握手失败的原因。

客户端接收到握手失败的应答消息之后,关闭客户端的 TCP 连接,等待 INTERVAL 时间之后,再次发起 TCP 连接,直到认证成功。

为了防止由服务端和客户端对链路状态理解不一致导致的客户端无法握手成功的问题,当服务端连续 N 次心跳超时之后需要主动关闭链路,清空该客户端的地址缓存信息,以保证后续该客户端可以重连成功,防止被重复登录保护机制拒绝掉。

4. 消息缓存重发

无论客户端还是服务端,当发生链路中断之后,在链路恢复之前,缓存在消息队列中待发送的消息不能丢失,等链路恢复之后,重新发送这些消息,保证链路中断期间消息不丢失。

考虑到内存溢出的风险,建议消息缓存队列设置上限,当达到上限之后,应该拒绝继续向该队列添加新的消息。

12.2.10 安全性设计

为了保证整个集群环境的安全,内部长连接采用基于 IP 地址的安全认证机制,服务端对握手请求消息的 IP 地址进行合法性校验:如果在白名单之内,则校验通过;否则,拒绝对方连接。

如果将 Netty 协议栈放到公网中使用,需要采用更加严格的安全认证机制,例如基于密钥和 AES 加密的用户名+密码认证机制,也可以采用 SSL/TSL 安全传输。

作为示例程序,Netty 协议栈采用最简单的基于 IP 地址的白名单安全认证机制。

12.2.11 可扩展性设计

Netty 协议需要具备一定的扩展能力,业务可以在消息头中自定义业务域字段,例如消息流水号、业务自定义消息头等。通过 Netty 消息头中的可选附件 attachment 字段,业务可以方便地进行自定义扩展。

Netty 协议栈架构需要具备一定的扩展能力,例如统一的消息拦截、接口日志、安全、加解密等可以被方便地添加和删除,不需要修改之前的逻辑代码,类似 Servlet 的 Filter Chain 和 AOP,但考虑到性能因素,不推荐通过 AOP 来实现功能的扩展。

12.3 Netty 协议栈开发

12.3.1 数据结构定义

首先,对 Netty 协议栈使用到的数据结构进行定义,Netty 消息定义如下。

代码清单 12-1　NettyMessage 类定义

```
1. public final class NettyMessage {
2.     private Header header; //消息头
3.     private Object body;//消息体
4.
5.     /**
6.      * @return the header
7.      */
8.     public final Header getHeader() {
9. return header;
10.     }
11.
12.     /**
13.      * @param header
14.      *            the header to set
15.      */
16.     public final void setHeader(Header header) {
17. this.header = header;
18.     }
19.
20.     /**
21.      * @return the body
22.      */
23.     public final Object getBody() {
24. return body;
25.     }
26.
```

```
27.    /**
28.     * @param body
29.     *             the body to set
30.     */
31.    public final void setBody(Object body) {
32. this.body = body;
33.    }
34.
35.    /*
36.     * (non-Javadoc)
37.     *
38.     * @see java.lang.Object#toString()
39.     */
40.    @Override
41.    public String toString() {
42. return "NettyMessage [header=" + header + "]";
43.    }
44. }
```

代码清单 12-2　消息头 Header 类定义

```
1. public final class Header {
2.     private int crcCode = 0xabef0101;
3.     private int length;// 消息长度
4.     private long sessionID;// 会话 ID
5.     private byte type;// 消息类型
6.     private byte priority;// 消息优先级
7.     private Map<String, Object> attachment = new HashMap<String, Object>();
// 附件
8.
9.     /**
10.     * @return the crcCode
11.     */
12.    public final int getCrcCode() {
13. return crcCode;
14.    }
15.
16.    /**
17.     * @param crcCode
18.     *             the crcCode to set
19.     */
20.    public final void setCrcCode(int crcCode) {
```

```
21.     this.crcCode = crcCode;
22.     }
23.
24.     /**
25.      * @return the length
26.      */
27.     public final int getLength() {
28. return length;
29.     }
30.
31.     /**
32.      * @param length
33.      *           the length to set
34.      */
35.     public final void setLength(int length) {
36. this.length = length;
37.     }
38.
39.     /**
40.      * @return the sessionID
41.      */
42.     public final long getSessionID() {
43. return sessionID;
44.     }
45.
46.     /**
47.      * @param sessionID
48.      *           the sessionID to set
49.      */
50.     public final void setSessionID(long sessionID) {
51. this.sessionID = sessionID;
52.     }
53.
54.     /**
55.      * @return the type
56.      */
57.     public final byte getType() {
58. return type;
59.     }
60.
61.     /**
```

```
62.     * @param type
63.     *           the type to set
64.     */
65.    public final void setType(byte type) {
66. this.type = type;
67.    }
68.
69.    /**
70.     * @return the priority
71.     */
72.    public final byte getPriority() {
73. return priority;
74.    }
75.
76.    /**
77.     * @param priority
78.     *           the priority to set
79.     */
80.    public final void setPriority(byte priority) {
81. this.priority = priority;
82.    }
83.
84.    /**
85.     * @return the attachment
86.     */
87.    public final Map<String, Object> getAttachment() {
88. return attachment;
89.    }
90.
91.    /**
92.     * @param attachment
93.     *           the attachment to set
94.     */
95.    public final void setAttachment(Map<String, Object> attachment) {
96. this.attachment = attachment;
97.    }
98.
99.    /*
100.    * (non-Javadoc)
101.    *
102.    * @see java.lang.Object#toString()
```

```
103.         */
104.        @Override
105.        public String toString() {
106.            return "Header [crcCode=" + crcCode + ", length=" + length
107.                    + ", sessionID=" + sessionID + ", type=" + type + ", priority="
108.                    + priority + ", attachment=" + attachment + "]";
109.        }
110.    }
```

由于心跳消息、握手请求和握手应答消息都可以统一由 NettyMessage 承载,所以不需要为这几类控制消息做单独的数据结构定义。

12.3.2 消息编解码

分别定义 NettyMessageDecoder 和 NettyMessageEncoder 用于 NettyMessage 消息的编解码,它们的具体实现如下。

代码清单 12-3　Netty 消息编码类:NettyMessageEncoder

```
1.  public final class NettyMessageEncoder extends
2.      MessageToMessageEncoder<NettyMessage> {
3.
4.      MarshallingEncoder marshallingEncoder;
5.
6.      public NettyMessageEncoder() throws IOException {
7.          this.marshallingEncoder = new MarshallingEncoder();
8.      }
9.
10.     @Override
11.     protected void encode(ChannelHandlerContext ctx, NettyMessage msg,
12.         List<Object> out) throws Exception {
13.
14.     if (msg == null || msg.getHeader() == null)
15.         throw new Exception("The encode message is null");
16.     ByteBuf sendBuf = Unpooled.buffer();
17.     sendBuf.writeInt((msg.getHeader().getCrcCode()));
18.     sendBuf.writeInt((msg.getHeader().getLength()));
19.     sendBuf.writeLong((msg.getHeader().getSessionID()));
20.     sendBuf.writeByte((msg.getHeader().getType()));
21.     sendBuf.writeByte((msg.getHeader().getPriority()));
22.     sendBuf.writeInt((msg.getHeader().getAttachment().size()));
```

```
23. String key = null;
24. byte[] keyArray = null;
25. Object value = null;
26. for (Map.Entry<String, Object> param : msg.getHeader().getAttachment()
27.     .entrySet()) {
28.     key = param.getKey();
29.     keyArray = key.getBytes("UTF-8");
30.     sendBuf.writeInt(keyArray.length);
31.     sendBuf.writeBytes(keyArray);
32.     value = param.getValue();
33.     marshallingEncoder.encode(value, sendBuf);
34. }
35. key = null;
36. keyArray = null;
37. value = null;
38. if (msg.getBody() != null) {
39.     marshallingEncoder.encode(msg.getBody(), sendBuf);
40. } else
41.     sendBuf.writeInt(0);
42.         sendBuf.setInt(4, sendBuf.readableBytes());
43.     }
44. }
```

代码清单 12-4　Netty 消息编码工具类：MarshallingEncoder

```
1. public class MarshallingEncoder {
2.     private static final byte[] LENGTH_PLACEHOLDER = new byte[4];
3.     Marshaller marshaller;
4.
5.     public MarshallingEncoder() throws IOException {
6. marshaller = MarshallingCodecFactory.buildMarshalling();
7.     }
8.
9.     protected void encode(Object msg, ByteBuf out) throws Exception {
10. try {
11.     int lengthPos = out.writerIndex();
12.     out.writeBytes(LENGTH_PLACEHOLDER);
13.     ChannelBufferByteOutput output = new ChannelBufferByteOutput(out);
14.     marshaller.start(output);
15.     marshaller.writeObject(msg);
16.     marshaller.finish();
17.     out.setInt(lengthPos, out.writerIndex() - lengthPos - 4);
```

```
18. } finally {
19.     marshaller.close();
20. }
21.   }
22. }
```

代码清单 12-5　Netty 消息解码类：NettyMessageDecoder

```
1. public class NettyMessageDecoder extends LengthFieldBasedFrameDecoder {
2.
3.     MarshallingDecoder marshallingDecoder;
4.
5.     public NettyMessageDecoder(int maxFrameLength, int lengthFieldOffset,
6.         int lengthFieldLength) throws IOException {
7.     super(maxFrameLength, lengthFieldOffset, lengthFieldLength);
8.     marshallingDecoder = new MarshallingDecoder();
9.     }
10.
11.    @Override
12.    protected Object decode(ChannelHandlerContext ctx, ByteBuf in)
13.        throws Exception {
14. ByteBuf frame = (ByteBuf) super.decode(ctx, in);
15. if (frame == null) {
16.     return null;
17. }
18.
19. NettyMessage message = new NettyMessage();
20. Header header = new Header();
21. header.setCrcCode(in.readInt());
22. header.setLength(in.readInt());
23. header.setSessionID(in.readLong());
24. header.setType(in.readByte());
25. header.setPriority(in.readByte());
26.
27. int size = in.readInt();
28. if (size > 0) {
29.     Map<String, Object> attch = new HashMap<String, Object>(size);
30.     int keySize = 0;
31.     byte[] keyArray = null;
32.     String key = null;
33.     for (int i = 0; i < size; i++) {
34.      keySize = in.readInt();
```

```
35.         keyArray = new byte[keySize];
36.         in.readBytes(keyArray);
37.             key = new String(keyArray, "UTF-8");
38.         attch.put(key, marshallingDecoder.decode(in));
39.     }
40.     keyArray = null;
41.     key = null;
42.     header.setAttachment(attch);
43. }
44. if (in.readableBytes() > 4) {
45.     message.setBody(marshallingDecoder.decode(in));
46. }
47. message.setHeader(header);
48. return message;
49.     }
50. }
```

在这里我们用到了 Netty 的 LengthFieldBasedFrameDecoder 解码器，它支持自动的 TCP 粘包和半包处理，只需要给出标识消息长度的字段偏移量和消息长度自身所占的字节数，Netty 就能自动实现对半包的处理。对于业务解码器来说，调用父类 LengthFieldBasedFrameDecoder 的解码方法后，返回的就是整包消息或者为空。如果为空则说明是个半包消息，直接返回继续由 I/O 线程读取后续的码流，代码如图 12-3 所示。

```
@Override
protected Object decode(ChannelHandlerContext ctx, ByteBuf in)
        throws Exception {
    ByteBuf frame = (ByteBuf) super.decode(ctx, in);
    if (frame == null) {
        return null;
    }
```

图 12-3　半包解码代码

代码清单 12-6　Netty 消息解码工具类：MarshallingDecoder

```
1. public class MarshallingDecoder {
2.     private final Unmarshaller unmarshaller;
3.
4.     /**
5.      * Creates a new decoder whose maximum object size is {@code 1048576} bytes.
6.      * If the size of the received object is greater than {@code 1048576} bytes,
7.      * a {@link StreamCorruptedException} will be raised.
8.      *
```

```
9.     * @throws IOException
10.    *
11.    */
12.   public MarshallingDecoder() throws IOException {
13. unmarshaller = MarshallingCodecFactory.buildUnMarshalling();
14.   }
15.
16.   protected Object decode(ByteBuf in) throws Exception {
17. int objectSize = in.readInt();
18. ByteBuf buf = in.slice(in.readerIndex(), objectSize);
19. ByteInput input = new ChannelBufferByteInput(buf);
20. try {
21.     unmarshaller.start(input);
22.     Object obj = unmarshaller.readObject();
23.     unmarshaller.finish();
24.     in.readerIndex(in.readerIndex() + objectSize);
25.     return obj;
26. } finally {
27.     unmarshaller.close();
28. }
29.   }
30. }
```

消息的编解码类按照 12.2.6 章节的消息编解码模块设计实现即可,如果读者对二进制编解码比较熟悉,结合第 9 章对 JBoss Marshall 序列化框架的介绍,相信可以比较轻松地读懂本章节的代码。如果对本章节的代码阅读起来比较吃力,建议补充下 JDK 的 ByteBuffer 和 Jboss Marshall 框架的相关知识,然后再学习本章。

12.3.3　握手和安全认证

握手的发起是在客户端和服务端 TCP 链路建立成功通道激活时,握手消息的接入和安全认证在服务端处理。下面看下具体实现。

首先开发一个握手认证的客户端 ChannelHandler,用于在通道激活时发起握手请求,具体代码实现如下。

代码清单 12-7　LoginAuthReqHandler

```
1. public class LoginAuthReqHandler extends ChannelHandlerAdapter {
2.
3.     /**
```

```
4.      * Calls {@link ChannelHandlerContext#fireChannelActive()} to forward to the
5.      * next {@link ChannelHandler} in the {@link ChannelPipeline}.
6.      *
7.      * Sub-classes may override this method to change behavior.
8.      */
9.     @Override
10.    public void channelActive(ChannelHandlerContext ctx) throws Exception {
11.        ctx.writeAndFlush(buildLoginReq());
12.    }
13.
14.    /**
15.     * Calls {@link ChannelHandlerContext#fireChannelRead(Object)} to forward to
16.     * the next {@link ChannelHandler} in the {@link ChannelPipeline}.
17.     *
18.     * Sub-classes may override this method to change behavior.
19.     */
20.    @Override
21.    public void channelRead(ChannelHandlerContext ctx, Object msg)
22.        throws Exception {
23.        NettyMessage message = (NettyMessage) msg;
24.
25.        // 如果是握手应答消息，需要判断是否认证成功
26.        if (message.getHeader() != null
27.            && message.getHeader().getType() == MessageType.LOGIN_RESP
28.                .value()) {
29.            byte loginResult = (byte) message.getBody();
30.            if (loginResult != (byte) 0) {
31.                // 握手失败，关闭连接
32.                ctx.close();
33.            } else {
34.                System.out.println("Login is ok : " + message);
35.                ctx.fireChannelRead(msg);
36.            }
37.        } else
38.            ctx.fireChannelRead(msg);
39.    }
40.
41.    private NettyMessage buildLoginReq() {
```

```
42.    NettyMessage message = new NettyMessage();
43.    Header header = new Header();
44.    header.setType(MessageType.LOGIN_REQ.value());
45.    message.setHeader(header);
46.    return message;
47.    }
48.
49.     public void exceptionCaught(ChannelHandlerContext ctx, Throwable cause)
50.        throws Exception {
51.    ctx.fireExceptionCaught(cause);
52.    }
53. }
```

第 10~12 行,当客户端跟服务端 TCP 三次握手成功之后,由客户端构造握手请求消息发送给服务端,由于采用 IP 白名单认证机制,因此,不需要携带消息体,消息体为空,消息类型为 "3:握手请求消息"。握手请求发送之后,按照协议规范,服务端需要返回握手应答消息。

第 21~39 行对握手应答消息进行处理,首先判断消息是否是握手应答消息,如果不是,直接透传给后面的 ChannelHandler 进行处理;如果是握手应答消息,则对应答结果进行判断,如果非 0,说明认证失败,关闭链路,重新发起连接。

接着看服务端的握手接入和安全认证代码。

代码清单 12-8　LoginAuthRespHandler

```
1. public class LoginAuthRespHandler extends ChannelHandlerAdapter {
2.     private Map<String, Boolean> nodeCheck = new ConcurrentHashMap<String, Boolean>();
3.     private String[] whitekList = { "127.0.0.1", "192.168.1.104" };
4.
5.     /**
6.      * Calls {@link ChannelHandlerContext#fireChannelRead(Object)} to forward to
7.      * the next {@link ChannelHandler} in the {@link ChannelPipeline}.
8.      *
9.      * Sub-classes may override this method to change behavior.
10.     */
11.    @Override
12.    public void channelRead(ChannelHandlerContext ctx, Object msg)
```

```
13.        throws Exception {
14. NettyMessage message = (NettyMessage) msg;
15.
16.     // 如果是握手请求消息，处理，其他消息透传
17.     if (message.getHeader() != null
18.         && message.getHeader().getType() == MessageType.LOGIN_REQ
19.             .value()) {
20.         String nodeIndex = ctx.channel().remoteAddress().toString();
21.         NettyMessage loginResp = null;
22.         // 重复登录，拒绝
23.         if (nodeCheck.containsKey(nodeIndex)) {
24.          loginResp = buildResponse((byte) -1);
25.         } else {
26.          InetSocketAddress address = (InetSocketAddress) ctx.channel()
27.             .remoteAddress();
28.          String ip = address.getAddress().getHostAddress();
29.          boolean isOK = false;
30.          for (String WIP : whitekList) {
31.             if (WIP.equals(ip)) {
32.              isOK = true;
33.              break;
34.             }
35.         }
36.         loginResp = isOK ? buildResponse((byte) 0)
37.             : buildResponse((byte) -1);
38.         if (isOK)
39.             nodeCheck.put(nodeIndex, true);
40.         }
41.         System.out.println("The login response is : " + loginResp
42.             + " body [" + loginResp.getBody() + "]");
43.         ctx.writeAndFlush(loginResp);
44.     } else {
45.         ctx.fireChannelRead(msg);
46.     }
47.     }
48.
49.     private NettyMessage buildResponse(byte result) {
50. NettyMessage message = new NettyMessage();
51. Header header = new Header();
52. header.setType(MessageType.LOGIN_RESP.value());
53. message.setHeader(header);
```

```
54.       message.setBody(result);
55.       return message;
56.    }
57.
58.    public void exceptionCaught(ChannelHandlerContext ctx, Throwable cause)
59.       throws Exception {
60. nodeCheck.remove(ctx.channel().remoteAddress().toString());//删除缓存
61.    ctx.close();
62.    ctx.fireExceptionCaught(cause);
63.    }
64. }
```

第 2、3 行分别定义了重复登录保护和 IP 认证白名单列表，主要用于提升握手的可靠性。第 17～47 行用于接入认证，首先根据客户端的源地址（/127.0.0.1:12088）进行重复登录判断，如果客户端已经登录成功，拒绝重复登录，以防止由于客户端重复登录导致的句柄泄漏。随后通过 ChannelHandlerContext 的 Channel 接口获取客户端的 InetSocketAddress 地址，从中取得发送方的源地址信息，通过源地址进行白名单校验，校验通过握手成功，否则握手失败。最后通过 buildResponse 构造握手应答消息返回给客户端。

当发生异常关闭链路的时候，需要将客户端的信息从登录注册表中去注册，以保证后续客户端可以重连成功。

12.3.4 心跳检测机制

握手成功之后，由客户端主动发送心跳消息，服务端接收到心跳消息之后，返回心跳应答消息。由于心跳消息的目的是为了检测链路的可用性，因此不需要携带消息体。

客户端发送心跳请求消息的代码如下。

代码清单 12-9　HeartBeatReqHandler

```
1. public class HeartBeatReqHandler extends ChannelHandlerAdapter {
2.    private volatile ScheduledFuture<?> heartBeat;
3.
4.    @Override
5.    public void channelRead(ChannelHandlerContext ctx, Object msg)
6.       throws Exception {
7. NettyMessage message = (NettyMessage) msg;
8.    // 握手成功，主动发送心跳消息
9.    if (message.getHeader() != null
```

```
10.            && message.getHeader().getType() == MessageType.LOGIN_RESP
11.                .value()) {
12.         heartBeat = ctx.executor().scheduleAtFixedRate(
13.             new HeartBeatReqHandler.HeartBeatTask(ctx), 0, 5000,
14.             TimeUnit.MILLISECONDS);
15.    } else if (message.getHeader() != null
16.            && message.getHeader().getType() == MessageType.HEARTBEAT_RESP
17.                .value()) {
18.         System.out
19.             .println("Client receive server heart beat message : ---> "
20.                 + message);
21.    } else
22.         ctx.fireChannelRead(msg);
23.     }
24.
25.     private class HeartBeatTask implements Runnable {
26.  private final ChannelHandlerContext ctx;
27.
28.  public HeartBeatTask(final ChannelHandlerContext ctx) {
29.      this.ctx = ctx;
30.  }
31.
32.  @Override
33.  public void run() {
34.      NettyMessage heatBeat = buildHeatBeat();
35.      System.out
36.          .println("Client send heart beat messsage to server : ---> "
37.              + heatBeat);
38.      ctx.writeAndFlush(heatBeat);
39.  }
40.
41.  private NettyMessage buildHeatBeat() {
42.      NettyMessage message = new NettyMessage();
43.      Header header = new Header();
44.      header.setType(MessageType.HEARTBEAT_REQ.value());
45.      message.setHeader(header);
46.      return message;
47.  }
48.     }
49.
50.     @Override
```

```
51.     public void exceptionCaught(ChannelHandlerContext ctx, Throwable cause)
52.         throws Exception {
53.      if (heartBeat != null) {
54.         heartBeat.cancel(true);
55.         heartBeat = null;
56.      }
57.     ctx.fireExceptionCaught(cause);
58.         }
59.     }
```

首先看第 9 行,当握手成功之后,握手请求 Handler 会继续将握手成功消息向下透传,HeartBeatReqHandler 接收到之后对消息进行判断,如果是握手成功消息,则启动无限循环定时器用于定期发送心跳消息。由于 NioEventLoop 是一个 Schedule,因此它支持定时器的执行。心跳定时器的单位是毫秒,默认为 5000,即每 5 秒发送一条心跳消息。

为了统一在一个 Handler 中处理所有的心跳消息,因此第 15~20 行用于接收服务端发送的心跳应答消息,并打印客户端接收和发送的心跳消息。

心跳定时器 HeartBeatTask 的实现很简单,通过构造函数获取 ChannelHandlerContext,构造心跳消息并发送。

服务端的心跳应答 Handler 代码如下。

代码清单 12-10 HeartBeatRespHandler

```
1.  public class HeartBeatRespHandler extends ChannelHandlerAdapter {
2.     @Override
3.     public void channelRead(ChannelHandlerContext ctx, Object msg)
4.         throws Exception {
5.  NettyMessage message = (NettyMessage) msg;
6.  // 返回心跳应答消息
7.  if (message.getHeader() != null
8.       && message.getHeader().getType() == MessageType.HEARTBEAT_REQ
9.          .value()) {
10.     System.out.println("Receive client heart beat message : ---> "
11.         + message);
12.     NettyMessage heartBeat = buildHeatBeat();
13.     System.out
14.         .println("Send heart beat response message to client : ---> "
15.            + heartBeat);
16.     ctx.writeAndFlush(heartBeat);
17.  } else
```

```
18.         ctx.fireChannelRead(msg);
19.     }
20.
21.     private NettyMessage buildHeatBeat() {
22. NettyMessage message = new NettyMessage();
23. Header header = new Header();
24. header.setType(MessageType.HEARTBEAT_RESP.value());
25. message.setHeader(header);
26. return message;
27.     }
28. }
```

服务端的心跳 Handler 非常简单,接收到心跳请求消息之后,构造心跳应答消息返回,并打印接收和发送的心跳消息。

心跳超时的实现非常简单,直接利用 Netty 的 ReadTimeoutHandler 机制,当一定周期内(默认值 50s)没有读取到对方任何消息时,需要主动关闭链路。如果是客户端,重新发起连接;如果是服务端,释放资源,清除客户端登录缓存信息,等待服务端重连。

具体代码实现在下面的小节中会进行说明。

12.3.5 断连重连

当客户端感知断连事件之后,释放资源,重新发起连接,具体代码实现如图 12-4 所示。

```
// 发起异步连接操作
            ChannelFuture future = b.connect(
                new InetSocketAddress(host, port),
                new InetSocketAddress(NettyConstant.LOCALIP,
                    NettyConstant.LOCAL_PORT)).sync();
            future.channel().closeFuture().sync();
} finally {
            // 所有资源释放完成之后,清空资源,再次发起重连操作
executor.execute(new Runnable() {
                @Override
                public void run() {
                    try {
                        TimeUnit.SECONDS.sleep(1);
                        try {
                            connect(NettyConstant.PORT,
NettyConstant.REMOTEIP);// 发起重连操作
                        } catch (Exception e) {
                            e.printStackTrace();
                        }
```

图 12-4 客户端重连代码

首先监听网络断连事件，如果 Channel 关闭，则执行后续的重连任务，通过 Bootstrap 重新发起连接，客户端挂在 closeFuture 上监听链路关闭信号，一旦关闭，则创建重连定时器，5s 之后重新发起连接，直到重连成功。

服务端感知到断连事件之后，需要清空缓存的登录认证注册信息，以保证后续客户端能够正常重连。

12.3.6　客户端代码

客户端主要用于初始化系统资源，根据配置信息发起连接，代码如下。

代码清单 12-11　NettyClient

```
1.  public class NettyClient {
2.      private ScheduledExecutorService executor = Executors
3.       .newScheduledThreadPool(1);
4.      EventLoopGroup group = new NioEventLoopGroup();
5.      public void connect(int port, String host) throws Exception {
6.  // 配置客户端NIO线程组
7.      try {
8.          Bootstrap b = new Bootstrap();
9.          b.group(group).channel(NioSocketChannel.class)
10.          .option(ChannelOption.TCP_NODELAY, true)
11.          .handler(new ChannelInitializer<SocketChannel>() {
12.          @Override
13.          public void initChannel(SocketChannel ch)
14.              throws Exception {
15.              ch.pipeline().addLast(
16.                  new NettyMessageDecoder(1024 * 1024, 4, 4));
17.              ch.pipeline().addLast("MessageEncoder",
18.                  new NettyMessageEncoder());
19.          ch.pipeline().addLast("readTimeoutHandler",
20.                  new ReadTimeoutHandler(50));
21.              ch.pipeline().addLast("LoginAuthHandler",
22.                  new LoginAuthReqHandler());
23.              ch.pipeline().addLast("HeartBeatHandler",
24.                  new HeartBeatReqHandler());
25.          }
26.          });
27.      // 发起异步连接操作
```

```
28.     ChannelFuture future = b.connect(
29.         new InetSocketAddress(host, port),
30.         new InetSocketAddress(NettyConstant.LOCALIP,
31.             NettyConstant.LOCAL_PORT)).sync();
32.     future.channel().closeFuture().sync();
33. } finally {
34.     // 所有资源释放完成之后,清空资源,再次发起重连操作
35.     executor.execute(new Runnable() {
36.      @Override
37.      public void run() {
38.         try {
39.          TimeUnit.SECONDS.sleep(5);
40.          try {
41.             connect(NettyConstant.PORT, NettyConstant.REMOTEIP);// 发起重连操作
42.          } catch (Exception e) {
43.             e.printStackTrace();
44.          }
45.         } catch (InterruptedException e) {
46.          e.printStackTrace();
47.         }
48.      }
49.     });
50. }
51.    }
52.
53.    /**
54.     * @param args
55.     * @throws Exception
56.     */
57.    public static void main(String[] args) throws Exception {
58. new              NettyClient().connect(NettyConstant.PORT, NettyConstant.REMOTEIP);
59.    }
60. }
```

第15和16行增加了 NettyMessageDecoder 用于 Netty 消息解码,为了防止由于单条消息过大导致的内存溢出或者畸形码流导致解码错位引起内存分配失败,我们对单条消息最大长度进行了上限限制。第17和18行新增了 Netty 消息编码器,用于协议消息的自动编码。随后依次增加了读超时 Handler、握手请求 Handler 和心跳消息 Handler。

第 28 行发起 TCP 连接的代码与之前的不同，这次我们绑定了本地端口，主要用于服务端重复登录保护，另外，从产品管理角度看，一般情况下不允许系统随便使用随机端口。

利用 Netty 的 ChannelPipeline 和 ChannelHandler 机制，可以非常方便地实现功能解耦和业务产品的定制。例如本例程中的心跳定时器、握手请求和后端的业务处理可以通过不同的 Handler 来实现，类似于 AOP。通过 Handler Chain 的机制可以方便地实现切面拦截和定制，相比于 AOP 它的性能更高。

12.3.7 服务端代码

相对于客户端，服务端的代码更简单一些，主要的工作就是握手的接入认证等，不用关心断连重连等事件。

服务端的代码如下。

代码清单 12-12　NettyServer

```
1. public class NettyServer {
2.     public void bind() throws Exception {
3.     // 配置服务端的NIO线程组
4.     EventLoopGroup bossGroup = new NioEventLoopGroup();
5.     EventLoopGroup workerGroup = new NioEventLoopGroup();
6.     ServerBootstrap b = new ServerBootstrap();
7.     b.group(bossGroup, workerGroup).channel(NioServerSocketChannel.class)
8.          .option(ChannelOption.SO_BACKLOG, 100)
9.          .handler(new LoggingHandler(LogLevel.INFO))
10.         .childHandler(new ChannelInitializer<SocketChannel>() {
11.            @Override
12.            public void initChannel(SocketChannel ch)
13.                throws IOException {
14.             ch.pipeline().addLast(
15.                 new NettyMessageDecoder(1024 * 1024, 4, 4));
16.             ch.pipeline().addLast(new NettyMessageEncoder());
17. ch.pipeline().addLast("readTimeoutHandler",
18.                 new ReadTimeoutHandler(50));
19.             ch.pipeline().addLast(new LoginAuthRespHandler());
20.             ch.pipeline().addLast("HeartBeatHandler",
21.                 new HeartBeatRespHandler());
22.           }
```

```
23.             });
24.
25.     // 绑定端口，同步等待成功
26.     b.bind(NettyConstant.REMOTEIP, NettyConstant.PORT).sync();
27.     System.out.println("Netty server start ok : "
28.         + (NettyConstant.REMOTEIP + " : " + NettyConstant.PORT));
29.     }
30.
31.     public static void main(String[] args) throws Exception {
32. new NettyServer().bind();
33.     }
34. }
```

与客户端不同的是，服务端 ChannelPipeline 中除了 Netty 编码器和解码器以外，还有握手和接入认证的 LoginAuthRespHandler 和心跳应答 HeartBeatRespHandler。

12.4 运行协议栈

12.4.1 正常场景

启动服务端，待服务端启动成功之后启动客户端，检查链路是否建立成功，是否每隔 5s 互发一次心跳请求和应答，运行结果如图 12-5 所示。

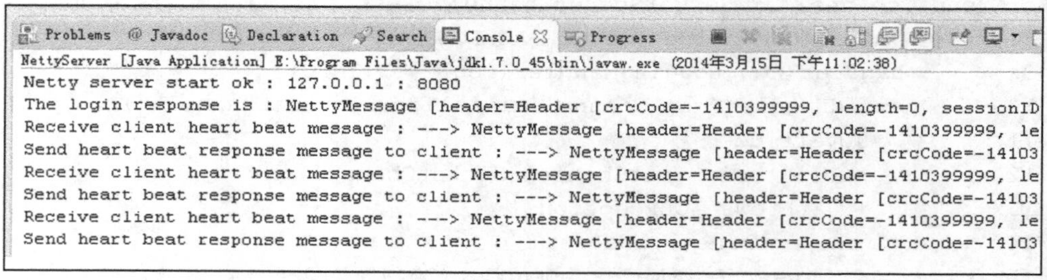

图 12-5 服务端运行结果

客户端运行结果如图 12-6 所示。

从上面的运行结果可以看出，客户端和服务端握手成功，双方可以互发心跳，链路正常，如图 12-7 所示。

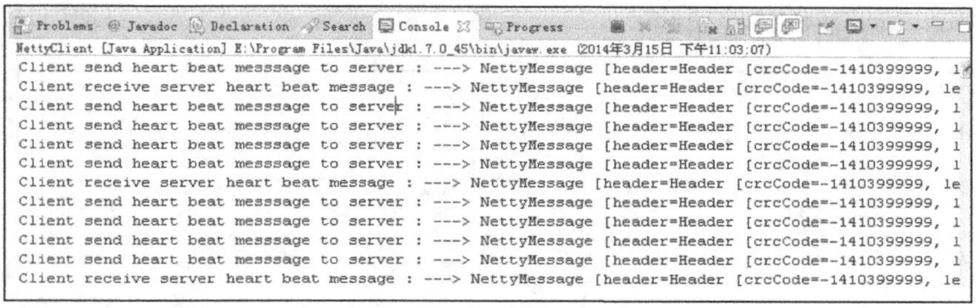

图 12-6　客户端运行结果

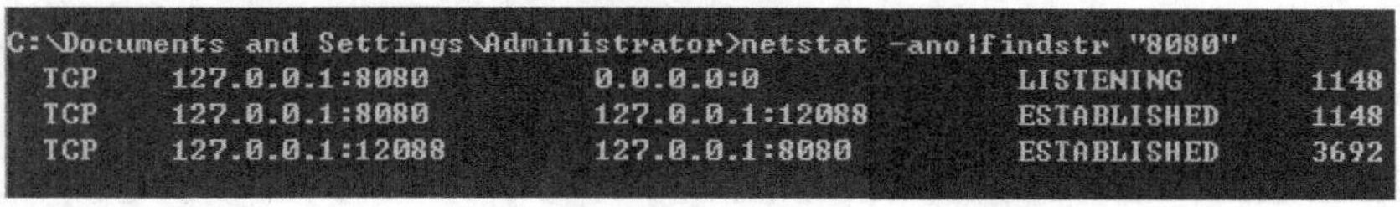

图 12-7　TCP 链接正常

12.4.2　异常场景：服务端宕机重启

假设服务端宕机一段时间重启，检验如下功能是否正常。

（1）客户端是否能够正常发起重连；

（2）重连成功之后，不再重连；

（3）断连期间，心跳定时器停止工作，不再发送心跳请求消息；

（4）服务端重启成功之后，允许客户端重新登录；

（5）服务端重启成功之后，客户端能够重连和握手成功；

（6）重连成功之后，双方的心跳能够正常互发；

（7）性能指标：重连期间，客户端资源得到了正常回收，不会导致句柄等资源泄漏。

服务端重启之前的客户端资源占用如图 12-8 所示。

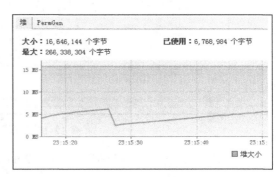

图 12-8　客户端堆内存占用

线程资源占用如图 12-9 所示。

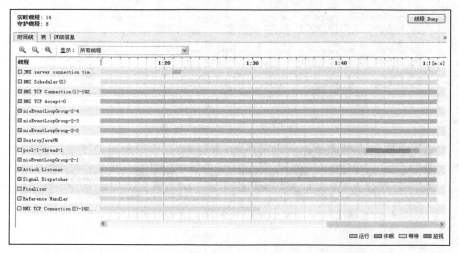

图 12-9　客户端线程资源信息列表

服务端宕机之后,重启之前,客户端周期性重连失败。如图 12-10 所示。

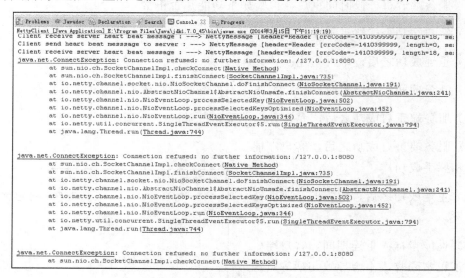

图 12-10　客户端重连失败

重连期间线程资源占用正常,如图 12-11 所示。

重连期间内存占用正常,如图 12-12 所示。

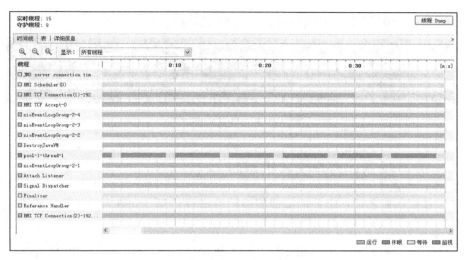

图 12-11　重连期间线程资源占用正常

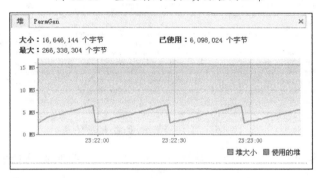

图 12-12　重连期间内存占用正常

服务端重启成功，握手成功，链路重新恢复，如图 12-13 所示。

```
        at java.lang.Thread.run(Thread.java:744)
java.net.ConnectException: Connection refused: no further information: /127.0.0.1:8080
        at sun.nio.ch.SocketChannelImpl.checkConnect(Native Method)
        at sun.nio.ch.SocketChannelImpl.finishConnect(SocketChannelImpl.java:735)
        at io.netty.channel.socket.nio.NioSocketChannel.doFinishConnect(NioSocketChannel.java:191)
        at io.netty.channel.nio.AbstractNioChannel$AbstractNioUnsafe.finishConnect(AbstractNioChannel.java:241)
        at io.netty.channel.nio.NioEventLoop.processSelectedKey(NioEventLoop.java:502)
        at io.netty.channel.nio.NioEventLoop.processSelectedKeysOptimized(NioEventLoop.java:452)
        at io.netty.channel.nio.NioEventLoop.run(NioEventLoop.java:346)
        at io.netty.util.concurrent.SingleThreadEventExecutor$5.run(SingleThreadEventExecutor.java:794)
        at java.lang.Thread.run(Thread.java:744)
Login is ok : NettyMessage [header=Header [crcCode=-1410399999, length=93, sessionID=0, type=4, priority=0, att
Client send heart beat messsage to server : ---> NettyMessage [header=Header [crcCode=-1410399999, length=0, se
Client receive server heart beat message : ---> NettyMessage [header=Header [crcCode=-1410399999, length=18, se
Client send heart beat messsage to server : ---> NettyMessage [header=Header [crcCode=-1410399999, length=0, se
Client receive server heart beat message : ---> NettyMessage [header=Header [crcCode=-1410399999, length=18, se
Client send heart beat messsage to server : ---> NettyMessage [header=Header [crcCode=-1410399999, length=0, se
Client receive server heart beat message : ---> NettyMessage [header=Header [crcCode=-1410399999, length=18, se
```

图 12-13　服务端重启成功后链路恢复

通过 netstat 命令查看 TCP 连接状态，如图 12-14 所示。

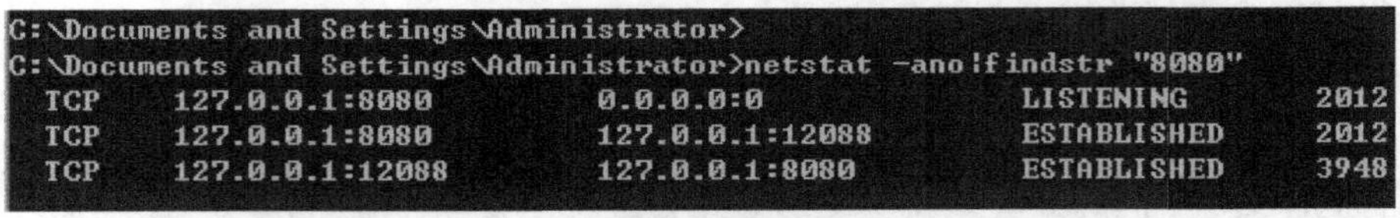

```
C:\Documents and Settings\Administrator>
C:\Documents and Settings\Administrator>netstat -ano|findstr "8080"
  TCP    127.0.0.1:8080         0.0.0.0:0              LISTENING       2012
  TCP    127.0.0.1:8080         127.0.0.1:12088        ESTABLISHED     2012
  TCP    127.0.0.1:12088        127.0.0.1:8080         ESTABLISHED     3948
```

图 12-14　TCP 连接正常

12.4.3　异常场景：客户端宕机重启

客户端宕机重启之后，服务端需要能够清除缓存信息，允许客户端重新登录。下面看测试结果。

客户端停机，然后重启，结果如图 12-15 所示。

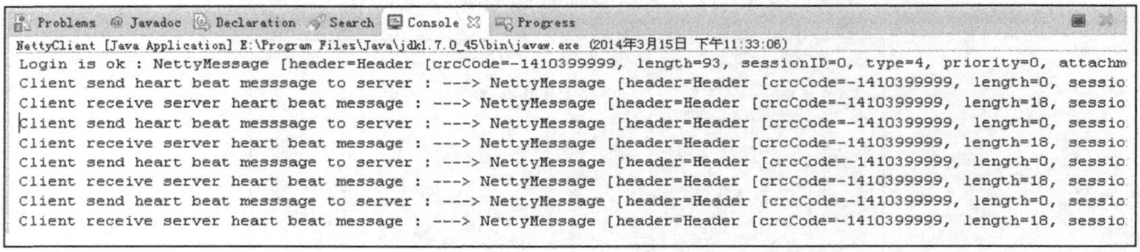

```
Problems  @ Javadoc  Declaration  Search  Console  Progress
NettyClient [Java Application] E:\Program Files\Java\jdk1.7.0_45\bin\javaw.exe (2014年3月15日 下午11:33:06)
Login is ok : NettyMessage [header=Header [crcCode=-1410399999, length=93, sessionID=0, type=4, priority=0, attachm
Client send heart beat messsage to server : ---> NettyMessage [header=Header [crcCode=-1410399999, length=0, sessio
Client receive server heart beat message : ---> NettyMessage [header=Header [crcCode=-1410399999, length=18, sessio
Client send heart beat messsage to server : ---> NettyMessage [header=Header [crcCode=-1410399999, length=0, sessio
Client receive server heart beat message : ---> NettyMessage [header=Header [crcCode=-1410399999, length=18, sessio
Client send heart beat messsage to server : ---> NettyMessage [header=Header [crcCode=-1410399999, length=0, sessio
Client receive server heart beat message : ---> NettyMessage [header=Header [crcCode=-1410399999, length=18, sessio
Client send heart beat messsage to server : ---> NettyMessage [header=Header [crcCode=-1410399999, length=0, sessio
Client receive server heart beat message : ---> NettyMessage [header=Header [crcCode=-1410399999, length=18, sessio
```

图 12-15　客户端宕机重启重新登录

运行结果表明客户端重启之后可以重新登录成功，说明服务端功能正常。

12.5　总结

本章首先介绍了私有协议栈的相关概念，然后通过一个模拟私有协议栈——Netty 协议栈的设计和开发，让读者掌握私有协议栈的功能和开发要点，为后续在实际工作中进行私有协议栈的设计和开发提供帮助。

尽管本章节在设计 Netty 协议栈的时候，已经考虑了很多可靠性方面的功能，但是对于实际商用协议栈而言，仍然是不足的。例如当链路断连的时候，已经放入发送队列中的消息不能丢失，更加通用的做法是提供通知机制，将发送失败的消息通知给业务侧，由业务做决定：是丢弃还是缓存重发。

本章综合了之前所学的 Netty 知识，还涉及到了通用半包解码器、读超时、自定义定时任务、安全认证等方面的知识，当读者能够综合运用所学知识进行灵活设计和开发时，说明对 Netty 的掌握程度更上了一层楼。

需要指出的是，本例程仅仅是个简单 Demo，限于篇幅，一些实现未必是最优的，读者在学习过程中也可以思考下哪些地方还可以进一步优化。

第 13 章

服务端创建

对于想要深入学习 Netty 原理的人而言，通过阅读源码是最有效的学习方式之一。尽管 Netty 使用起来并不复杂，但是通过对源码的分析和学习，掌握一些必备的基础知识还是很有必要的。

Netty 服务端创建需要的必备知识如下：

（1）熟悉 JDK NIO 主要类库的使用，例如 ByteBuffer、Selector、ServerSocketChannel 等；

（2）熟悉 JDK 的多线程编程；

（3）了解 Reactor 模式。

本文首先对 Java NIO 服务端的创建进行简单介绍，然后对 Netty 服务端的创建进行原理讲解和源码分析，以期让更多希望了解 Netty 底层原理的读者可以快速入门。

本章主要内容包括：

◎ 原生 NIO 类库的复杂性

◎ Netty 服务端创建源码分析

◎ 客户端接入源码分析

13.1 原生 NIO 类库的复杂性

在开始本文之前,我先讲一件自己亲身经历的事。大约在 2011 年的时候,身边有两个业务团队同时进行新版本开发,他们都需要基于 NIO 非阻塞特性构建高性能、异步和高可靠性的底层通信框架。

当时两个项目组的设计师都咨询了我的意见,在了解了两个项目团队的 NIO 编程经验和现状之后,我建议他们都使用 Netty 构建业务通信框架。令人遗憾的是其中 1 个项目组并没有按照我的建议做,而是选择直接基于 JDK 的 NIO 类库构建自己的通信框架。在他们看来,构建业务层的 NIO 通信框架并不是件难事,即便当前他们还缺乏相关经验。

两个多月过去之后,自研 NIO 框架团队的通信框架始终无法稳定地工作,他们频繁遭遇客户端断连、句柄泄露和消息丢失等问题,项目的进度出现了严重的延迟。形成鲜明对比的是,另一个团队由于基于 Netty 研发,在通信框架上节省了大量的人力和时间,加之 Netty 自身的可靠性和稳定性非常好,他们的项目进展非常顺利。

这两个项目组的不同遭遇告诉我们:开发高质量的 NIO 程序并不是一件简单的事情,除去 NIO 类库的固有复杂性和 Bug,作为 NIO 服务端,需要能够处理网络的闪断、客户端的重连、安全认证和消息的编解码、半包处理等。如果没有足够的 NIO 编程经验积累,自研 NIO 框架往往需要半年甚至数年的时间才能最终稳定下来,这种成本即便对一个大公司而言也是个严重的挑战。

13.2 Netty 服务端创建源码分析

当我们直接使用 JDK NIO 的类库开发基于 NIO 的异步服务端时,需要使用到多路复用器 Selector、ServerSocketChannel、SocketChannel、ByteBuffer、SelectionKey 等,相比于传统的 BIO 开发,NIO 的开发要复杂很多,开发出稳定、高性能的异步通信框架,一直是个难题。

Netty 为了向使用者屏蔽 NIO 通信的底层细节,在和用户交互的边界做了封装,目的就是为了减少用户开发工作量,降低开发难度。ServerBootstrap 是 Socket 服务端的启动辅助类,用户通过 ServerBootstrap 可以方便地创建 Netty 的服务端。时序图如图 13-1 所示。

13.2.1 Netty 服务端创建时序图

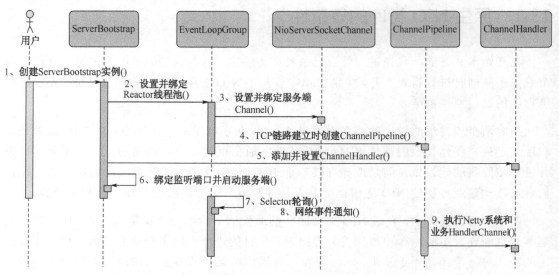

图 13-1 Netty 服务端创建时序图

下面我们对 Netty 服务端创建的关键步骤和原理进行讲解。

步骤 1：创建 ServerBootstrap 实例。ServerBootstrap 是 Netty 服务端的启动辅助类，它提供了一系列的方法用于设置服务端启动相关的参数。底层通过门面模式对各种能力进行抽象和封装，尽量不需要用户跟过多的底层 API 打交道，以降低用户的开发难度。

我们在创建 ServerBootstrap 实例时，会惊讶地发现 ServerBootstrap 只有一个无参的构造函数，作为启动辅助类这让人不可思议，因为它需要与多个其他组件或者类交互。ServerBootstrap 构造函数没有参数的根本原因是因为它的参数太多了，而且未来也可能会发生变化，为了解决这个问题，就需要引入 Builder 模式。《Effective Java》第二版第 2 条建议遇到多个构造器参数时要考虑用构建器，关于多个参数构造函数的缺点和使用构建器的优点大家可以查阅《Effective Java》，在此不再详述。

步骤 2：设置并绑定 Reactor 线程池。Netty 的 Reactor 线程池是 EventLoopGroup，它实际就是 EventLoop 的数组。EventLoop 的职责是处理所有注册到本线程多路复用器 Selector 上的 Channel，Selector 的轮询操作由绑定的 EventLoop 线程 run 方法驱动，在一个循环体内循环执行。值得说明的是，EventLoop 的职责不仅仅是处理网络 I/O 事件，用户自定义的 Task 和定时任务 Task 也统一由 EventLoop 负责处理,这样线程模型就实现了统一。从调度层面看,也不存在从 EventLoop 线程中再启动其他类型的线程用于异步执行另外的任务,这样就避免了多线程并发操作和锁竞争,提升了 I/O 线程的处理和调度性能。

步骤 3:设置并绑定服务端 Channel。作为 NIO 服务端,需要创建 ServerSocketChannel,Netty 对原生的 NIO 类库进行了封装,对应实现是 NioServerSocketChannel。对于用户而言,不需要关心服务端 Channel 的底层实现细节和工作原理,只需要指定具体使用哪种服务端 Channel 即可。因此,Netty 的 ServerBootstrap 方法提供了 channel 方法用于指定服务端 Channel 的类型。Netty 通过工厂类,利用反射创建 NioServerSocketChannel 对象。由于服务端监听端口往往只需要在系统启动时才会调用,因此反射对性能的影响并不大。相关代码如下。

```
public ServerBootstrap channel(Class<? extends ServerChannel> channelClass) {
        if (channelClass == null) {
            throw new NullPointerException("channelClass");
        }
        return channelFactory(new ServerBootstrapChannelFactory<ServerChannel>(channelClass));
    }
```

步骤 4:链路建立的时候创建并初始化 ChannelPipeline。ChannelPipeline 并不是 NIO 服务端必需的,它本质就是一个负责处理网络事件的职责链,负责管理和执行 ChannelHandler。网络事件以事件流的形式在 ChannelPipeline 中流转,由 ChannelPipeline 根据 ChannelHandler 的执行策略调度 ChannelHandler 的执行。典型的网络事件如下。

(1) 链路注册;

(2) 链路激活;

(3) 链路断开;

(4) 接收到请求消息;

(5) 请求消息接收并处理完毕;

(6) 发送应答消息;

(7) 链路发生异常;

(8) 发生用户自定义事件。

步骤 5：初始化 ChannelPipeline 完成之后，添加并设置 ChannelHandler。ChannelHandler 是 Netty 提供给用户定制和扩展的关键接口。利用 ChannelHandler 用户可以完成大多数的功能定制，例如消息编解码、心跳、安全认证、TSL/SSL 认证、流量控制和流量整形等。Netty 同时也提供了大量的系统 ChannelHandler 供用户使用，比较实用的系统 ChannelHandler 总结如下。

（1）系统编解码框架——ByteToMessageCodec；

（2）通用基于长度的半包解码器——LengthFieldBasedFrameDecoder；

（3）码流日志打印 Handler——LoggingHandler；

（4）SSL 安全认证 Handler——SslHandler；

（5）链路空闲检测 Handler——IdleStateHandler；

（6）流量整形 Handler——ChannelTrafficShapingHandler；

（7）Base64 编解码——Base64Decoder 和 Base64Encoder。

创建和添加 ChannelHandler 的代码示例如下。

```
.childHandler(new ChannelInitializer<SocketChannel>() {
    @Override
    public void initChannel(SocketChannel ch)
        throws Exception {
        ch.pipeline().addLast(
            new EchoServerHandler());
    }
});
```

步骤 6：绑定并启动监听端口。在绑定监听端口之前系统会做一系列的初始化和检测工作，完成之后，会启动监听端口，并将 ServerSocketChannel 注册到 Selector 上监听客户端连接，相关代码如下。

```
protected void doBind(SocketAddress localAddress) throws Exception {
    javaChannel().socket().bind(localAddress, config.getBacklog());
}
```

步骤 7：Selector 轮询。由 Reactor 线程 NioEventLoop 负责调度和执行 Selector 轮询操作，选择准备就绪的 Channel 集合，相关代码如下。

```
private void select() throws IOException {
    Selector selector = this.selector;
    try {
        //此处代码省略...
        int selectedKeys = selector.select(timeoutMillis);
        selectCnt ++;
        //此处代码省略...
}
```

步骤 8：当轮询到准备就绪的 Channel 之后，就由 Reactor 线程 NioEventLoop 执行 ChannelPipeline 的相应方法，最终调度并执行 ChannelHandler，接口如图 13-2 所示。

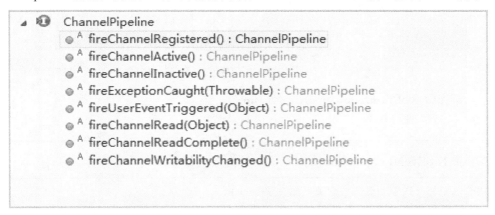

图 13-2　调度相关方法

步骤 9：执行 Netty 系统 ChannelHandler 和用户添加定制的 ChannelHandler。ChannelPipeline 根据网络事件的类型，调度并执行 ChannelHandler，相关代码如下。

```
public ChannelHandlerContext fireChannelRead(Object msg) {
    DefaultChannelHandlerContext next = findContextInbound(MASK_CHANNEL_READ);
    next.invoker.invokeChannelRead(next, msg);
    return this;
}
```

13.2.2　Netty 服务端创建源码分析

首先通过构造函数创建 ServerBootstrap 实例，随后，通常会创建两个 EventLoopGroup

（并不是必须要创建两个不同的 EventLoopGroup，也可以只创建一个并共享），代码如下。

```
EventLoopGroup acceptorGroup = new NioEventLoopGroup();
EventLoopGroup IOGroup = new NioEventLoopGroup();
```

NioEventLoopGroup 实际就是 Reactor 线程池，负责调度和执行客户端的接入、网络读写事件的处理、用户自定义任务和定时任务的执行。通过 ServerBootstrap 的 group 方法将两个 EventLoopGroup 实例传入，代码如下。

```
public ServerBootstrap group(EventLoopGroup parentGroup, EventLoopGroup childGroup) {
    super.group(parentGroup);
    if (childGroup == null) {
        throw new NullPointerException("childGroup");
    }
    if (this.childGroup != null) {
        throw new IllegalStateException("childGroup set already");
    }
    this.childGroup = childGroup;
    return this;
}
```

其中父 NioEventLoopGroup 被传入了父类构造函数中，代码如下。

```
public B group(EventLoopGroup group) {
    if (group == null) {
        throw new NullPointerException("group");
    }
    if (this.group != null) {
        throw new IllegalStateException("group set already");
    }
    this.group = group;
    return (B) this;
}
```

该方法会被客户端和服务端重用，用于设置工作 I/O 线程，执行和调度网络事件的读写。

线程组和线程类型设置完成后，需要设置服务端 Channel 用于端口监听和客户端链路接入。Netty 通过 Channel 工厂类来创建不同类型的 Channel，对于服务端，需要创建 NioServerSocketChannel。所以，通过指定 Channel 类型的方式创建 Channel 工厂。

ServerBootstrapChannelFactory 是 ServerBootstrap 的内部静态类，职责是根据 Channel 的类型通过反射创建 Channel 的实例，服务端需要创建的是 NioServerSocketChannel 实例，代码如下。

```
public T newChannel(EventLoop eventLoop, EventLoopGroup childGroup) {
    try {
        Constructor<? extends T> constructor = clazz.getConstructor(EventLoop.class, EventLoopGroup.class);
        return constructor.newInstance(eventLoop, childGroup);
    } catch (Throwable t) {
        throw new ChannelException("Unable to create Channel from class " + clazz, t);
    }
}
```

指定 NioServerSocketChannel 后，需要设置 TCP 的一些参数，作为服务端，主要是要设置 TCP 的 backlog 参数，底层 C 的对应接口定义如下。

```
int listen(int fd, int backlog);
```

backlog 指定了内核为此套接口排队的最大连接个数，对于给定的监听套接口，内核要维护两个队列：未链接队列和已连接队列，根据 TCP 三路握手过程中三个分节来分隔这两个队列。服务器处于 listen 状态时，收到客户端 syn 分节（connect）时在未完成队列中创建一个新的条目，然后用三路握手的第二个分节即服务器的 syn 响应客户端，此条目在第三个分节到达前（客户端对服务器 syn 的 ack）一直保留在未完成连接队列中，如果三路握手完成，该条目将从未完成连接队列搬到已完成连接队列尾部。当进程调用 accept 时，从已完成队列中的头部取出一个条目给进程，当已完成队列为空时进程将睡眠，直到有条目在已完成连接队列中才唤醒。backlog 被规定为两个队列总和的最大值，大多数实现默认值为 5，但在高并发 Web 服务器中此值显然不够，Lighttpd 中此值达到 128×8。需要设置此值更大一些的原因是未完成连接队列的长度可能因为客户端 syn 的到达及等待三路握手第三个分节的到达延时而增大。Netty 默认的 backlog 为 100，当然，用户可以修改默认值，这需要根据实际场景和网络状况进行灵活设置。

TCP 参数设置完成后，用户可以为启动辅助类和其父类分别指定 Handler。两类 Handler 的用途不同：子类中的 Handler 是 NioServerSocketChannel 对应的 ChannelPipeline 的 Handler；父类中的 Handler 是客户端新接入的连接 SocketChannel 对应的 ChannelPipeline 的 Handler。两者的区别可以通过图 13-3 来展示。

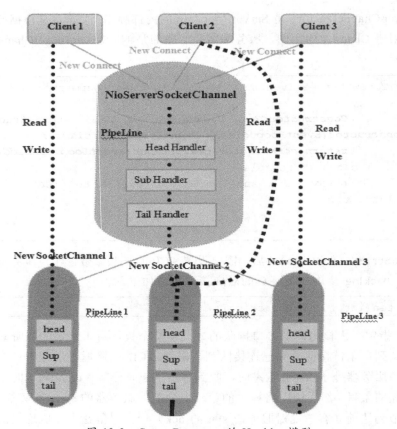

图 13-3　ServerBootstrap 的 Hanlder 模型

　　本质区别就是：ServerBootstrap 中的 Handler 是 NioServerSocketChannel 使用的，所有连接该监听端口的客户端都会执行它；父类 AbstractBootstrap 中的 Handler 是个工厂类，它为每个新接入的客户端都创建一个新的 Handler。

　　服务端启动的最后一步，就是绑定本地端口，启动服务，下面我们来分析下这部分代码。

```
private ChannelFuture doBind(final SocketAddress localAddress) {
    final ChannelFuture regFuture = initAndRegister();NO. 1
    final Channel channel = regFuture.channel();
    if (regFuture.cause() != null) {
        return regFuture;
    }
```

```
        final ChannelPromise promise;
        if (regFuture.isDone()) { NO. 2
            promise = channel.newPromise();
            doBind0(regFuture, channel, localAddress, promise);
        } else {
            promise       =       new       DefaultChannelPromise(channel,
GlobalEventExecutor.INSTANCE);
            regFuture.addListener(new ChannelFutureListener() {
                @Override NO. 3
                public void operationComplete(ChannelFuture future) throws
Exception {
                    doBind0(regFuture, channel, localAddress, promise);
                }
            });
        }
        return promise;
    }
```

先看下 NO.1。首先创建 Channel，createChannel 由子类 ServerBootstrap 实现，创建新的 NioServerSocketChannel。它有两个参数：参数 1 是从父类的 NIO 线程池中顺序获取一个 NioEventLoop，它就是服务端用于监听和接收客户端连接的 Reactor 线程；参数 2 是所谓的 workerGroup 线程池，它就是处理 I/O 读写的 Reactor 线程组，相关代码如下。

```
    final ChannelFuture initAndRegister() {
        Channel channel;
        try {
            channel = createChannel();
        } catch (Throwable t) {
            return VoidChannel.INSTANCE.newFailedFuture(t);
        }
        try {
            init(channel);
        } catch (Throwable t) {
            channel.unsafe().closeForcibly();
            return channel.newFailedFuture(t);
        }
        //后续代码省略.....
    }
```

NioServerSocketChannel 创建成功后，对它进行初始化，初始化工作主要有以下三点。

（1）设置 Socket 参数和 NioServerSocketChannel 的附加属性，代码如下。

```
void init(Channel channel) throws Exception {
    final Map<ChannelOption<?>, Object> options = options();
    synchronized (options) {
        channel.config().setOptions(options);
    }

    final Map<AttributeKey<?>, Object> attrs = attrs();
    synchronized (attrs) {
        for (Entry<AttributeKey<?>, Object> e: attrs.entrySet()) {
            AttributeKey<Object> key = (AttributeKey<Object>) e.getKey();
            channel.attr(key).set(e.getValue());
        }
    }
}
```

（2）将 AbstractBootstrap 的 Handler 添加到 NioServerSocketChannel 的 ChannelPipeline 中，代码如下。

```
ChannelPipeline p = channel.pipeline();
    if (handler() != null) {
        p.addLast(handler());
    }
```

（3）将用于服务端注册的 Handler ServerBootstrapAcceptor 添加到 ChannelPipeline 中，代码如下。

```
p.addLast(new ChannelInitializer<Channel>() {
            @Override
            public void initChannel(Channel ch) throws Exception {
                ch.pipeline().addLast(new
ServerBootstrapAcceptor(currentChildHandler, currentChildOptions,
                    currentChildAttrs));
            }
        });
```

到此，Netty 服务端监听的相关资源已经初始化完毕，就剩下最后一步——注册 NioServerSocketChannel 到 Reactor 线程的多路复用器上，然后轮询客户端连接事件。在分析注册代码之前，我们先通过图 13-4 看看目前 NioServerSocketChannel 的 ChannelPipeline 的组成。

图 13-4　NioServerSocketChannel 的 ChannelPipeline

最后，我们看下 NioServerSocketChannel 的注册。当 NioServerSocketChannel 初始化完成之后，需要将它注册到 Reactor 线程的多路复用器上监听新客户端的接入，代码如下。

```
public final void register(final ChannelPromise promise) {
    if (eventLoop.inEventLoop()) {
        register0(promise);
    } else {
        try {
            eventLoop.execute(new Runnable() {
                public void run() {
                    register0(promise);
                }
            });
        } catch (Throwable t) {
            //此处代码省略......
    }
}
```

首先判断是否是 NioEventLoop 自身发起的操作。如果是，则不存在并发操作，直接执行 Channel 注册；如果由其他线程发起，则封装成一个 Task 放入消息队列中异步执行。此处，由于是由 ServerBootstrap 所在线程执行的注册操作，所以会将其封装成 Task 投递到 NioEventLoop 中执行，代码如下。

```
private void register0(ChannelPromise promise) {
    try {
        if (!ensureOpen(promise)) {
            return;
        }
        doRegister();
        registered = true;
        promise.setSuccess();
        pipeline.fireChannelRegistered();
        if (isActive()) {
            pipeline.fireChannelActive();
        }
    } catch (Throwable t) {
```

```
            //此处代码省略...
    }
```

将 NioServerSocketChannel 注册到 NioEventLoop 的 Selector 上，代码如下：

```
protected void doRegister() throws Exception {
    boolean selected = false;
    for (;;) {
        try {
            selectionKey = javaChannel().register(eventLoop().selector, 0, this);
            //此处代码省略...
    }
```

大家可能会很诧异，应该注册 OP_ACCEPT（16）到多路复用器上，怎么注册 0 呢？0 表示只注册，不监听任何网络操作。这样做的原因如下。

（1）注册方法是多态的，它既可以被 NioServerSocketChannel 用来监听客户端的连接接入，也可以注册 SocketChannel 用来监听网络读或者写操作；

（2）通过 SelectionKey 的 interestOps(int ops)方法可以方便地修改监听操作位。所以，此处注册需要获取 SelectionKey 并给 AbstractNioChannel 的成员变量 selectionKey 赋值。

注册成功之后，触发 ChannelRegistered 事件，方法如下。

```
            promise.setSuccess();
            pipeline.fireChannelRegistered();
```

当 ChannelRegistered 事件传递到 TailHandler 后结束，TailHandler 也不关心 ChannelRegistered 事件，因此是空实现，代码如下。

```
@Override
    public void channelRegistered(ChannelHandlerContext ctx) throws Exception { }
```

ChannelRegistered 事件传递完成后，判断 ServerSocketChannel 监听是否成功，如果成功，需要出发 NioServerSocketChannel 的 ChannelActive 事件，代码如下。

```
    if (isActive()) {
            pipeline.fireChannelActive();
        }
```

isActive()也是个多态方法。如果是服务端，判断监听是否启动；如果是客户端，判断

TCP 连接是否完成。ChannelActive 事件在 ChannelPipeline 中传递，完成之后根据配置决定是否自动触发 Channel 的读操作，代码如下。

```
public ChannelPipeline fireChannelActive() {
    head.fireChannelActive();
    if (channel.config().isAutoRead()) {
        channel.read();
    }
    return this;
}
```

AbstractChannel 的读操作触发 ChannelPipeline 的读操作，最终调用到 HeadHandler 的读方法，代码如下。

```
public void read(ChannelHandlerContext ctx) {
    unsafe.beginRead();
}
```

继续看 AbstractUnsafe 的 beginRead 方法，代码如下。

```
public void beginRead() {
    if (!isActive()) {
        return;
    }
    try {
        doBeginRead();
//后续代码省略……
}
```

由于不同类型的 Channel 对读操作的准备工作不同，因此，beginRead 也是个多态方法，对于 NIO 通信，无论是客户端还是服务端，都是要修改网络监听操作位为自身感兴趣的，对于 NioServerSocketChannel 感兴趣的操作是 OP_ACCEPT（16），于是重新修改注册的操作位为 OP_ACCEPT，代码如下。

```
protected void doBeginRead() throws Exception {
    //以上代码省略……
    final int interestOps = selectionKey.interestOps();
    if ((interestOps & readInterestOp) == 0) {
        selectionKey.interestOps(interestOps | readInterestOp);
    }
}
```

在某些场景下，当前监听的操作类型和 Chanel 关心的网络事件是一致的，不需要重复注册，所以增加了&操作的判断，只有两者不一致，才需要重新注册操作位。

JDK SelectionKey 有 4 种操作类型，分别为：

（1）OP_READ = 1 << 0；

（2）OP_WRITE = 1 << 2；

（3）OP_CONNECT = 1 << 3；

（4）OP_ACCEPT = 1 << 4。

由于只有 4 种网络操作类型，所以用 4 bit 就可以表示所有的网络操作位，由于 Java 语言没有 bit 类型，所以使用了整型来表示，每个操作位代表一种网络操作类型，分别为：0001、0010、0100、1000，这样做的好处是可以非常方便地通过位操作来进行网络操作位的状态判断和状态修改，从而提升操作性能。

由于创建 NioServerSocketChannel 将 readInterestOp 设置成了 OP_ACCEPT，所以，在服务端链路注册成功之后重新将操作位设置为监听客户端的网络连接操作，初始化 NioServerSocketChannel 的代码如下。

```
    public NioServerSocketChannel(EventLoop eventLoop, EventLoopGroup childGroup) {
        super(null, eventLoop, childGroup, newSocket(),
SelectionKey.OP_ACCEPT);
        config = new DefaultServerSocketChannelConfig(this,
javaChannel().socket());
    }
```

到此，服务端监听启动部分源码已经分析完成，下一章节，让我们继续分析一个新的客户端是如何接入的。

13.3 客户端接入源码分析

负责处理网络读写、连接和客户端请求接入的 Reactor 线程就是 NioEventLoop，下面我们分析下 NioEventLoop 是如何处理新的客户端连接接入的。当多路复用器检测到新的准备就绪的 Channel 时，默认执行 processSelectedKeysOptimized 方法，代码如下。

```
        if (selectedKeys != null) {
            processSelectedKeysOptimized(selectedKeys.flip());
        } else {
            processSelectedKeysPlain(selector.selectedKeys());
        }
```

由于 Channel 的 Attachment 是 NioServerSocketChannel，所以执行 processSelectedKey 方法，根据就绪的操作位，执行不同的操作。此处，由于监听的是连接操作，所以执行 unsafe.read()方法。由于不同的 Channel 执行不同的操作，所以 NioUnsafe 被设计成接口，由不同的 Channel 内部的 NioUnsafe 实现类负责具体实现。我们发现 read()方法的实现有两个，分别是 NioByteUnsafe 和 NioMessageUnsafe。对于 NioServerSocketChannel，它使用的是 NioMessageUnsafe，它的 read 方法代码如下。

```
public void read() {
        //代码省略......
        final ChannelConfig config = config();
        final int maxMessagesPerRead = config.getMaxMessagesPerRead();
        final boolean autoRead = config.isAutoRead();
        final ChannelPipeline pipeline = pipeline();
        boolean closed = false;
        Throwable exception = null;
        try {
            for (;;) {
                int localRead = doReadMessages(readBuf);
                if (localRead == 0) {
                    break;
                }
                if (localRead < 0) {
                    closed = true;
                    break;
                }
                if (readBuf.size() >= maxMessagesPerRead | !autoRead) {
                    break;
                }
            //此处代码省略......
        }
```

对 doReadMessages 方法进行分析，发现它实际就是接收新的客户端连接并创建 NioSocketChannel，代码如下。

```
protected int doReadMessages(List<Object> buf) throws Exception {
    SocketChannel ch = javaChannel().accept();
    try {
        if (ch != null) {
            buf.add(new NioSocketChannel(this, childEventLoopGroup().next(), ch));
            return 1;
        }
    } catch (Throwable t) {
        //后续代码省略...
    }
```

接收到新的客户端连接后，触发 ChannelPipeline 的 ChannelRead 方法，代码如下。

```
int size = readBuf.size();
for (int i = 0; i < size; i ++) {
    pipeline.fireChannelRead(readBuf.get(i));
}
```

执行 headChannelHandlerContext 的 fireChannelRead 方法，事件在 ChannelPipeline 中传递，执行 ServerBootstrapAcceptor 的 channelRead 方法，代码如下。

```
public void channelRead(ChannelHandlerContext ctx, Object msg) {
    Channel child = (Channel) msg;
    child.pipeline().addLast(childHandler);
    //代码省略...
    child.unsafe().register(child.newPromise());
}
```

该方法主要分为如下三个步骤。

第一步：将启动时传入的 childHandler 加入到客户端 SocketChannel 的 ChannelPipeline 中；

第二步：设置客户端 SocketChannel 的 TCP 参数；

第三步：注册 SocketChannel 到多路复用器。

以上三个步骤执行完成之后，下面我们展开看下 NioSocketChannel 的 register 方法，代码如图 13-5 所示。

图 13-5　NioSocketChannel 的 register 方法实现

NioSocketChannel 的注册方法与 ServerSocketChannel 的一致，也是将 Channel 注册到 Reactor 线程的多路复用器上。由于注册的操作位是 0，所以，此时 NioSocketChannel 还不能读取客户端发送的消息，那什么时候修改监听操作位为 OP_READ 呢，别着急，继续看代码。

执行完注册操作之后，紧接着会触发 ChannelReadComplete 事件。我们继续分析 ChannelReadComplete 在 ChannelPipeline 中的处理流程：Netty 的 Header 和 Tail 本身不关注 ChannelReadComplete 事件就直接透传，执行完 ChannelReadComplete 后，接着执行 PipeLine 的 read()方法，最终执行 HeadHandler 的 read()方法。

HeadHandler read()方法的代码已经在之前的小节介绍过，用来将网络操作位修改为读操作。创建 NioSocketChannel 的时候已经将 AbstractNioChannel 的 readInterestOp 设置为 OP_READ，这样，执行 selectionKey.interestOps(interestOps | readInterestOp)操作时就会把操作位设置为 OP_READ。代码如下。

```
    protected AbstractNioByteChannel(Channel parent, EventLoop eventLoop,
SelectableChannel ch) {
        super(parent, eventLoop, ch, SelectionKey.OP_READ);
    }
```

到此，新接入的客户端连接处理完成，可以进行网络读写等 I/O 操作。

13.4　总结

本章首先对原生 NIO 类库的使用复杂性进行了讲解，然后对 Netty 服务端创建的时序图和步骤进行了详细地说明，随后结合 Netty 的源码对服务端创建进行剖析，最后对新的客户端的接入进行了源码层面的分析和讲解。

通过本章的学习，希望广大读者能够掌握 Netty 服务端创建的要点，并能够在实际的工作中正确的使用服务端相关的类库，编写出高效的业务代码。

第 14 章
客户端创建

相对于服务端，Netty 客户端的创建更加复杂，除了要考虑线程模型、异步连接、客户端连接超时等因素外，还需要对连接过程中的各种异常进行考虑。本章将对 Netty 客户端创建的关键流程和源码进行分析，以期读者能够了解客户端创建的细节。

本章主要内容包括：

◎ 客户端创建流程分析

◎ 客户端创建源码分析

14.1 Netty 客户端创建流程分析

Netty 为了向使用者屏蔽 NIO 通信的底层细节，在和用户交互的边界做了封装，目的就是为了减少用户开发工作量，降低开发难度。Bootstrap 是 Socket 客户端创建工具类，用户通过 Bootstrap 可以方便地创建 Netty 的客户端并发起异步 TCP 连接操作。

14.2.1 Netty 客户端创建时序图

Netty 客户端创建时序图如图 14-1 所示。

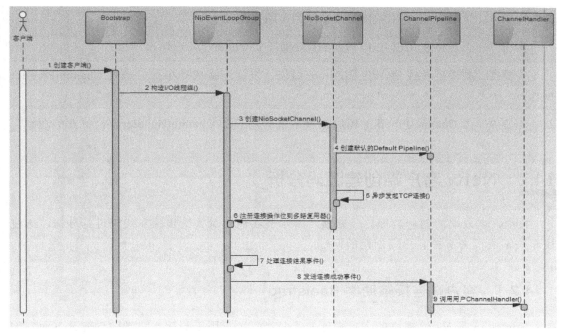

图 14-1 Netty 客户端创建时序图

14.2.2 Netty 客户端创建流程分析

步骤 1：用户线程创建 Bootstrap 实例，通过 API 设置创建客户端相关的参数，异步发起客户端连接。

步骤 2：创建处理客户端连接、I/O 读写的 Reactor 线程组 NioEventLoopGroup。可以通过构造函数指定 I/O 线程的个数，默认为 CPU 内核数的 2 倍；

步骤 3：通过 Bootstrap 的 ChannelFactory 和用户指定的 Channel 类型创建用于客户端连接的 NioSocketChannel，它的功能类似于 JDK NIO 类库提供的 SocketChannel；

步骤 4：创建默认的 Channel Handler Pipeline，用于调度和执行网络事件；

步骤 5：异步发起 TCP 连接，判断连接是否成功。如果成功，则直接将 NioSocketChannel 注册到多路复用器上，监听读操作位，用于数据报读取和消息发送；如果没有立即连接成功，则注册连接监听位到多路复用器，等待连接结果；

步骤 6：注册对应的网络监听状态位到多路复用器；

步骤 7：由多路复用器在 I/O 现场中轮询各 Channel，处理连接结果；

步骤 8：如果连接成功，设置 Future 结果，发送连接成功事件，触发 ChannelPipeline 执行；

步骤 9：由 ChannelPipeline 调度执行系统和用户的 ChannelHandler，执行业务逻辑。

14.2 Netty 客户端创建源码分析

Netty 客户端的创建流程比较繁琐，本章节我们针对关键步骤和代码进行分析，通过梳理关键流程来掌握客户端创建的原理。

14.2.1 客户端连接辅助类 Bootstrap

Bootstrap 是 Netty 提供的客户端连接工具类，主要用于简化客户端的创建，下面我们对它的主要 API 进行讲解。

设置 I/O 线程组：在前面的章节我们介绍过，非阻塞 I/O 的特点就是一个多路复用器可以同时处理成百上千条链路，这就意味着使用 NIO 模式一个线程可以处理多个 TCP 连接。考虑到 I/O 线程的处理性能，大多数 NIO 框架都采用线程池的方式处理 I/O 读写，Netty 也不例外。客户端相对于服务端，只需要一个处理 I/O 读写的线程组即可，因为 Bootstrap 提供了设置 I/O 线程组的接口，代码如下。

```
public B group(EventLoopGroup group) {
    if (group == null) {
        throw new NullPointerException("group");
    }
    if (this.group != null) {
        throw new IllegalStateException("group set already");
    }
    this.group = group;
    return (B) this;
}
```

由于 Netty 的 NIO 线程组默认采用 EventLoopGroup 接口，因此线程组参数使用

EventLoopGroup。

TCP 参数设置接口：无论是异步 NIO，还是同步 BIO，创建客户端套接字的时候通常都会设置连接参数，例如接收和发送缓冲区大小、连接超时时间等。Bootstrap 也提供了客户端 TCP 参数设置接口，代码如下。

```
public <T> B option(ChannelOption<T> option, T value) {
    if (option == null) {
        throw new NullPointerException("option");
    }
    if (value == null) {
        synchronized (options) {
            options.remove(option);
        }
    } else {
        synchronized (options) {
            options.put(option, value);
        }
    }
    return (B) this;
}
```

Netty 提供的主要 TCP 参数如下。

（1）SO_TIMEOUT：控制读取操作将阻塞多少毫秒。如果返回值为 0，计时器就被禁止了，该线程将无限期阻塞；

（2）SO_SNDBUF：套接字使用的发送缓冲区大小；

（3）SO_RCVBUF：套接字使用的接收缓冲区大小；

（4）SO_REUSEADDR：用于决定如果网络上仍然有数据向旧的 ServerSocket 传输数据，是否允许新的 ServerSocket 绑定到与旧的 ServerSocket 同样的端口上。SO_REUSEADDR 选项的默认值与操作系统有关，在某些操作系统中，允许重用端口，而在某些操作系统中不允许重用端口；

（5）CONNECT_TIMEOUT_MILLIS：客户端连接超时时间，由于 NIO 原生的客户端并不提供设置连接超时的接口，因此，Netty 采用的是自定义连接超时定时器负责检测和超时控制；

（6）TCP_NODELAY：激活或禁止 TCP_NODELAY 套接字选项，它决定是否使用 Nagle

算法。如果是时延敏感型的应用，建议关闭 Nagle 算法。

channel 接口：用于指定客户端使用的 channel 接口，对于 TCP 客户端连接，默认使用 NioSocketChannel，代码如下。

```
public Bootstrap channel(Class<? extends Channel> channelClass) {
    if (channelClass == null) {
        throw new NullPointerException("channelClass");
    }
    return channelFactory(new BootstrapChannelFactory<Channel>(channelClass));
}
```

BootstrapChannelFactory 利用 channelClass 类型信息，通过反射机制创建 NioSocketChannel 对象。

设置 Handler 接口：Bootstrap 为了简化 Handler 的编排，提供了 ChannelInitializer，它继承了 ChannelHandlerAdapter，当 TCP 链路注册成功之后，调用 initChannel 接口，用于设置用户 ChannelHandler。它的代码如下。

```
public final void channelRegistered(ChannelHandlerContext ctx) throws Exception {
    ChannelPipeline pipeline = ctx.pipeline();
    boolean success = false;
    try {
        initChannel((C) ctx.channel());
        pipeline.remove(this);
        ctx.fireChannelRegistered();
        success = true;
        //后续代码省略...
}
```

其中 initChannel 为抽象接口，用户可以在此方法中设置 ChannelHandler，代码如下。

```
.handler(new ChannelInitializer<SocketChannel>() {
    @Override
    public void initChannel(SocketChannel ch)
        throws Exception {
        ch.pipeline().addLast(
            new EchoClientHandler(firstMessageSize));
    }
});
```

最后一个比较重要的接口就是发起客户端连接,代码如下。

```
ChannelFuture f = b.connect(host, port).sync();
```

由于客户端连接方法比较复杂,下个小节对此进行详细讲解。

14.2.2 客户端连接操作

首先要创建和初始化 NioSocketChannel,代码如下。

```
private ChannelFuture doConnect(final SocketAddress remoteAddress, final SocketAddress localAddress) {
    final ChannelFuture regFuture = initAndRegister();
    final Channel channel = regFuture.channel();
    if (regFuture.cause() != null) {
        return regFuture;
    }
    //代码省略…
}
```

从 NioEventLoopGroup 中获取 NioEventLoop,然后使用其作为参数创建 NioSocketChannel,代码如下。

```
Channel createChannel() {
    EventLoop eventLoop = group().next();
    return channelFactory().newChannel(eventLoop);
}
```

初始化 Channel 之后,将其注册到 Selector 上,代码如下。

```
ChannelPromise regFuture = channel.newPromise();
channel.unsafe().register(regFuture);
```

链路创建成功之后,发起异步的 TCP 连接,代码如下。

```
private static void doConnect0(
        final ChannelFuture regFuture, final Channel channel,
        final SocketAddress remoteAddress, final SocketAddress localAddress, final ChannelPromise promise) {
    channel.eventLoop().execute(new Runnable() {
        @Override
        public void run() {
```

```
            if (regFuture.isSuccess()) {
            if (localAddress == null) {
                channel.connect(remoteAddress, promise);
            } else {
                channel.connect(remoteAddress, localAddress, promise);
            }
            //后续代码省略...
    }
```

由上述代码可以看出,从 doConnect0 操作开始,连接操作切换到了 Netty 的 NIO 线程 NioEventLoop 中进行,此时客户端返回,连接操作异步执行。

doConnect0 最终调用 HeadHandler 的 connect 方法,代码如下。

```
public void connect(
        ChannelHandlerContext ctx,
        SocketAddress remoteAddress, SocketAddress localAddress,
        ChannelPromise promise) throws Exception {
    unsafe.connect(remoteAddress, localAddress, promise);
}
```

AbstractNioUnsafe 的 connect 操作如下。

```
if (doConnect(remoteAddress, localAddress)) {
            fulfillConnectPromise(promise, wasActive);
        } else {
//后续代码省略...
}
```

需要注意的是,SocketChannel 执行 connect()操作后有以下三种结果。

(1)连接成功,返回 True;

(2)暂时没有连接上,服务端没有返回 ACK 应答,连接结果不确定,返回 False;

(3)连接失败,直接抛出 I/O 异常。

如果是第二种结果,需要将 NioSocketChannel 中的 selectionKey 设置为 OP_CONNECT,监听连接结果。

异步连接返回之后,需要判断连接结果,如果连接成功,则触发 ChannelActive 事件,代码如下。

```
if (!wasActive && isActive()) {
            pipeline().fireChannelActive();
        }
```

ChannelActive 事件处理在前面章节已经详细说明,最终会将 NioSocketChannel 中的 selectionKey 设置为 SelectionKey.OP_READ,用于监听网络读操作。

如果没有立即连接上服务端,则注册 SelectionKey.OP_CONNECT 到多路复用器,代码如下。

```
boolean success = false;
    try {
        boolean connected = javaChannel().connect(remoteAddress);
        if (!connected) {
            selectionKey().interestOps(SelectionKey.OP_CONNECT);
        }
        success = true;
        return connected;
```

如果连接过程发生异常,则关闭链路,进入连接失败处理流程,代码如下。

```
finally {
        if (!success) {
            doClose();
        }
```

14.2.3 异步连接结果通知

NioEventLoop 的 Selector 轮询客户端连接 Channel,当服务端返回握手应答之后,对连接结果进行判断,代码如下。

```
if ((readyOps & SelectionKey.OP_CONNECT) != 0) {
            int ops = k.interestOps();
            ops &= ~SelectionKey.OP_CONNECT;
            k.interestOps(ops);
            unsafe.finishConnect();
        }
```

下面对 finishConnect 方法进行分析,代码如下。

```
    try {
        boolean wasActive = isActive();
        doFinishConnect();
        fulfillConnectPromise(connectPromise, wasActive);
    }
```

doFinishConnect 用于判断 JDK 的 SocketChannel 的连接结果，如果返回 true 表示连接成功，其他值或者发生异常表示连接失败。

```
protected void doFinishConnect() throws Exception {
    if (!javaChannel().finishConnect()) {
        throw new Error();
    }
}
```

连接成功之后，调用 fulfillConnectPromise 方法，触发链路激活事件，该事件由 ChannelPipeline 进行传播，代码如下。

```
private void fulfillConnectPromise(ChannelPromise promise, boolean wasActive) {
    boolean promiseSet = promise.trySuccess();
    if (!wasActive && isActive()) {
        pipeline().fireChannelActive();
    }
    if (!promiseSet) {
        close(voidPromise());
    }
}
```

前面章节已经对 fireChannelActive 方法进行过讲解，主要用于修改网络监听位为读操作。

14.2.4　客户端连接超时机制

对于 SocketChannel 接口，JDK 并没有提供连接超时机制，需要 NIO 框架或者用户自己扩展实现。Netty 利用定时器提供了客户端连接超时控制功能，下面我们对该功能进行详细讲解。

首先，用户在创建 Netty 客户端的时候，可以通过 ChannelOption.CONNECT_TIMEOUT_MILLIS 配置项设置连接超时时间，代码如下。

```
b.group(group).channel(NioSocketChannel.class)
        .option(ChannelOption.TCP_NODELAY, true)
        .option(ChannelOption.CONNECT_TIMEOUT_MILLIS, 3000)
//后续代码省略...
```

发起连接的同时，启动连接超时检测定时器，代码如下。

```
// Schedule connect timeout.
                int connectTimeoutMillis = config().getConnectTimeoutMillis();
                if (connectTimeoutMillis > 0) {
                    connectTimeoutFuture = eventLoop().schedule(new Runnable() {
                        @Override
                        public void run() {
                            ChannelPromise connectPromise = AbstractNioChannel.this.connectPromise;
                            ConnectTimeoutException cause =
                                    new ConnectTimeoutException("connection timed out: " + remoteAddress);
                            if (connectPromise != null && connectPromise.tryFailure(cause)) {
                                close(voidPromise());
                            }
                        }
                    }, connectTimeoutMillis, TimeUnit.MILLISECONDS);
                }
```

一旦超时定时器执行，说明客户端连接超时，构造连接超时异常，将异常结果设置到 connectPromise 中，同时关闭客户端连接，释放句柄。

如果在连接超时之前获取到连接结果，则删除连接超时定时器，防止其被触发，代码如下。

```
public void finishConnect() {
} finally {
            if (connectTimeoutFuture != null) {
                connectTimeoutFuture.cancel(false);
            }
            connectPromise = null;
```

```
        }
//后续代码省略...
    }
```

无论连接是否成功,只要获取到连接结果,之后就删除连接超时定时器。

14.3 总结

本章首先对 Netty 客户端创建的主要流程进行说明,然后对客户端创建的关键步骤进行详细分析和讲解。希望通过本章的学习,广大读者能够掌握 Netty 客户端创建的原理,熟悉相关流程,编写出更加高效和可靠的业务代码。

源码分析篇

Netty 功能介绍和源码分析

第 15 章　ByteBuf 和相关辅助类

第 16 章　Channel 和 Unsafe

第 17 章　ChannelPipeline 和 ChannelHandler

第 18 章　EventLoop 和 EventLoopGroup

第 19 章　Future 和 Promise

第 15 章

ByteBuf 和相关辅助类

从本章开始，我们将学习 Netty NIO 相关的主要接口和模块的 API 功能，并对其源码实现进行分析。希望读者通过对功能和 API 的学习，能够更加熟练地掌握和应用这些类库。对源码的学习不仅能够帮助读者从源码的层面掌握 Netty 框架，方便日后的维护、扩展和定制，更能够起到触类旁通的作用，拓展读者的知识面，提升编程技能。

本章主要内容包括：

◎ ByteBuf 功能说明

◎ ByteBuf 源码分析

◎ ByteBuf 相关辅助类功能说明

15.1 ByteBuf 功能说明

当我们进行数据传输的时候，往往需要使用到缓冲区，常用的缓冲区就是 JDK NIO 类库提供的 java.nio.Buffer，它的实现类如图 15-1 所示。

实际上，7 种基础类型（Boolean 除外）都有自己的缓冲区实现。对于 NIO 编程而言，我们主要使用的是 ByteBuffer。从功能角度而言，ByteBuffer 完全可以满足 NIO 编程的需要，但是由于 NIO 编程的复杂性，ByteBuffer 也有其局限性，它的主要缺点如下：

第 15 章 ByteBuf 和相关辅助类

图 15-1 java.nio.Buffer 继承关系图

（1）ByteBuffer 长度固定，一旦分配完成，它的容量不能动态扩展和收缩，当需要编码的 POJO 对象大于 ByteBuffer 的容量时，会发生索引越界异常；

（2）ByteBuffer 只有一个标识位置的指针 position，读写的时候需要手工调用 flip() 和 rewind() 等，使用者必须小心谨慎地处理这些 API，否则很容易导致程序处理失败；

（3）ByteBuffer 的 API 功能有限，一些高级和实用的特性它不支持，需要使用者自己编程实现。

为了弥补这些不足，Netty 提供了自己的 ByteBuffer 实现——ByteBuf，下面我们一起学习 ByteBuf 的原理和主要功能。

15.1.1 ByteBuf 的工作原理

不同 ByteBuf 实现类的工作原理不尽相同，本小节我们从 ByteBuf 的设计原理出发，一起探寻 Netty ByteBuf 的设计理念。

首先，ByteBuf 依然是个 Byte 数组的缓冲区，它的基本功能应该与 JDK 的 ByteBuffer

一致，提供以下几类基本功能。

- ◎ 7种Java基础类型、byte数组、ByteBuffer（ByteBuf）等的读写；
- ◎ 缓冲区自身的copy和slice等；
- ◎ 设置网络字节序；
- ◎ 构造缓冲区实例；
- ◎ 操作位置指针等方法。

由于JDK的ByteBuffer已经提供了这些基础能力的实现，因此，Netty ByteBuf的实现可以有两种策略。

- ◎ 参考JDK ByteBuffer的实现，增加额外的功能，解决原ByteBuffer的缺点；
- ◎ 聚合JDK ByteBuffer，通过Facade模式对其进行包装，可以减少自身的代码量，降低实现成本。

JDK ByteBuffer由于只有一个位置指针用于处理读写操作，因此每次读写的时候都需要额外调用flip()和clear()等方法，否则功能将出错，它的典型用法如下。

```
ByteBuffer buffer = ByteBuffer.allocate(88);
String value = "Netty权威指南";
buffer.put(value.getBytes());
buffer.flip();
byte[] vArray = new byte[buffer.remaining()];
buffer.get(vArray);
String decodeValue = new String(vArray);
```

我们看下调用flip()操作前后的对比。

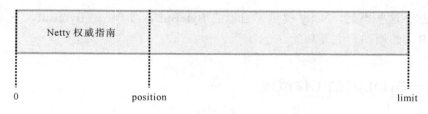

图15-2　ByteBuffer flip()操作之前

如图15-2所示，如果不做flip操作，读取到的将是position到capacity之间的错误内容。

当执行 flip() 操作之后，它的 limit 被设置为 position，position 设置为 0，capacity 不变。由于读取的内容是从 position 到 limit 之间，因此，它能够正确地读取到之前写入缓冲区的内容。如图 15-3 所示。

图 15-3　ByteBuffer flip() 操作之后

ByteBuf 通过两个位置指针来协助缓冲区的读写操作，读操作使用 readerIndex，写操作使用 writerIndex。

readerIndex 和 writerIndex 的取值一开始都是 0，随着数据的写入 writerIndex 会增加，读取数据会使 readerIndex 增加，但是它不会超过 writerIndex。在读取之后，0～readerIndex 就被视为 discard 的，调用 discardReadBytes 方法，可以释放这部分空间，它的作用类似 ByteBuffer 的 compact 方法。ReaderIndex 和 writerIndex 之间的数据是可读取的，等价于 ByteBuffer position 和 limit 之间的数据。WriterIndex 和 capacity 之间的空间是可写的，等价于 ByteBuffer limit 和 capacity 之间的可用空间。

由于写操作不修改 readerIndex 指针，读操作不修改 writerIndex 指针，因此读写之间不再需要调整位置指针，这极大地简化了缓冲区的读写操作，避免了由于遗漏或者不熟悉 flip() 操作导致的功能异常。

初始分配的 ByteBuf 如图 15-4 所示。

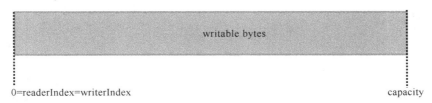

图 15-4　初始分配的 ByteBuf

写入 N 个字节之后的 ByteBuf 如图 15-5 所示。

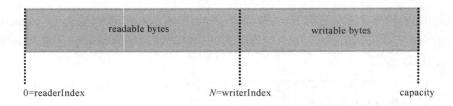

图 15-5 写入 N 个字节后的 ByteBuf

读取 M（<N）个字节之后的 ByteBuf 如图 15-6 所示。

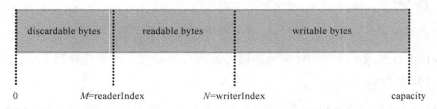

图 15-6 读取 M 个字节后的 ByteBuf

调用 discardReadBytes 操作之后的 ByteBuf 如图 15-7 所示。

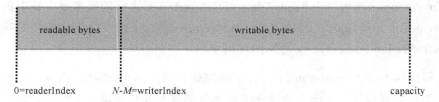

图 15-7 discardReadBytes 操作之后的 ByteBuf

调用 clear 操作之后的 ByteBuf 如图 15-8 所示。

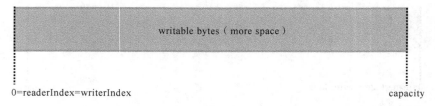

图 15-8 clear 操作之后的 ByteBuf

下面我们继续分析 ByteBuf 是如何实现动态扩展的。通常情况下，当我们对 ByteBuffer 进行 put 操作的时候，如果缓冲区剩余可写空间不够，就会发生 BufferOverflowException

异常。为了避免发生这个问题，通常在进行 put 操作的时候会对剩余可用空间进行校验。如果剩余空间不足，需要重新创建一个新的 ByteBuffer，并将之前的 ByteBuffer 复制到新创建的 ByteBuffer 中，最后释放老的 ByteBuffer，代码示例如下。

```
if (this.buffer.remaining() < needSize)
    {
    int toBeExtSize = needSize > 128 ? needSize : 128;
    ByteBuffer tmpBuffer = ByteBuffer.allocate(this.buffer.capacity() + toBeExtSize);
    this.buffer.flip();
    tmpBuffer.put(this.buffer);
    this.buffer = tmpBuffer;
    }
```

从示例代码可以看出，为了防止 ByteBuffer 溢出，每进行一次 put 操作，都需要对可用空间进行校验，这导致了代码冗余，稍有不慎，就可能引入其他问题。为了解决这个问题，ByteBuf 对 write 操作进行了封装，由 ByteBuf 的 write 操作负责进行剩余可用空间的校验。如果可用缓冲区不足，ByteBuf 会自动进行动态扩展。对于使用者而言，不需要关心底层的校验和扩展细节，只要不超过设置的最大缓冲区容量即可。当可用空间不足时，ByteBuf 会帮助我们实现自动扩展，这极大地降低了 ByteBuf 的学习和使用成本，提升了开发效率。校验和扩展的相关代码如图 15-9、15-10 所示。

```
@Override
public ByteBuf writeByte(int value) {
    ensureWritable(1);
    setByte(writerIndex++, value);
    return this;
}
```

图 15-9　ByteBuf 写入字节

通过源码分析，我们发现当进行 write 操作时，会对需要 write 的字节进行校验。如果可写的字节数小于需要写入的字节数，并且需要写入的字节数小于可写的最大字节数，就对缓冲区进行动态扩展。无论缓冲区是否进行了动态扩展，从功能角度看使用者并不感知，这样就简化了上层的应用。

由于 NIO 的 Channel 读写的参数都是 ByteBuffer，因此，Netty 的 ByteBuf 接口必须提供 API，以方便地将 ByteBuf 转换成 ByteBuffer，或者将 ByteBuffer 包装成 ByteBuf。考虑到性能，应该尽量避免缓冲区的复制，内部实现的时候可以考虑聚合一个 ByteBuffer 的私有指针用来代表 ByteBuffer。在后面的源码分析章节我们将详细介绍它的实现原理。

```
    @Override
    public ByteBuf ensureWritable(int minWritableBytes) {
        if (minWritableBytes < 0) {
            throw new IllegalArgumentException(String.format(
                "minWritableBytes: %d (expected: >= 0)", minWritableBytes));
        }
        if (minWritableBytes <= writableBytes()) {
            return this;
        }
        if (minWritableBytes > maxCapacity - writerIndex) {
            throw new IndexOutOfBoundsException(String.format(
                "writerIndex(%d) + minWritableBytes(%d) exceeds maxCapacity(%d): %s",
                writerIndex, minWritableBytes, maxCapacity, this));
        }
        // Normalize the current capacity to the power of 2.
        int newCapacity = calculateNewCapacity(writerIndex + minWritableBytes);
        // Adjust to the new capacity.
        capacity(newCapacity);
        return this;
    }
```

图 15-10　ByteBuf 写入字节

学习完 ByteBuf 的原理之后，下面我们继续学习它的主要 API 功能。

15.1.2　ByteBuf 的功能介绍

本小节我们将对 ByteBuf 的常用 API 进行分类说明，讲解它的主要功能，后面的章节将给出重要 API 的一些典型用法。

1. 顺序读操作（read）

ByteBuf 的 read 操作类似于 ByteBuffer 的 get 操作，主要的 API 功能说明如表 15-1 所示。

表 15-1　ByteBuf 的读操作 API 列表

方法名称	返 回 值	功能说明
readBoolean	boolean	从 readerIndex 开始获取 boolean 值，readerIndex 增加 1
readByte	byte	从 readerIndex 开始获取字节值，readerIndex 增加 1
readUnsignedByte	byte	从 readerIndex 开始获取无符号字节值，readerIndex 增加 1
readShort	short	从 readerIndex 开始获取短整型值，readerIndex 增加 2
readUnsignedShort	short	从 readerIndex 开始获取无符号短整型值，readerIndex 增加 2

续表

方法名称	返回值	功能说明
readMedium	int	从 readerIndex 开始获取 24 位整型值，readerIndex 增加 3（注意：该类型并非 Java 的基本类型，大多数场景使用不到）
readUnsignedMedium	int	从 readerIndex 开始获取 24 位无符号整型值，readerIndex 增加 3（注意：该类型并非 Java 的基本类型，大多数场景使用不到）
readInt	int	从 readerIndex 开始获取整型值，readerIndex 增加 4
readUnsignedInt	int	从 readerIndex 开始获取无符号整型值，readerIndex 增加 4
readLong	long	从 readerIndex 开始获取长整型值，readerIndex 增加 8
readChar	char	从 readerIndex 开始获取字符值，readerIndex 增加 2
readFloat	float	从 readerIndex 开始获取浮点值，readerIndex 增加 4
readDouble	double	从 readerIndex 开始获取双精度浮点值，readerIndex 增加 8
readBytes(int length)	ByteBuf	将当前 ByteBuf 中的数据读取到新创建的 ByteBuf 中，读取的长度为 length。操作成功完成之后，返回的 ByteBuf 的 readerIndex 为 0，writerIndex 为 length 如果读取的长度 length 大于当前操作的 ByteBuf 的可写字节数，将抛出 IndexOutOfBoundsException，操作失败
readSlice(int length)	ByteBuf	返回当前 ByteBuf 新创建的子区域，子区域与原 ByteBuf 共享缓冲区，但是独立维护自己的 readerIndex 和 writerIndex 新创建的子区域 readerIndex 为 0，writerIndex 为 length 如果读取的长度 length 大于当前操作的 ByteBuf 的可写字节数，将抛出 IndexOutOfBoundsException，操作失败
readBytes(ByteBuf dst)	ByteBuf	将当前 ByteBuf 的数据读取到目标 ByteBuf 中，直到目标 ByteBuf 没有剩余的空间可写 操作完成之后，当前 ByteBuf 的 readerIndex += 读取的字节数 如果目标 ByteBuf 可写的字节数大于当前 ByteBuf 可读取的字节数，则抛出 IndexOutOfBoundsException，操作失败
readBytes(ByteBuf dst, int length)	ByteBuf	将当前 ByteBuf 的数据读取到目标 ByteBuf 中，读取的字节数长度为 length 操作完成之后，当前 ByteBuf 的 readerIndex += length 如果需要读取的字节数长度 length 大于当前 ByteBuf 可读的字节数或者目标 ByteBuf 可写的字节数，则抛出 IndexOutOfBoundsException，操作失败

续表

方法名称	返 回 值	功能说明
readBytes(ByteBuf dst, int dstIndex, int length)	ByteBuf	将当前 ByteBuf 的数据读取到目标 ByteBuf 中，读取的字节数长度为 length 目标 ByteBuf 的起始索引为 dstIndex，非 writerIndex 操作完成之后，当前 ByteBuf 的 readerIndex += length 如果需要读取的字节数长度 length 大于当前 ByteBuf 可读的字节数，或者 dstIndex 小于 0，或者 dstIndex+length 大于目标 ByteBuf 的 capacity，则抛出 IndexOutOfBoundsException，操作失败
readBytes(byte[] dst)	ByteBuf	将当前 ByteBuf 的数据读取到目标 byte 数组中，读取的字节数长度为 dst.length 操作完成之后，当前 ByteBuf 的 readerIndex += dst.length 如果目标字节数组的长度大于当前 ByteBuf 可读的字节数，则抛出 IndexOutOfBoundsException，操作失败
readBytes(byte[] dst, int dstIndex, int length)	ByteBuf	将当前 ByteBuf 的数据读取到目标 byte 数组中，读取的字节数长度为 length，目标字节数组的起始索引为 dstIndex 如果 dstIndex 小于 0，或者 length 大于当前 ByteBuf 的可读字节数，或者 dstIndex+length 大于 dst.length，则抛出 IndexOutOfBoundsException，操作失败
readBytes(ByteBuffer dst)	ByteBuf	将当前 ByteBuf 的数据读取到目标 ByteBuffer 中，直到位置指针到达 ByteBuffer 的 limit 操作成功完成之后，当前 ByteBuf 的 readerIndex += dest.remaining() 如果目标 ByteBuffer 的可写字节数大于当前 ByteBuf 可读字节数，则抛出 IndexOutOfBoundsException，操作失败
readBytes(OutputStream out, int length)	ByteBuf	将当前 ByteBuf 的数据读取到目标输出流中，读取的字节数长度为 length 如果操作成功，当前 ByteBuf 的 readerIndex+=length 如果 length 大于当前 ByteBuf 可读取的字节数，则抛出 IndexOutOfBoundsException，操作失败 如果读取过程中 OutputStream 自身发生了 I/O 异常，则抛出 IOException

续表

方法名称	返 回 值	功能说明
readBytes(GatheringByteChannel out, int length)	int	将当前 ByteBuf 的数据写入到目标 GatheringByteChannel 中，写入的最大字节数长度为 length 注意：由于 GatheringByteChannel 是非阻塞 Channel，调用它的 write 操作并不能保证一次能够将所有需要写入的字节数都写入成功，即存在"写半包"问题。因此，它写入的字节数范围为[0,length] 如果操作成功，当前 ByteBuf 的 readerIndex+=实际写入的字节数 如果需要写入的 length 大于当前 ByteBuf 的可读字节数，则抛出 IndexOutOfBoundsException 异常；如果操作过程中 GatheringByteChannel 发生了 I/O 异常，则抛出 IOException，无论抛出何种异常，操作都将失败 与其他 read 方法不同的是，本方法的返回值不是当前的 ByteBuf，而是写入 GatheringByteChannel 的实际字节数

2. 顺序写操作（write）

ByteBuf 的 write 操作类似于 ByteBuffer 的 put 操作，主要的 API 功能说明如表 15-2 所示。

表 15-2 ByteBuf 的写操作 API 列表

方法名称	返 回 值	功能说明
writeBoolean(boolean value)	ByteBuf	将参数 value 写入到当前的 ByteBuf 中 操作成功之后 writerIndex+=1 如果当前 ByteBuf 可写的字节数小于 1，则抛出 IndexOutOfBoundsException，操作失败
writeByte(int value)	ByteBuf	将参数 value 写入到当前的 ByteBuf 中 操作成功之后 writerIndex+=1 如果当前 ByteBuf 可写的字节数小于 1，则抛出 IndexOutOfBoundsException，操作失败

续表

方法名称	返回值	功能说明
writeShort(int value)	ByteBuf	将参数 value 写入到当前的 ByteBuf 中 操作成功之后 writerIndex+=2 如果当前 ByteBuf 可写的字节数小于 2，则抛出 IndexOutOfBoundsException，操作失败
writeMedium(int value)	ByteBuf	将参数 value 写入到当前的 ByteBuf 中 操作成功之后 writerIndex+=3 如果当前 ByteBuf 可写的字节数小于 3，则抛出 IndexOutOfBoundsException，操作失败
writeInt(int value)	ByteBuf	将参数 value 写入到当前的 ByteBuf 中 操作成功之后 writerIndex+=4 如果当前 ByteBuf 可写的字节数小于 4，则抛出 IndexOutOfBoundsException，操作失败
writeLong(long value)	ByteBuf	将参数 value 写入到当前的 ByteBuf 中 操作成功之后 writerIndex+=8 如果当前 ByteBuf 可写的字节数小于 8，则抛出 IndexOutOfBoundsException，操作失败
writeChar(int value)	ByteBuf	将参数 value 写入到当前的 ByteBuf 中 操作成功之后 writerIndex+=2 如果当前 ByteBuf 可写的字节数小于 2，则抛出 IndexOutOfBoundsException，操作失败
writeBytes(ByteBuf src)	ByteBuf	将源 ByteBuf src 中的所有可读字节写入到当前 ByteBuf 中 操作成功之后当前 ByteBuf 的 writerIndex+= src.readableBytes 如果源 ByteBuf src 可读的字节数大于当前 ByteBuf 的可写字节数，则抛出 IndexOutOfBoundsException，操作失败
writeBytes(ByteBuf src, int length)	ByteBuf	将源 ByteBuf src 中的可读字节写入到当前 ByteBuf 中，写入的字节数长度为 length 操作成功之后 当前 ByteBuf 的 writerIndex+= length 如果 length 大于源 ByteBuf 的可读字节数或者当前 ByteBuf 的可写字节数，则抛出 IndexOutOfBoundsException，操作失败

续表

方法名称	返回值	功能说明
writeBytes(ByteBuf src, int srcIndex, int length)	ByteBuf	将源 ByteBuf src 中的可读字节写入到当前 ByteBuf 中，写入的字节数长度为 length，起始索引为 srcIndex 操作成功之后 当前 ByteBuf 的 writerIndex+= length 如果 srcIndex 小于 0，或者 srcIndex + length 大于源 src 的容量；或者写入长度 length 大于当前 ByteBuf 的可写字节数，则抛出 IndexOutOfBoundsException，操作失败
writeBytes(byte[] src)	ByteBuf	将源字节数组 src 中的所有字节写入到当前 ByteBuf 中 操作成功之后 当前 ByteBuf 的 writerIndex+= src.length 如果源字节数组 src 的长度大于当前 ByteBuf 的可写字节数，则抛出 IndexOutOfBoundsException，操作失败
writeBytes(byte[] src, int srcIndex, int length)	ByteBuf	将源字节数组 src 中的字节写入到当前 ByteBuf 中，写入的字节数长度为 length，起始索引为 srcIndex 操作成功之后 当前 ByteBuf 的 writerIndex+= length 如果 srcIndex 小于 0，或者 srcIndex + length 大于源 src 的容量；或者写入长度 length 大于当前 ByteBuf 的可写字节数，则抛出 IndexOutOfBoundsException，操作失败
writeBytes(ByteBuffer src)	ByteBuf	将源 ByteBuffer src 中所有可读字节写入到当前 ByteBuf 中，写入的长度为 src.remaining() 操作成功之后 当前 ByteBuf 的 writerIndex+= src.remaining() 如果源 ByteBuffer src 的可读字节数大于当前 ByteBuf 的可写字节数，则抛出 IndexOutOfBoundsException，操作失败
writeBytes(InputStream in, int length)	int	将源 InputStream src 中的内容写入到当前 ByteBuf 中，写入的最大字节数长度为 length 实际写入的字节数可能小于 length 操作成功之后当前 ByteBuf 的 writerIndex+= 实际写入的字节数 如果 length 大于源 ByteBuf 的可读字节数或者当前 ByteBuf 的可写字节数，则抛出 IndexOutOfBoundsException，操作失败 如果从 InputStream 读取的时候发生了 I/O 异常，则抛出 IOException

续表

方法名称	返回值	功能说明
writeBytes(ScatteringByteChannel in, int length)	int	将源 ScatteringByteChannel src 中的内容写入到当前 ByteBuf 中，写入的最大字节数长度为 length 实际写入的字节数可能小于 length 操作成功之后当前 ByteBuf 的 writerIndex+=实际写入的字节数 如果 length 大于源 src 的可读字节数或者当前 ByteBuf 的可写字节数，则抛出 IndexOutOfBoundsException，操作失败 如果从 ScatteringByteChannel 读取的时候发生了 I/O 异常，则抛出 IOException
writeZero(int length)	ByteBuf	将当前的缓冲区内容填充为 NUL (0x00)，起始位置为 writerIndex，填充的长度为 length 填充成功之后 writerIndex+= length 如果 length 大于当前 ByteBuf 的可写字节数则抛出 IndexOutOfBoundsException 操作失败

3. readerIndex 和 writerIndex

Netty 提供了两个指针变量用于支持顺序读取和写入操作：readerIndex 用于标识读取索引，writerIndex 用于标识写入索引。两个位置指针将 ByteBuf 缓冲区分割成三个区域，如图 15-11 所示。

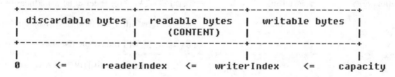

图 15-11 ByteBuf 的 readerIndex 和 writerIndex

调用 ByteBuf 的 read 操作时，从 readerIndex 处开始读取。readerIndex 到 writerIndex 之间的空间为可读的字节缓冲区；从 writerIndex 到 capacity 之间为可写的字节缓冲区；0 到 readerIndex 之间是已经读取过的缓冲区，可以调用 discardReadBytes 操作来重用这部分空间，以节约内存，防止 ByteBuf 的动态扩张。这在私有协议栈消息解码的时候非常有用，因为 TCP 底层可能粘包，几百个整包消息被 TCP 粘包后作为一个整包发送。这样，通过 discardReadBytes 操作可以重用之前已经解码过的缓冲区，从而防止接收缓冲区因为容量不足导致的扩张。但是，discardReadBytes 操作是把双刃剑，不能滥用，关于这一点在后续章节会进行详细说明。

4. Discardable bytes

相比于其他的 Java 对象，缓冲区的分配和释放是个耗时的操作，因此，我们需要尽量重用它们。由于缓冲区的动态扩张需要进行字节数组的复制，它是个耗时的操作，因此，为了最大程度地提升性能，往往需要尽最大努力提升缓冲区的重用率。

假如缓冲区包含了 N 个整包消息，每个消息的长度为 L，消息的可写字节数为 R。当读取 M 个整包消息后，如果不对 ByteBuf 做压缩或者 discardReadBytes 操作，则可写的缓冲区长度依然为 R。如果调用 discardReadBytes 操作，则可写字节数会变为 $R=（R+M\times L）$，之前已经读取的 M 个整包的空间会被重用。假如此时 ByteBuf 需要写入 $R+1$ 个字节，则不需要动态扩张 ByteBuf。

ByteBuf 的 discardReadBytes 操作效果图如下。

操作之前如图 15-12 所示。

图 15-12　discardReadBytes 操作之前的 ByteBuf

操作之后如图 15-13 所示。

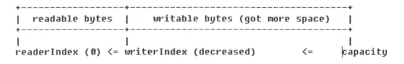

图 15-13　discardReadBytes 操作之后的 ByteBuf

需要指出的是，调用 discardReadBytes 会发生字节数组的内存复制，所以，频繁调用将会导致性能下降，因此在调用它之前要确认你确实需要这样做，例如牺牲性能来换取更多的可用内存。discardReadBytes 的相关源码如图 15-14 所示。

需要指出的是，调用 discardReadBytes 操作之后的 writable bytes 内容处理策略跟 ByteBuf 接口的具体实现有关。

```java
@Override
public ByteBuf discardReadBytes() {
    ensureAccessible();
    if (readerIndex == 0) {
        return this;
    }

    if (readerIndex != writerIndex) {
        setBytes(0, this, readerIndex, writerIndex - readerIndex);
        writerIndex -= readerIndex;
        adjustMarkers(readerIndex);
        readerIndex = 0;
    } else {
        adjustMarkers(readerIndex);
        writerIndex = readerIndex = 0;
    }
    return this;
}
```

图 15-14 discardReadBytes 操作会导致内存复制

5. Readable bytes 和 Writable bytes

可读空间段是数据实际存储的区域，以 read 或者 skip 开头的任何操作都将会从 readerIndex 开始读取或者跳过指定的数据，操作完成之后 readerIndex 增加了读取或者跳过的字节数长度。如果读取的字节数长度大于实际可读的字节数，则抛出 IndexOutOfBoundsException。当新分配、包装或者复制一个新的 ByteBuf 对象时，它的 readerIndex 为 0。

可写空间段是尚未被使用可以填充的空闲空间，任何以 write 开头的操作都会从 writerIndex 开始向空闲空间写入字节，操作完成之后 writerIndex 增加了写入的字节数长度。如果写入的字节数大于可写的字节数，则会抛出 IndexOutOfBoundsException 异常。新分配一个 ByteBuf 对象时，它的 readerIndex 为 0。通过包装或者复制的方式创建一个新的 ByteBuf 对象时，它的 writerIndex 是 ByteBuf 的容量。

6. Clear 操作

正如 JDK ByteBuffer 的 clear 操作，它并不会清空缓冲区内容本身，例如填充为 NUL（0x00）。它主要用来操作位置指针，例如 position、limit 和 mark。对于 ByteBuf，它也是用来操作 readerIndex 和 writerIndex，将它们还原为初始分配值。具体的处理示例图如下。

Clear()操作之前如图 15-15 所示。

Clear()操作之后如图 15-16 所示。

```
+-------------------+------------------+------------------+
| discardable bytes |  readable bytes  |  writable bytes  |
+-------------------+------------------+------------------+
|                   |                  |                  |
|0      <=     readerIndex   <=    writerIndex   <=   capacity
```

图 15-15　clear 操作之前的缓冲区

```
+--------------------------------------------------------+
|              writable bytes (got more space)           |
+--------------------------------------------------------+
|                                                        |
|0 = readerIndex = writerIndex             <=    capacity|
```

图 15-16　clear 操作之后的缓冲区

7. Mark 和 Rest

当对缓冲区进行读操作时，由于某种原因，可能需要对之前的操作进行回滚。读操作并不会改变缓冲区的内容，回滚操作主要就是重新设置索引信息。

对于 JDK 的 ByteBuffer，调用 mark 操作会将当前的位置指针备份到 mark 变量中，当调用 rest 操作之后，重新将指针的当前位置恢复为备份在 mark 中的值，源码代码如图 15-17 所示。

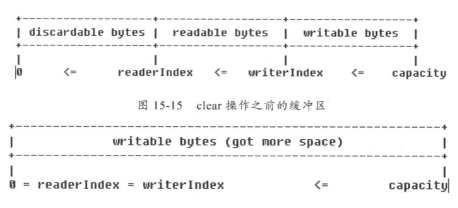

图 15-17　mark 操作之后的缓冲区位置指针

调用 reset 操作之后如图 15-18 所示。

Netty 的 ByteBuf 也有类似的 rest 和 mark 接口，因为 ByteBuf 有读索引和写索引，因此，它总共有 4 个相关的方法，分别如下。

- markReaderIndex：将当前的 readerIndex 备份到 markedReaderIndex 中；
- resetReaderIndex：将当前的 readerIndex 设置为 markedReaderIndex；
- markWriterIndex：将当前的 writerIndex 备份到 markedWriterIndex 中；
- resetWriterIndex：将当前的 writerIndex 设置为 markedWriterIndex。

```
public final Buffer reset() {
  int m = mark;
  if (m < 0)
    throw new InvalidMarkException();
  position = m;
  return this;
}
```

图 15-18 rest 操作之后的缓冲区位置指针

相关的代码实现如图 15-19 所示。

```
@Override
public ByteBuf markReaderIndex() {
  markedReaderIndex = readerIndex;
  return this;
}

@Override
public ByteBuf resetReaderIndex() {
  readerIndex(markedReaderIndex);
  return this;
}

@Override
public ByteBuf markWriterIndex() {
  markedWriterIndex = writerIndex;
  return this;
}

@Override
public ByteBuf resetWriterIndex() {
  writerIndex = markedWriterIndex;
  return this;
}
```

图 15-19 mark 和 rest 操作前后的缓冲区位置指针

8. 查找操作

很多时候，需要从 ByteBuf 中查找某个字符，例如通过 "\r\n" 作为文本字符串的换行符，利用 "NUL(0x00)" 作为分隔符。

ByteBuf 提供了多种查找方法用于满足不同的应用场景，详细分类如下。

（1）indexOf(int fromIndex, int toIndex, byte value)：从当前 ByteBuf 中定位出首次出现 value 的位置。起始索引为 fromIndex，终点是 toIndex。如果没有查找到则返回-1，否则返回第一条满足搜索条件的位置索引。

（2）bytesBefore(byte value)：从当前 ByteBuf 中定位出首次出现 value 的位置。起始

索引为 readerIndex，终点是 writerIndex。如果没有查找到则返回-1，否则返回第一条满足搜索条件的位置索引。该方法不会修改 readerIndex 和 writerIndex。

（3）bytesBefore(int length, byte value)：从当前 ByteBuf 中定位出首次出现 value 的位置。起始索引为 readerIndex，终点是 readerIndex+length。如果没有查找到则返回-1，否则返回第一条满足搜索条件的位置索引。如果 length 大于当前字节缓冲区的可读字节数，则抛出 IndexOutOfBoundsException 异常。

（4）bytesBefore(int index, int length, byte value)：从当前 ByteBuf 中定位出首次出现 value 的位置。起始索引为 index，终点是 index+length。如果没有查找到则返回-1，否则返回第一条满足搜索条件的位置索引。如果 index+length 大于当前字节缓冲区的容量，则抛出 IndexOutOfBoundsException 异常。

（5）forEachByte(ByteBufProcessor processor)：遍历当前 ByteBuf 的可读字节数组，与 ByteBufProcessor 设置的查找条件进行对比。如果满足条件，则返回位置索引，否则返回-1。

（6）forEachByte(int index, int length, ByteBufProcessor processor)：以 index 为起始位置，index + length 为终止位置进行遍历，与 ByteBufProcessor 设置的查找条件进行对比。如果满足条件，则返回位置索引，否则返回-1。

（7）forEachByteDesc(ByteBufProcessor processor)：遍历当前 ByteBuf 的可读字节数组，与 ByteBufProcessor 设置的查找条件进行对比。如果满足条件，则返回位置索引，否则返回-1。注意对字节数组进行迭代的时候采用逆序的方式，也就是从 writerIndex-1 开始迭代，直到 readerIndex。

（8）forEachByteDesc(int index, int length, ByteBufProcessor processor)：以 index 为起始位置，index +length 为终止位置进行遍历，与 ByteBufProcessor 设置的查找条件进行对比。如果满足条件，则返回位置索引，否则返回-1。采用逆序查找的方式，从 index + length - 1 开始，直到 index。

对于查找的字节而言，存在一些常用值，例如回车换行符、常用的分隔符等，Netty 为了减少业务的重复定义，在 ByteBufProcessor 接口中对这些常用的查找字节进行了抽象，定义如下。

（1）FIND_NUL：NUL（0x00）；

（2）FIND_CR：CR ('\r')；

（3）FIND_LF：LF ('\n')；

（4）FIND_CRLF：CR ('\r')或者 LF ('\n')；

（5）FIND_LINEAR_WHITESPACE：' '或者'\t'。

使用者也可以自定义查找规则，实现如 15-20 所示接口即可。

```
/**
 * @return {@code true} if the processor wants to continue the loop and
 *         handle the next byte in the buffer.
 *         {@code false} if the processor wants to stop handling bytes and
 abort the loop.
 */
boolean process(byte value) throws Exception;
```

图 15-20　字节查找接口

9. Derived buffers

类似于数据库的视图，ByteBuf 提供了多个接口用于创建某个 ByteBuf 的视图或者复制 ByteBuf，具体方法如下。

（1）duplicate：返回当前 ByteBuf 的复制对象，复制后返回的 ByteBuf 与操作的 ByteBuf 共享缓冲区内容，但是维护自己独立的读写索引。当修改复制后的 ByteBuf 内容后，之前原 ByteBuf 的内容也随之改变，双方持有的是同一个内容指针引用。

（2）copy：复制一个新的 ByteBuf 对象，它的内容和索引都是独立的，复制操作本身并不修改原 ByteBuf 的读写索引。

（3）copy(int index, int length)：从指定的索引开始复制，复制的字节长度为 length，复制后的 ByteBuf 内容和读写索引都与之前的独立。

（4）slice：返回当前 ByteBuf 的可读子缓冲区，起始位置从 readerIndex 到 writerIndex，返回后的 ByteBuf 与原 ByteBuf 共享内容，但是读写索引独立维护。该操作并不修改原 ByteBuf 的 readerIndex 和 writerIndex。

（5）slice(int index, int length)：返回当前 ByteBuf 的可读子缓冲区，起始位置从 index 到 index + length，返回后的 ByteBuf 与原 ByteBuf 共享内容，但是读写索引独立维护。该操作并不修改原 ByteBuf 的 readerIndex 和 writerIndex。

10. 转换成标准的 ByteBuffer

我们知道，当通过 NIO 的 SocketChannel 进行网络读写时，操作的对象是 JDK 标准的

java.nio.ByteBuffer，由于 Netty 统一使用 ByteBuf 替代 JDK 原生的 java.nio.ByteBuffer，所以必须从接口层面支持两者的相互转换，下面就一起看下如何将 ByteBuf 转换成 java.nio.ByteBuffer。

将 ByteBuf 转换成 java.nio.ByteBuffer 的方法有两个，详细说明如下。

（1）ByteBuffer nioBuffer()：将当前 ByteBuf 可读的缓冲区转换成 ByteBuffer，两者共享同一个缓冲区内容引用，对 ByteBuffer 的读写操作并不会修改原 ByteBuf 的读写索引。需要指出的是，返回后的 ByteBuffer 无法感知原 ByteBuf 的动态扩展操作。

（2）ByteBuffer nioBuffer(int index, int length)：将当前 ByteBuf 从 index 开始长度为 length 的缓冲区转换成 ByteBuffer，两者共享同一个缓冲区内容引用，对 ByteBuffer 的读写操作并不会修改原 ByteBuf 的读写索引。需要指出的是，返回后的 ByteBuffer 无法感知原 ByteBuf 的动态扩展操作。

11. 随机读写（set 和 get）

除了顺序读写之外，ByteBuf 还支持随机读写，它与顺序读写的最大差别在于可以随机指定读写的索引位置。

读取操作的 API 列表如图 15-21 所示。

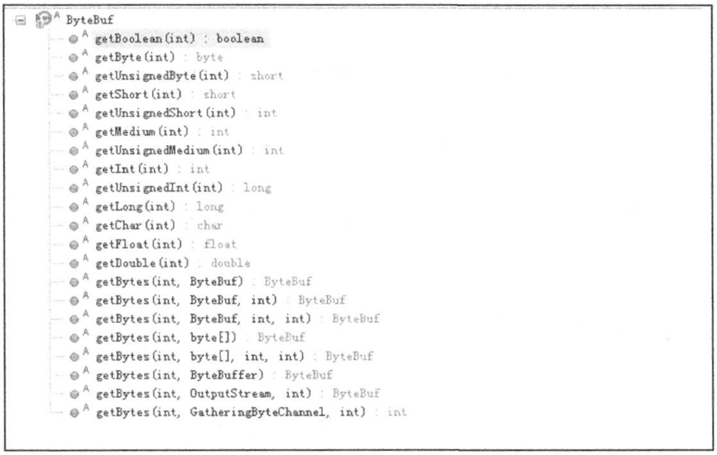

图 15-21　ByteBuf 随机读操作 API 列表

随机写操作的 API 列表如图 15-22 所示。

```
ByteBuf
  setIndex(int, int) : ByteBuf
  setBoolean(int, boolean) : ByteBuf
  setByte(int, int) : ByteBuf
  setShort(int, int) : ByteBuf
  setMedium(int, int) : ByteBuf
  setInt(int, int) : ByteBuf
  setLong(int, long) : ByteBuf
  setChar(int, int) : ByteBuf
  setFloat(int, float) : ByteBuf
  setDouble(int, double) : ByteBuf
  setBytes(int, ByteBuf) : ByteBuf
  setBytes(int, ByteBuf, int) : ByteBuf
  setBytes(int, ByteBuf, int, int) : ByteBuf
  setBytes(int, byte[]) : ByteBuf
  setBytes(int, byte[], int, int) : ByteBuf
  setBytes(int, ByteBuffer) : ByteBuf
  setBytes(int, InputStream, int) : int
  setBytes(int, ScatteringByteChannel, int) : int
  setZero(int, int) : ByteBuf
```

图 15-22　ByteBuf 随机写操作 API 列表

无论是 get 还是 set 操作，ByteBuf 都会对其索引和长度等进行合法性校验，与顺序读写一致。但是，set 操作与 write 操作不同的是它不支持动态扩展缓冲区，所以使用者必须保证当前的缓冲区可写的字节数大于需要写入的字节长度，否则会抛出数组或者缓冲区越界异常。相关代码如图 15-23 所示。

```
@Override
public ByteBuf setByte(int index, int value) {
    checkIndex(index);
    _setByte(index, value);
    return this;
}
```

图 15-23　ByteBuf 随机写操作不支持动态扩展缓冲区

15.2　ByteBuf 源码分析

由于 ByteBuf 的实现非常繁杂，因此本书不会对其所有子类都进行穷举分析，我们挑选 ByteBuf 的主要接口实现类和主要方法进行分析说明。相信理解了这些主要功能之后，再去阅读和分析其他辅助类会更加简单。

15.2.1 ByteBuf 的主要类继承关系

首先,我们通过主要功能类库的继承关系图(见图 15-24),来看下 ByteBuf 接口的不同实现。

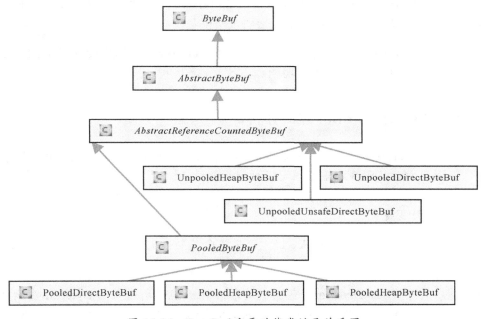

图 15-24 ByteBuf 主要功能类继承关系图

从内存分配的角度看,ByteBuf 可以分为两类。

(1)堆内存(HeapByteBuf)字节缓冲区:特点是内存的分配和回收速度快,可以被 JVM 自动回收;缺点就是如果进行 Socket 的 I/O 读写,需要额外做一次内存复制,将堆内存对应的缓冲区复制到内核 Channel 中,性能会有一定程度的下降。

(2)直接内存(DirectByteBuf)字节缓冲区:非堆内存,它在堆外进行内存分配,相比于堆内存,它的分配和回收速度会慢一些,但是将它写入或者从 Socket Channel 中读取时,由于少了一次内存复制,速度比堆内存快。

正是因为各有利弊,所以 Netty 提供了多种 ByteBuf 供开发者使用,经验表明,ByteBuf 的最佳实践是在 I/O 通信线程的读写缓冲区使用 DirectByteBuf,后端业务消息的编解码模块使用 HeapByteBuf,这样组合可以达到性能最优。

从内存回收角度看，ByteBuf 也分为两类：基于对象池的 ByteBuf 和普通 ByteBuf。两者的主要区别就是基于对象池的 ByteBuf 可以重用 ByteBuf 对象，它自己维护了一个内存池，可以循环利用创建的 ByteBuf，提升内存的使用效率，降低由于高负载导致的频繁 GC。测试表明使用内存池后的 Netty 在高负载、大并发的冲击下内存和 GC 更加平稳。

尽管推荐使用基于内存池的 ByteBuf，但是内存池的管理和维护更加复杂，使用起来也需要更加谨慎，因此，Netty 提供了灵活的策略供使用者来做选择。

下面我们对主要的功能类和方法的源码进行分析和解读，以便能够更加深刻地理解 ByteBuf 的实现，掌握其更加高级的功能。

15.2.2 AbstractByteBuf 源码分析

AbstractByteBuf 继承自 ByteBuf，ByteBuf 的一些公共属性和功能会在 AbstractByteBuf 中实现，下面我们对其属性和重要代码进行分析解读。

1. 主要成员变量

首先，像读索引、写索引、mark、最大容量等公共属性需要定义，具体定义如图 15-25 所示。

我们重点关注下 leakDetector，它被定义为 static，意味着所有的 ByteBuf 实例共享同一个 ResourceLeakDetector 对象。ResourceLeakDetector 用于检测对象是否泄漏，后面有专门章节进行讲解。

```
static final ResourceLeakDetector<ByteBuf> leakDetector = new
ResourceLeakDetector<ByteBuf>(ByteBuf.class);

    int readerIndex;
    private int writerIndex;
    private int markedReaderIndex;
    private int markedWriterIndex;

    private int maxCapacity;

    private SwappedByteBuf swappedBuf;
```

图 15-25　AbstractByteBuf 成员变量定义

我们发现，在 AbstractByteBuf 中并没有定义 ByteBuf 的缓冲区实现，例如 byte 数组或者 DirectByteBuffer。原因显而易见，因为 AbstractByteBuf 并不清楚子类到底是基于堆内存还是直接内存，因此无法提前定义。

2. 读操作簇

无论子类如何实现 ByteBuf，例如 UnpooledHeapByteBuf 使用 byte 数组表示字节缓冲区，UnpooledDirectByteBuf 直接使用 ByteBuffer，它们的功能都是相同的，操作的结果是等价的。

因此，读操作以及其他的一些公共功能都由父类实现，差异化功能由子类实现，这也就是抽象和继承的价值所在。

read 类操作的方法如图 15-26 所示。

```
● readByte() : byte
● readBoolean() : boolean
● readUnsignedByte() : short
● readShort() : short
● readUnsignedShort() : int
● readMedium() : int
● readUnsignedMedium() : int
● readInt() : int
● readUnsignedInt() : long
● readLong() : long
● readChar() : char
● readFloat() : float
● readDouble() : double
● readBytes(int) : ByteBuf
● readSlice(int) : ByteBuf
● readBytes(byte[], int, int) : ByteBuf
● readBytes(byte[]) : ByteBuf
● readBytes(ByteBuf) : ByteBuf
● readBytes(ByteBuf, int) : ByteBuf
● readBytes(ByteBuf, int, int) : ByteBuf
● readBytes(ByteBuffer) : ByteBuf
● readBytes(GatheringByteChannel, int) : int
● readBytes(OutputStream, int) : ByteBuf
```

图 15-26　读操作方法一览表

限于篇幅，我们不能一一枚举，挑选其中框线所示的 readBytes(byte[] dst, int dstIndex, int length)方法进行分析，首先看源码实现，如图 15-27 所示。

```
@Override
public ByteBuf readBytes(byte[] dst, int dstIndex, int length) {
    checkReadableBytes(length);
    getBytes(readerIndex, dst, dstIndex, length);
    readerIndex += length;
    return this;
}
```

图 15-27　readBytes(byte[] dst, int dstIndex, int length)方法源码

在读之前，首先对缓冲区的可用空间进行校验，校验的代码如图 15-28 所示。

```
protected final void checkReadableBytes(int minimumReadableBytes) {
    ensureAccessible();
    if (minimumReadableBytes < 0) {
        throw new IllegalArgumentException("minimumReadableBytes: " +
minimumReadableBytes + " (expected: >= 0)");
    }
    if (readerIndex > writerIndex - minimumReadableBytes) {
        throw new IndexOutOfBoundsException(String.format(
            "readerIndex(%d) + length(%d) exceeds writerIndex(%d): %s",
            readerIndex, minimumReadableBytes, writerIndex, this));
    }
}
```

图 15-28　readBytes(byte[] dst, int dstIndex, int length)方法源码

如果读取的长度小于 0，则抛出 IllegalArgumentException 异常提示参数非法；如果可写的字节数小于需要读取的长度，则抛出 IndexOutOfBoundsException 异常。由于异常中封装了详细的异常信息，所以使用者可以非常方便地进行问题定位。

校验通过之后，调用 getBytes 方法，从当前的读索引开始，复制 length 个字节到目标 byte 数组中。由于不同的子类复制操作的技术实现细节不同，因此该方法由子类实现，如图 15-29 所示。

```
/**
 * Transfers this buffer's data to the specified destination starting at
 * the specified absolute {@code index}.
 * This method does not modify {@code readerIndex} or {@code writerIndex}
 * of both the source (i.e. {@code this}) and the destination.
 *
 * @param dstIndex the first index of the destination
 * @param length   the number of bytes to transfer
 *
 * @throws IndexOutOfBoundsException
 *         if the specified {@code index} is less than {@code 0},
 *         if the specified {@code dstIndex} is less than {@code 0},
 *         if {@code index + length} is greater than
 *            {@code this.capacity}, or
 *         if {@code dstIndex + length} is greater than
 *            {@code dst.capacity}
 */
public abstract ByteBuf getBytes(int index, ByteBuf dst, int dstIndex, int length);
```

图 15-29　字节数组读取操作

如果读取成功，需要对读索引进行递增：readerIndex += length。其他类型的读取操作与之类似，不再展开介绍，感兴趣的读者可以自行阅读相关代码。

3. 写操作簇

与读取操作类似，写操作的公共行为在 AbstractByteBuf 中实现，它的 API 列表如图

15-30 所示。

我们选择与读取配套的 writeBytes(byte[] src, int srcIndex, int length)进行分析，它的功能是将源字节数组中从 srcIndex 开始，到 srcIndex + length 截止的字节数组写入到当前的 ByteBuf 中。下面具体看源码实现，如图 15-31 所示。

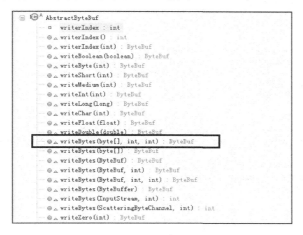

图 15-30　写操作方法一览

```
@Override
public ByteBuf writeBytes(ByteBuf src, int srcIndex, int length) {
    ensureWritable(length);
    setBytes(writerIndex, src, srcIndex, length);
    writerIndex += length;
    return this;
}
```

图 15-31　指定的字节数组写入缓冲区源码

首先对写入字节数组的长度进行合法性校验，校验代码如图 15-32 所示。

如果写入的字节数组长度小于 0，则抛出 IllegalArgumentException 异常；如果写入的字节数组长度小于当前 ByteBuf 可写的字节数，说明可以写入成功，直接返回；如果写入的字节数组长度大于可以动态扩展的最大可写字节数，说明缓冲区无法写入超过其最大容量的字节数组，抛出 IndexOutOfBoundsException 异常。

如果当前写入的字节数组长度虽然大于目前 ByteBuf 的可写字节数，但是通过自身的动态扩展可以满足新的写入请求，则进行动态扩展。可能有读者会产生疑问，既然需要写入的字节数组长度大于当前缓冲区可写的空间，为什么不像 JDK 的 ByteBuffer 那样抛出缓冲区越界异常呢？

```java
    @Override
    public ByteBuf ensureWritable(int minWritableBytes) {
      if (minWritableBytes < 0) {
        throw new IllegalArgumentException(String.format(
            "minWritableBytes: %d (expected: >= 0)", minWritableBytes));
      }

      if (minWritableBytes <= writableBytes()) {
        return this;
      }

      if (minWritableBytes > maxCapacity - writerIndex) {
        throw new IndexOutOfBoundsException(String.format(
            "writerIndex(%d) + minWritableBytes(%d) exceeds maxCapacity(%d): %s",
            writerIndex, minWritableBytes, maxCapacity, this));
      }

      // Normalize the current capacity to the power of 2.
      int newCapacity = calculateNewCapacity(writerIndex + minWritableBytes);

      // Adjust to the new capacity.
      capacity(newCapacity);
      return this;
    }
```

图 15-32　对写入的字节数组长度进行校验

在前面我们分析 JDK ByteBuffer 缺点的时候已经有过介绍，ByteBuffer 的一个最大的缺点就是一旦完成分配之后不能动态调整其容量。由于很多场景下我们无法预先判断需要编码和解码的 POJO 对象长度，因此只能根据经验数据给个估计值。如果这个值偏大，就会导致内存的浪费；如果这个值偏小，遇到大消息编码的时候就会发生缓冲区溢出异常。使用者需要自己捕获这个异常，并重新计算缓冲区的大小，将原来的内容复制到新的缓冲区中，然后重置指针。这种处理策略对用户非常不友好，而且稍有不慎，就会引入新的问题。

Netty 的 ByteBuffer 可以动态扩展，为了保证安全性，允许使用者指定最大的容量，在容量范围内，可以先分配个较小的初始容量，后面不够用再动态扩展，这样可以达到功能和性能的最优组合。

我们继续看 calculateNewCapacity 方法的实现。首先需要重新计算下扩展后的容量，它有一个参数，等于 writerIndex + minWritableBytes，也就是满足要求的最小容量。如图 15-33 所示。

首先设置门限阈值为 4MB，当需要的新容量正好等于门限阈值时，使用阈值作为新的缓冲区容量。如果新申请的内存空间大于阈值，不能采用倍增的方式（防止内存膨胀和浪费）扩张内存，而采用每次步进 4MB 的方式进行内存扩张。扩张的时候需要对扩张后的内存和最大内存（maxCapacity）进行比较，如果大于缓冲区的最大长度，则使用 maxCapacity 作为扩容后的缓冲区容量。

```
private int calculateNewCapacity(int minNewCapacity) {
    final int maxCapacity = this.maxCapacity;
    final int threshold = 1048576 * 4; // 4 MiB page

    if (minNewCapacity == threshold) {
        return threshold;
    }

    // If over threshold, do not double but just increase by threshold.
    if (minNewCapacity > threshold) {
        int newCapacity = minNewCapacity / threshold * threshold;
        if (newCapacity > maxCapacity - threshold) {
            newCapacity = maxCapacity;
        } else {
            newCapacity += threshold;
        }
        return newCapacity;
    }

    // Not over threshold. Double up to 4 MiB, starting from 64.
    int newCapacity = 64;
    while (newCapacity < minNewCapacity) {
        newCapacity <<= 1;
    }

    return Math.min(newCapacity, maxCapacity);
}
```

图 15-33　重新计算缓冲区的容量

如果扩容后的新容量小于阈值，则以 64 为计数进行倍增，直到倍增后的结果大于或等于需要的容量值。

采用倍增或者步进算法的原因如下：如果以 minNewCapacity 作为目标容量，则本次扩容后的可写字节数刚好够本次写入使用。写入完成后，它的可写字节数会变为 0，下次做写入操作的时候，需要再次动态扩张。这样就会形成第一次动态扩张后，每次写入操作都会进行动态扩张，由于动态扩张需要进行内存复制，频繁的内存复制会导致性能下降。

采用先倍增后步进的原因如下：当内存比较小的情况下，倍增操作并不会带来太多的内存浪费，例如 64 字节-->128 字节-->256 字节，这样的内存扩张方式对于大多数应用系统是可以接受的。但是，当内存增长到一定阈值后，再进行倍增就可能会带来额外的内存浪费，例如 10MB，采用倍增后变为 20MB。但很有可能系统只需要 12MB，则扩张到 20MB 后会带来 8MB 的内存浪费。由于每个客户端连接都可能维护自己独立的接收和发送缓冲区，这样随着客户读的线性增长，内存浪费也会成比例地增加，因此，达到某个阈值后就需要以步进的方式对内存进行平滑的扩张。

这个阈值是个经验值，不同的应用场景，这个值可能不同，此处 ByteBuf 取值为 4MB。

重新计算完动态扩张后的目标容量后，需要重新创建个新的缓冲区，将原缓冲区的内容复制到新创建的 ByteBuf 中，最后设置读写索引和 mark 标签等。由于不同的子类会对应不同的复制操作，所以该方法依然是个抽象方法，由子类负责实现。如图 15-34 所示。

```
public void initChannel(SocketChannel ch)
        throws Exception {
    ByteBuf delimiter = Unpooled.copiedBuffer("$_"
            .getBytes());
    // ch.pipeline().addLast(
    // new DelimiterBasedFrameDecoder(1024,
    // delimiter));
    ch.pipeline().addLast(new StringDecoder());
    ch.pipeline().addLast(new EchoServerHandler());
}
```

图 15-34　重新分配缓冲区的容量

4. 操作索引

与索引相关的操作主要涉及设置读写索引、mark 和 rest 等。如图 15-35 所示。

```
● ▲ maxWritableBytes() : int
● ▲ markReaderIndex() : ByteBuf
● ▲ resetReaderIndex() : ByteBuf
● ▲ markWriterIndex() : ByteBuf
● ▲ resetWriterIndex() : ByteBuf
● ▲ writerIndex() : int
● ▲ writerIndex(int) : ByteBuf
● ▲ readerIndex() : int
● ▲ readerIndex(int) : ByteBuf
```

图 15-35　索引操作相关 API 列表

由于这部分代码非常简单，我们就以设置读索引为例进行分析，相关代码如图 15-36 所示。

```
@Override
public ByteBuf readerIndex(int readerIndex) {
    if (readerIndex < 0 || readerIndex > writerIndex) {
        throw new IndexOutOfBoundsException(String.format(
            "readerIndex: %d (expected: 0 <= readerIndex <= writerIndex(%d))", readerIndex, writerIndex));
    }
    this.readerIndex = readerIndex;
    return this;
}
```

图 15-36　重设读索引

在重新设置读索引之前需要对索引进行合法性判断，如果它小于 0 或者大于写索引，则抛出 IndexOutOfBoundsException 异常，设置失败。校验通过之后，将索引设置为新的值，然后返回当前的 ByteBuf 对象。

5. 重用缓冲区

前面介绍功能的时候已经简单讲解了如何通过 discardReadBytes 和 discardSomeReadBytes 方法重用已经读取过的缓冲区，下面结合 discardReadBytes 方法的实现进行分析，源码如图 15-37 所示。

首先对读索引进行判断，如果为 0 则说明没有可重用的缓冲区，直接返回。如果读索引大于 0 且读索引不等于写索引，说明缓冲区中既有已经读取过的被丢弃的缓冲区，也有尚未读取的可读缓冲区。调用 setBytes(0, this, readerIndex, writerIndex - readerIndex) 方法进行字节数组复制。将尚未读取的字节数组复制到缓冲区的起始位置，然后重新设置读写索引，读索引设置为 0，写索引设置为之前的写索引减去读索引（重用的缓冲区长度）。

```
@Override
public ByteBuf discardReadBytes() {
    ensureAccessible();
    if (readerIndex == 0) {
        return this;
    }

    if (readerIndex != writerIndex) {
        setBytes(0, this, readerIndex, writerIndex - readerIndex);
        writerIndex -= readerIndex;
        adjustMarkers(readerIndex);
        readerIndex = 0;
    } else {
        adjustMarkers(readerIndex);
        writerIndex = readerIndex = 0;
    }
    return this;
}
```

图 15-37　重用读取过的缓冲区

在设置读写索引的同时，需要同时调整 markedReaderIndex 和 markedWriterIndex，调整 mark 的代码如图 15-38 所示。

首先对备份的 markedReaderIndex 和需要减少的 decrement 进行判断，如果小于需要减少的值，则将 markedReaderIndex 设置为 0。注意，无论是 markedReaderIndex 还是 markedWriterIndex，它的取值都不能小于 0。如果 markedWriterIndex 也小于需要减少的值，则 markedWriterIndex 置为 0，否则，markedWriterIndex 减去 decrement 之后的值就是新的 markedWriterIndex。

```
protected final void adjustMarkers(int decrement) {
  int markedReaderIndex = this.markedReaderIndex;
  if (markedReaderIndex <= decrement) {
    this.markedReaderIndex = 0;
    int markedWriterIndex = this.markedWriterIndex;
    if (markedWriterIndex <= decrement) {
      this.markedWriterIndex = 0;
    } else {
      this.markedWriterIndex = markedWriterIndex - decrement;
    }
  } else {
    this.markedReaderIndex = markedReaderIndex - decrement;
    markedWriterIndex -= decrement;
  }
}
```

图 15-38　调整 mark

如果需要减小的值小于 markedReaderIndex，则它也一定也小于 markedWriterIndex，markedReaderIndex 和 markedWriterIndex 的新值就是减去 decrement 之后的取值。

如果 readerIndex 等于 writerIndex，则说明没有可读的字节数组，那就不需要进行内存复制，直接调整 mark，将读写索引设置为 0 即可完成缓冲区的重用，代码如图 15-39 所示。

```
} else {
  adjustMarkers(readerIndex);
  writerIndex = readerIndex = 0;
}
```

图 15-39　没有可读的字节数组，不需要内存复制

6. skipBytes

在解码的时候，有时候需要丢弃非法的数据报，或者跳跃过不需要读取的字节或字节数组，此时，使用 skipBytes 方法就非常方便。它可以忽略指定长度的字节数组，读操作时直接跳过这些数据读取后面的可读缓冲区，详细的代码实现如图 15-40 所示。

```
@Override
public ByteBuf skipBytes(int length) {
  checkReadableBytes(length);

  int newReaderIndex = readerIndex + length;
  if (newReaderIndex > writerIndex) {
    throw new IndexOutOfBoundsException(String.format(
      "length: %d (expected: readerIndex(%d) + length <= writerIndex(%d))",
      length, readerIndex, writerIndex));
  }
  readerIndex = newReaderIndex;
  return this;
}
```

图 15-40　跳过指定长度的缓冲区

首先判断跳过的长度是否大于当前缓冲区可读的字节数组长度，如果大于可读字节数组长度，则抛出 IndexOutOfBoundsException；如果参数本身为负数，则抛出 IllegalArgumentException 异常。

如果校验通过，则设置新的读索引为旧的索引值与跳跃的长度之和，然后对新的读索引进行判断。如果大于写索引，则抛出 IndexOutOfBoundsException 异常；如果合法，则将读索引设置为新的读索引。这样后续读操作的时候就会从新的读索引开始，跳过 length 个字节。

15.2.3　AbstractReferenceCountedByteBuf 源码分析

从类的名字就可以看出该类主要是对引用进行计数，类似于 JVM 内存回收的对象引用计数器，用于跟踪对象的分配和销毁，做自动内存回收。

下面通过源码来看它的具体实现。

1. 成员变量

AbstractReferenceCountedByteBuf 成员变量列表如图 15-41 所示。

```
    private static final
AtomicIntegerFieldUpdater<AbstractReferenceCountedByteBuf>
refCntUpdater =

AtomicIntegerFieldUpdater.newUpdater(AbstractReferenceCountedByteBuf.class, "refCnt");

    private static final long REFCNT_FIELD_OFFSET;

    static {
        long refCntFieldOffset = -1;
        try {
            if (PlatformDependent.hasUnsafe()) {
                refCntFieldOffset = PlatformDependent.objectFieldOffset(
AbstractReferenceCountedByteBuf.class.getDeclaredField("refCnt"));
            }
        } catch (Throwable t) {
            // Ignored
        }

        REFCNT_FIELD_OFFSET = refCntFieldOffset;
    }

    @SuppressWarnings("FieldMayBeFinal")
    private volatile int refCnt = 1;
```

图 15-41　AbstractReferenceCountedByteBuf 成员变量列表

首先看第一个字段 refCntUpdater，它是 AtomicIntegerFieldUpdater 类型变量，通过原子的方式对成员变量进行更新等操作，以实现线程安全，消除锁。第二个字段是 REFCNT_FIELD_OFFSET，它用于标识 refCnt 字段在 AbstractReferenceCountedByteBuf 中的内存地址。该内存地址的获取是 JDK 实现强相关的，如果使用 SUN 的 JDK，它通过 sun.misc.Unsafe 的 objectFieldOffset 接口来获得，ByteBuf 的实现子类 UnpooledUnsafeDirectByteBuf 和 PooledUnsafeDirectByteBuf 会使用到这个偏移量。

最后定义了一个 volatile 修饰的 refCnt 字段用于跟踪对象的引用次数，使用 volatile 是为了解决多线程并发访问的可见性问题，此处不对 volatile 的用法展开说明，后续多线程章节会有详细介绍。

2. 对象引用计数器

每调用一次 retain 方法，引用计数器就会加一，由于可能存在多线程并发调用的场景，所以它的累加操作必须是线程安全的，下面我们一起看下它的具体实现细节，如图 15-42 所示。

```
@Override
public ByteBuf retain() {
    for (;;) {
        int refCnt = this.refCnt;
        if (refCnt == 0) {
            throw new IllegalReferenceCountException(0, 1);
        }
        if (refCnt == Integer.MAX_VALUE) {
            throw new IllegalReferenceCountException(Integer.MAX_VALUE, 1);
        }
        if (refCntUpdater.compareAndSet(this, refCnt, refCnt + 1)) {
            break;
        }
    }
    return this;
}
```

图 15-42 调用 retain 函数，引用计数器加一

通过自旋对引用计数器进行加一操作，由于引用计数器的初始值为 1，如果申请和释放操作能够保证正确使用，则它的最小值为 1。当被释放和被申请的次数相等时，就调用回收方法回收当前的 ByteBuf 对象。如果为 0，说明对象被意外、错误地引用，抛出 IllegalReferenceCountException。如果引用计数器达到整型数的最大值，抛出引用越界的异常 IllegalReferenceCountException。最后通过 compareAndSet 进行原子更新，它会使用自己获取的值跟期望值进行对比。如果其间已经被其他线程修改了，则比对失败，进行自旋，重新获取引用计数器的值再次比对；如果比对成功则对其加一。注意：compareAndSet 是

由操作系统层面提供的原子操作，这类原子操作被称为 CAS，感兴趣的读者可以看下 Java CAS 的原理。

下面看下释放引用计数器的代码，如图 15-43 所示。

与 retain 方法类似，它也是在一个自旋循环里面进行判断和更新的。需要注意的是：当 refCnt == 1 时意味着申请和释放相等，说明对象引用已经不可达，该对象需要被释放和垃圾回收掉，则通过调用 deallocate 方法来释放 ByteBuf 对象。

```
@Override
public final boolean release() {
    for (;;) {
        int refCnt = this.refCnt;
        if (refCnt == 0) {
            throw new IllegalReferenceCountException(0, -1);
        }
        if (refCntUpdater.compareAndSet(this, refCnt, refCnt - 1)) {
            if (refCnt == 1) {
                deallocate();
                return true;
            }
            return false;
        }
    }
}
```

图 15-43　调用释放函数，引用计数器减一

15.2.4　UnpooledHeapByteBuf 源码分析

UnpooledHeapByteBuf 是基于堆内存进行内存分配的字节缓冲区，它没有基于对象池技术实现，这就意味着每次 I/O 的读写都会创建一个新的 UnpooledHeapByteBuf，频繁进行大块内存的分配和回收对性能会造成一定影响，但是相比于堆外内存的申请和释放，它的成本还是会低一些。

相比于 PooledHeapByteBuf，UnpooledHeapByteBuf 的实现原理更加简单，也不容易出现内存管理方面的问题，因此在满足性能的情况下，推荐使用 UnpooledHeapByteBuf。

下面我们就一起来看下 UnpooledHeapByteBuf 的代码实现。

1. 成员变量

首先看下 UnpooledHeapByteBuf 的成员变量定义，如图 15-44 所示。

```
public class UnpooledHeapByteBuf extends
AbstractReferenceCountedByteBuf {

    private final ByteBufAllocator alloc;
    private byte[] array;
    private ByteBuffer tmpNioBuf;
}
```

图 15-44　UnpooledHeapByteBuf 成员变量定义

首先，它聚合了一个 ByteBufAllocator，用于 UnpooledHeapByteBuf 的内存分配，紧接着定义了一个 byte 数组作为缓冲区，最后定义了一个 ByteBuffer 类型的 tmpNioBuf 变量用于实现 Netty ByteBuf 到 JDK NIO ByteBuffer 的转换。

事实上，如果使用 JDK 的 ByteBuffer 替换 byte 数组也是可行的，直接使用 byte 数组的根本原因就是提升性能和更加便捷地进行位操作。JDK 的 ByteBuffer 底层实现也是 byte 数组，代码如图 15-45 所示。

```
public abstract class ByteBuffer
    extends Buffer
    implements Comparable<ByteBuffer>
{

    // These fields are declared here rather than in Heap-X-Buffer in order to
    // reduce the number of virtual method invocations needed to access these
    // values, which is especially costly when coding small buffers.
    //
    final byte[] hb;                  // Non-null only for heap buffers
    final int offset;
    boolean isReadOnly;               // Valid only for heap buffers
```

图 15-45　JDK ByteBuffer 内部实现源码

2. 动态扩展缓冲区

在前一章介绍 AbstractByteBuf 的时候，我们讲到 ByteBuf 在最大容量范围内能够实现自动扩张，下面我们一起看下缓冲区的自动扩展在 UnpooledHeapByteBuf 中的实现，如图 15-46 所示。

方法入口首先对新容量进行合法性校验，如果大于容量上限或者小于 0，则抛出 IllegalArgumentException 异常。

判断新的容量值是否大于当前的缓冲区容量，如果大于则需要进行动态扩展，通过 byte[] newArray = new byte[newCapacity] 创建新的缓冲区字节数组，然后通过 System.arraycopy 进行内存复制，将旧的字节数组复制到新创建的字节数组中，最后调用 setArray 替换旧的字节数组。如图 15-47 所示。

需要指出的是，当动态扩容完成后，需要将原来的视图 tmpNioBuf 设置为空。

```
@Override
public ByteBuf capacity(int newCapacity) {
    ensureAccessible();
    if (newCapacity < 0 || newCapacity > maxCapacity()) {
        throw new IllegalArgumentException("newCapacity: " + newCapacity);
    }
    int oldCapacity = array.length;
    if (newCapacity > oldCapacity) {
        byte[] newArray = new byte[newCapacity];
        System.arraycopy(array, 0, newArray, 0, array.length);
        setArray(newArray);
    } else if (newCapacity < oldCapacity) {
        byte[] newArray = new byte[newCapacity];
        int readerIndex = readerIndex();
        if (readerIndex < newCapacity) {
            int writerIndex = writerIndex();
            if (writerIndex > newCapacity) {
                writerIndex(writerIndex = newCapacity);
            }
            System.arraycopy(array, readerIndex, newArray, readerIndex,
                writerIndex - readerIndex);
        } else {
            setIndex(newCapacity, newCapacity);
        }
        setArray(newArray);
    }
    return this;
}
```

图 15-46　缓冲区的动态扩展

```
private void setArray(byte[] initialArray) {
    array = initialArray;
    tmpNioBuf = null;
}
```

图 15-47　字节数组替换

如果新的容量小于当前的缓冲区容量，不需要动态扩展，但是需要截取当前缓冲区创建一个新的子缓冲区，具体的算法如下：首先判断下读索引是否小于新的容量值，如果小于进一步判断写索引是否大于新的容量值，如果大于则将写索引设置为新的容量值（防止越界）。更新完写索引之后通过内存复制 System.arraycopy 将当前可读的字节数组复制到新创建的子缓冲区中，代码如下。

```
System.arraycopy(array, readerIndex, newArray, readerIndex, writerIndex - readerIndex);
```

如果新的容量值小于读索引，说明没有可读的字节数组需要复制到新创建的缓冲区中，将读写索引设置为新的容量值即可。最后调用 setArray 方法替换原来的字节数组。

3. 字节数组复制

在前一章节里我们介绍 setBytes(int index, byte[] src, int srcIndex, int length)方法的时候说它有子类实现，下面我们看看 UnpooledHeapByteBuf 如何进行字节数组的复制。如图 15-48 所示。

```
@Override
public ByteBuf setBytes(int index, byte[] src, int srcIndex, int length) {
    checkSrcIndex(index, length, srcIndex, src.length);
    System.arraycopy(src, srcIndex, array, index, length);
    return this;
}
```

图 15-48　字节数组复制

首先仍然是合法性校验，我们看下校验代码。如图 15-49 所示。

```
protected final void checkSrcIndex(int index, int length, int srcIndex, int srcCapacity) {
    checkIndex(index, length);
    if (srcIndex < 0 || srcIndex > srcCapacity - length) {
        throw new IndexOutOfBoundsException(String.format(
            "srcIndex: %d, length: %d (expected: range(0, %d))", srcIndex,
            length, srcCapacity));
    }
}
```

图 15-49　字节数组复制前的校验

校验 index 和 length 的值，如果它们小于 0，则抛出 IllegalArgumentException，然后对两者之和进行判断；如果大于缓冲区的容量，则抛出 IndexOutOfBoundsException。srcIndex 和 srcCapacity 的校验与 index 类似，不再赘述。校验通过之后，调用 System.arraycopy(src, srcIndex, array, index, length)方法进行字节数组的复制。

需要指出的是，ByteBuf 以 set 和 get 开头读写缓冲区的方法并不会修改读写索引。

4. 转换成 JDK ByteBuffer

熟悉 JDK NIO ByteBuffer 的读者可能会想到转换非常简单，因为 ByteBuf 基于 byte 数组实现，NIO 的 ByteBuffer 提供了 wrap 方法，可以将 byte 数组转换成 ByteBuffer 对象，JDK 的相关源码实现如图 15-50 所示。

大家的猜想是对的，下面我们一起看下 UnpooledHeapByteBuf 的实现，如图 15-51 所示。

我们发现，唯一不同的是它还调用了 ByteBuffer 的 slice 方法，slice 的功能前面已经介绍过了，此处不再展开说明。由于每次调用 nioBuffer 都会创建一个新的 ByteBuffer，因此此处的 slice 方法起不到重用缓冲区内容的效果，只能保证读写索引的独立性。

```
public static ByteBuffer wrap(byte[] array,
                              int offset, int length)
{
  try {
    return new HeapByteBuffer(array, offset, length);
  } catch (IllegalArgumentException x) {
    throw new IndexOutOfBoundsException();
  }
}
```

图 15-50　JDK ByteBuffer 的 warp 方法源码

```
@Override
public ByteBuffer nioBuffer(int index, int length) {
  ensureAccessible();
  return ByteBuffer.wrap(array, index, length).slice();
}
```

图 15-51　UnpooledHeapByteBuf 的 warp 源码

5. 子类实现相关的方法

ByteBuf 中的一些接口是跟具体子类实现相关的，不同的子类功能是不同的，本小节我们将列出这些不同点。

- isDirect 方法：如果是基于堆内存实现的 ByteBuf，它返回 false，相关的代码实现如图 15-52 所示。

- hasArray 方法：由于 UnpooledHeapByteBuf 基于字节数组实现，所以它的返回值是 true。

- array 方法：由于 UnpooledHeapByteBuf 基于字节数组实现，所以它的返回值是内部的字节数组成员变量。如图 15-53 所示。

```
@Override
public boolean isDirect() {
  return false;
}
```

```
@Override
public byte[] array() {
  ensureAccessible();
  return array;
}
```

图 15-52　UnpooledHeapByteBuf 的 isDirect 方法　　图 15-53　UnpooledHeapByteBuf 的 array 方法

读者在调用 array 方法之前，可以先通过 hasArray 进行判断。如果返回 false 说明当前的 ByteBuf 不支持 array 方法。

◎ 其他本地相关的方法有：arrayOffset、hasMemoryAddress 和 memoryAddress，这些方法的实现如图 15-54 所示。

内存地址相关的接口主要由 UnsafeByteBuf 使用，它基于 SUN JDK 的 sun.misc.Unsafe 方法实现，本书的重点并不是介绍 sun.misc.Unsafe 的，如果读者对 sun.misc.Unsafe 的实现感兴趣，可以阅读 OPEN JDK 的相关源码实现，也可以通过其他的 DOC 文档进行深入学习。

```
@Override
public int arrayOffset() {
    return 0;
}

@Override
public boolean hasMemoryAddress() {
    return false;
}

@Override
public long memoryAddress() {
    throw new UnsupportedOperationException();
}
```

图 15-54　UnpooledHeapByteBuf 的 address 相关方法

由于 UnpooledDirectByteBuf 与 UnpooledHeapByteBuf 的实现原理相同，不同之处就是它内部缓冲区由 java.nio.DirectByteBuffer 实现。当掌握了 UnpooledHeapByteBuf 之后，阅读 UnpooledDirectByteBuf 的代码会非常容易，所以本书不再对 UnpooledDirectByteBuf 进行源码解读。

15.2.5　PooledByteBuf 内存池原理分析

由于 ByteBuf 内存池的实现涉及到的类和数据结构非常多，限于篇幅，本章节不对其源码进行展开说明，而是从设计原理角度来讲解内存池的实现。

1. PoolArena

Arena 本身是指一块区域，在内存管理中，Memory Arena 是指内存中的一大块连续的区域，PoolArena 就是 Netty 的内存池实现类。

为了集中管理内存的分配和释放，同时提高分配和释放内存时候的性能，很多框架和应用都会通过预先申请一大块内存，然后通过提供相应的分配和释放接口来使用内存。这样一来，对内存的管理就被集中到几个类或者函数中，由于不再频繁使用系统调用来申请和释放内存，应用或者系统的性能也会大大提高。在这种设计思路下，预先申请的那一大块内存就被称为 Memory Arena。

不同的框架，Memory Arena 的实现不同，Netty 的 PoolArena 是由多个 Chunk 组成的大块内存区域，而每个 Chunk 则由一个或者多个 Page 组成，因此，对内存的组织和管理也就主要集中在如何管理和组织 Chunk 和 Page 了。PoolArena 中的内存 Chunk 定义如图 15-55 所示。

```
abstract class PoolArena<T> {

final PooledByteBufAllocator parent;

private final int pageSize;
private final int maxOrder;
private final int pageShifts;
private final int chunkSize;
private final int subpageOverflowMask;

private final PoolSubpage<T>[] tinySubpagePools;
private final PoolSubpage<T>[] smallSubpagePools;

private final PoolChunkList<T> q050;
private final PoolChunkList<T> q025;
private final PoolChunkList<T> q000;
private final PoolChunkList<T> qInit;
private final PoolChunkList<T> q075;
private final PoolChunkList<T> q100;
```

图 15-55　Netty 的 Memory Arena 实现

2. PoolChunk

Chunk 主要用来组织和管理多个 Page 的内存分配和释放，在 Netty 中，Chunk 中的 Page 被构建成一棵二叉树。假设一个 Chunk 由 16 个 Page 组成，那么这些 Page 将会被按照图 15-56 所示的形式组织起来。

Page 的大小是 4 个字节，Chunk 的大小是 64 个字节(4×16)。整棵树有 5 层，第 1 层（也就是叶子节点所在的层）用来分配所有 Page 的内存，第 4 层用来分配 2 个 Page 的内存，依此类推。

每个节点都记录了自己在整个 Memory Arena 中的偏移地址，当一个节点代表的内存区域被分配出去之后，这个节点就会被标记为已分配，自这个节点以下的所有节点在后面的内存分配请求中都会被忽略。举例来说，当我们请求一个 16 字节的存储区域时，上面这个树中的第 3 层中的 4 个节点中的一个就会被标记为已分配，这就表示整个 Memroy Arena 中有 16 个字节被分配出去了，新的分配请求只能从剩下的 3 个节点及其子树中寻找合适的节点。

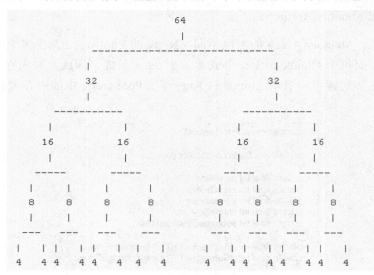

图 15-56　Chunk 的数据结构

对树的遍历采用深度优先的算法，但是在选择哪个子节点继续遍历时则是随机的，并不像通常的深度优先算法中那样总是访问左边的子节点。

3. PoolSubpage

对于小于一个 Page 的内存，Netty 在 Page 中完成分配。每个 Page 会被切分成大小相等的多个存储块，存储块的大小由第一次申请的内存块大小决定。假如一个 Page 是 8 个字节，如果第一次申请的块大小是 4 个字节，那么这个 Page 就包含 2 个存储块；如果第一次申请的是 8 个字节，那么这个 Page 就被分成 1 个存储块。

一个 Page 只能用于分配与第一次申请时大小相同的内存，比如，一个 4 字节的 Page，如果第一次分配了 1 字节的内存，那么后面这个 Page 只能继续分配 1 字节的内存，如果有一个申请 2 字节内存的请求，就需要在一个新的 Page 中进行分配。

Page 中存储区域的使用状态通过一个 long 数组来维护，数组中每个 long 的每一位表示一个块存储区域的占用情况：0 表示未占用，1 表示以占用。对于一个 4 字节的 Page 来说，如果这个 Page 用来分配 1 个字节的存储区域，那么 long 数组中就只有一个 long 类型的元素，这个数值的低 4 位用来指示各个存储区域的占用情况。对于一个 128 字节的 Page 来说，如果这个 Page 也是用来分配 1 个字节的存储区域，那么 long 数组中就会包含 2 个元素，总共 128 位，每一位代表一个区域的占用情况。

相关的代码实现如图 15-57 所示。

```
final class PoolSubpage<T> {

    final PoolChunk<T> chunk;
    final int memoryMapIdx;
    final int runOffset;
    final int pageSize;
    final long[] bitmap;

    PoolSubpage<T> prev;
    PoolSubpage<T> next;

    boolean doNotDestroy;
    int elemSize;
    int maxNumElems;
    int nextAvail;
    int bitmapLength;
    int numAvail;
}
```

图 15-57　PoolSubpage 的变量定义

4. 内存回收策略

无论是 Chunk 还是 Page，都通过状态位来标识内存是否可用，不同之处是 Chunk 通过在二叉树上对节点进行标识实现，Page 是通过维护块的使用状态标识来实现。

对于使用者来说，不需要关心内存池的实现细节，也不需要与这些类库打交道，只需要按照 API 说明正常使用即可。

15.2.6　PooledDirectByteBuf 源码分析

PooledDirectByteBuf 基于内存池实现，与 UnPooledDirectByteBuf 的唯一不同就是缓冲区的分配是销毁策略不同，其他功能都是等同的，也就是说，两者唯一的不同就是内存分配策略不同。

1. 创建字节缓冲区实例

由于采用内存池实现，所以新创建 PooledDirectByteBuf 对象时不能直接 new 一个实例，而是从内存池中获取，然后设置引用计数器的值，代码如图 15-58 所示。

```
static PooledDirectByteBuf newInstance(int maxCapacity) {
    PooledDirectByteBuf buf = RECYCLER.get();
    buf.setRefCnt(1);
    buf.maxCapacity(maxCapacity);
    return buf;
}
```

图 15-58　PooledDirectByteBuf 的创建

直接从内存池 Recycler<PooledDirectByteBuf>中获取 PooledDirectByteBuf 对象，然后设置它的引用计数器为 1，设置缓冲区最大容量后返回。

2. 复制新的字节缓冲区实例

如果使用者确实需要复制一个新的实例，与原来的 PooledDirectByteBuf 独立，则调用它的 copy（int index, int length）可以达到上述目标，代码如图 15-59 所示。

```
@Override
public ByteBuf copy(int index, int length) {
    checkIndex(index, length);
    ByteBuf copy = alloc().directBuffer(length, maxCapacity());
    copy.writeBytes(this, index, length);
    return copy;
}
```

图 15-59　PooledDirectByteBuf 的 copy 方法

首先对索引和长度进行合法性校验，通过之后调用 PooledByteBufAllocator 分配一个新的 ByteBuf，由于 PooledByteBufAllocator 没有实现 directBuffer 方法，所以最终会调用到 AbstractByteBufAllocator 的 directBuffer 方法，相关代码如图 15-60 所示。

```
@Override
public ByteBuf directBuffer(int initialCapacity, int maxCapacity) {
    if (initialCapacity == 0 && maxCapacity == 0) {
        return emptyBuf;
    }
    validate(initialCapacity, maxCapacity);
    return newDirectBuffer(initialCapacity, maxCapacity);
}
```

图 15-60　AbstractByteBufAllocator 的缓冲区分配

newDirectBuffer 方法对于不同的子类有不同的实现策略，如果是基于内存池的分配器，它会从内存池中获取可用的 ByteBuf，如果是非池，则直接创建新的 ByteBuf，相关代码实现如图 15-61、15-62 所示。

```
@Override
protected ByteBuf newDirectBuffer(int initialCapacity, int maxCapacity) {
    PoolThreadCache cache = threadCache.get();
    PoolArena<ByteBuffer> directArena = cache.directArena;

    ByteBuf buf;
    if (directArena != null) {
        buf = directArena.allocate(cache, initialCapacity, maxCapacity);
    } else {
        if (PlatformDependent.hasUnsafe()) {
            buf = new UnpooledUnsafeDirectByteBuf(this, initialCapacity, maxCapacity);
        } else {
            buf = new UnpooledDirectByteBuf(this, initialCapacity, maxCapacity);
        }
    }

    return toLeakAwareBuffer(buf);
}
```

图 15-61　基于内存池的缓冲区分配

```
@Override
protected ByteBuf newDirectBuffer(int initialCapacity, int maxCapacity) {
    ByteBuf buf;
    if (PlatformDependent.hasUnsafe()) {
        buf = new UnpooledUnsafeDirectByteBuf(this, initialCapacity, maxCapacity);
    } else {
        buf = new UnpooledDirectByteBuf(this, initialCapacity, maxCapacity);
    }

    return toLeakAwareBuffer(buf);
}
```

图 15-62　非内存池实现直接创建新的缓冲区

通过上述代码对比我们可以看出，基于内存池的实现直接从缓存中获取 ByteBuf 而不是创建一个新的对象。

3. 子类实现相关的方法

正如 UnpooledHeapByteBuf，PooledDirectByteBuf 也有子类实现相关的功能，这些方法如图 15-63 所示。

从上述代码可以看出，当我们操作子类实现相关的方法时，需要对是否支持这些操作进行判断，否则会导致异常。

```java
    @Override
    public boolean hasArray() {
        return false;
    }

    @Override
    public byte[] array() {
        throw new UnsupportedOperationException("direct buffer");
    }

    @Override
    public int arrayOffset() {
        throw new UnsupportedOperationException("direct buffer");
    }

    @Override
    public boolean hasMemoryAddress() {
        return false;
    }

    @Override
    public long memoryAddress() {
        throw new UnsupportedOperationException();
    }
```

图 15-63　PooledDirectByteBuf 实现相关的方法

15.3　ByteBuf 相关的辅助类功能介绍

学习完了核心的 ByteBuf 之后，下面一起继续学习它的一些常用辅助功能类。

15.3.1　ByteBufHolder

ByteBufHolder 是 ByteBuf 的容器，在 Netty 中，它非常有用。例如 HTTP 协议的请求消息和应答消息都可以携带消息体，这个消息体在 NIO ByteBuffer 中就是个 ByteBuffer 对象，在 Netty 中就是 ByteBuf 对象。由于不同的协议消息体可以包含不同的协议字段和功能，因此，需要对 ByteBuf 进行包装和抽象，不同的子类可以有不同的实现。

为了满足这些定制化的需求，Netty 抽象出了 ByteBufHolder 对象，它包含了一个 ByteBuf，另外还提供了一些其他实用的方法，使用者继承 ByteBufHolder 接口后可以按需封装自己的实现。相关类库的继承关系如图 15-64 所示。

第 15 章　ByteBuf 和相关辅助类

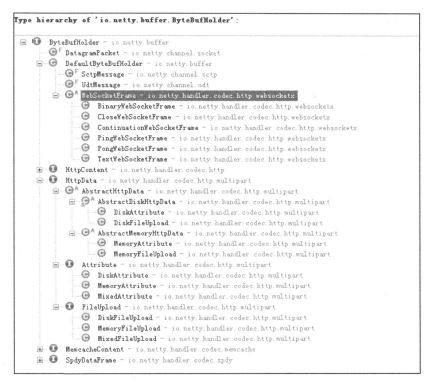

图 15-64　ByteBufHolder 继承关系图

15.3.2　ByteBufAllocator

ByteBufAllocator 是字节缓冲区分配器，按照 Netty 的缓冲区实现不同，共有两种不同的分配器：基于内存池的字节缓冲区分配器和普通的字节缓冲区分配器。接口的继承关系如图 15-65 所示。

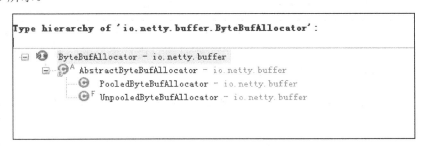

图 15-65　ByteBufAllocator 继承关系图

· 333 ·

下面我们给出 ByteBufAllocator 的主要 API 功能列表（表 15-3）。

表 15-3　ByteBufAllocator 主要 API 功能列表

方法名称	返回值说明	功能说明
buffer()	ByteBuf	分配一个字节缓冲区，缓冲区的类型由 ByteBufAllocator 的实现类决定
buffer(int initialCapacity)	ByteBuf	分配一个初始容量为 initialCapacity 的字节缓冲区，缓冲区的类型由 ByteBufAllocator 的实现类决定
buffer(int initialCapacity, int maxCapacity)	ByteBuf	分配一个初始容量为 initialCapacity，最大容量为 maxCapacity 的字节缓冲区，缓冲区的类型由 ByteBufAllocator 的实现类决定
ioBuffer(int initialCapacity, int maxCapacity)	ByteBuf	分配一个初始容量为 initialCapacity，最大容量为 maxCapacity 的 direct buffer，因为 direct buffer 的 I/O 操作性能更高
heapBuffer(int initialCapacity, int maxCapacity)	ByteBuf	分配一个初始容量为 initialCapacity，最大容量为 maxCapacity 的 heap buffer
directBuffer(int initialCapacity, int maxCapacity)	ByteBuf	分配一个初始容量为 initialCapacity，最大容量为 maxCapacity 的 direct buffer
compositeBuffer(int maxNumComponents)	CompositeByteBuf	分配一个最大容量为 maxCapacity 的 CompositeByteBuf，内存类型由 ByteBufAllocator 的实现类决定
isDirectBufferPooled()	boolean	是否使用了直接内存内存池

15.3.3　CompositeByteBuf

CompositeByteBuf 允许将多个 ByteBuf 的实例组装到一起，形成一个统一的视图，有点类似于数据库将多个表的字段组装到一起统一用视图展示。

CompositeByteBuf 在一些场景下非常有用，例如某个协议 POJO 对象包含两部分：消息头和消息体，它们都是 ByteBuf 对象。当需要对消息进行编码的时候需要进行整合，如果使用 JDK 的默认能力，有以下两种方式。

（1）将某个 ByteBuffer 复制到另一个 ByteBuffer 中，或者创建一个新的 ByteBuffer，将两者复制到新建的 ByteBuffer 中；

（2）通过 List 或数组等容器，将消息头和消息体放到容器中进行统一维护和处理。

上面的做法非常别扭，实际上我们遇到的问题跟数据库中视图解决的问题一致——缓冲区有多个，但是需要统一展示和处理，必须有存放它们的统一容器。为了解决这个问题，Netty 提供了 CompositeByteBuf。

我们一起简单看下它的实现，如图 15-66 所示。

```
public class CompositeByteBuf extends
AbstractReferenceCountedByteBuf {

  private final ResourceLeak leak;
  private final ByteBufAllocator alloc;
  private final boolean direct;
  private final List<Component> components = new ArrayList<Component>();
  private final int maxNumComponents;
  private static final ByteBuffer FULL_BYTEBUFFER = (ByteBuffer)
ByteBuffer.allocate(1).position(1);

  private boolean freed;
```

图 15-66　CompositeByteBuf 源码

它定义了一个 Component 类型的集合，实际上 Component 就是 ByteBuf 的包装实现类，它聚合了 ByteBuf 对象，维护了在集合中的位置偏移量信息等，它的实现如图 15-67 所示。

```
private final class Component {
  final ByteBuf buf;
  final int length;
  int offset;
  int endOffset;

  Component(ByteBuf buf) {
    this.buf = buf;
    length = buf.readableBytes();
  }

  void freeIfNecessary() {
    // Unwrap so that we can free slices, too.
    buf.release(); // We should not get a NPE here. If so, it must be a bug.
  }
}
```

图 15-67　Component 源码

向 CompositeByteBuf 中新增一个 ByteBuf 的代码，如图 15-68 所示。

```
public CompositeByteBuf addComponent(ByteBuf buffer) {
  addComponent0(components.size(), buffer);
  consolidateIfNeeded();
  return this;
}
```

图 15-68　CompositeByteBuf 中新增 ByteBuf 源码

删除增加的 ByteBuf 源码，如图 15-69 所示。

```
public CompositeByteBuf removeComponent(int cIndex) {
    checkComponentIndex(cIndex);
    components.remove(cIndex).freeIfNecessary();
    updateComponentOffsets(cIndex);
    return this;
}
```

图 15-69　CompositeByteBuf 中删除 ByteBuf 源码

注意：删除 ByteBuf 之后，需要更新各个 Component 的索引偏移量。

15.3.4　ByteBufUtil

ByteBufUtil 是一个非常有用的工具类，它提供了一系列静态方法用于操作 ByteBuf 对象。它的功能列表如图 15-70 所示。

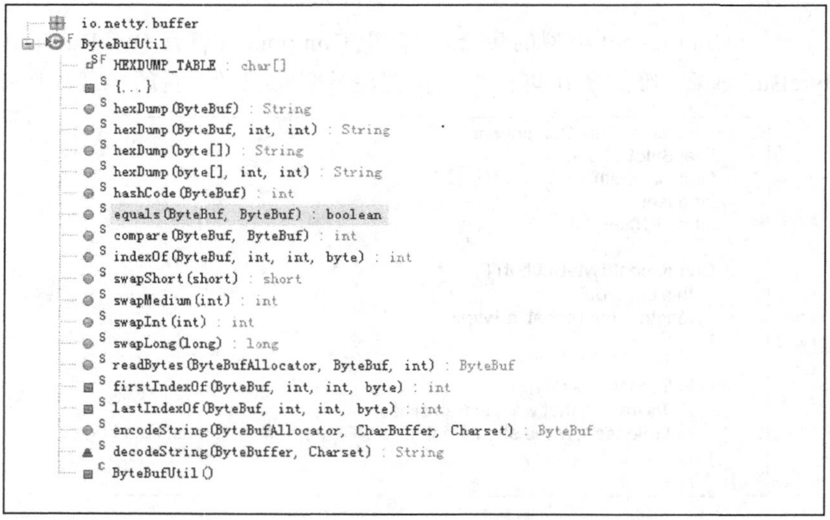

图 15-70　ByteBufUtil 工具类

其中最有用的方法就是对字符串的编码和解码，具体如下。

（1）encodeString(ByteBufAllocator alloc, CharBuffer src, Charset charset)：对需要编码的字符串 src 按照指定的字符集 charset 进行编码，利用指定的 ByteBufAllocator 生成一个新的 ByteBuf；

（2）decodeString(ByteBuffer src, Charset charset)：使用指定的 ByteBuffer 和 charset 进行对 ByteBuffer 进行解码，获取解码后的字符串。

还有一个非常有用的方法就是 hexDump，它能够将参数 ByteBuf 的内容以十六进制字符串的方式打印出来，用于输出日志或者打印码流，方便问题定位，提升系统的可维护性。

hexDump 包含了一系列的方法，参数不同，输出的结果也不同，如图 15-71 所示。

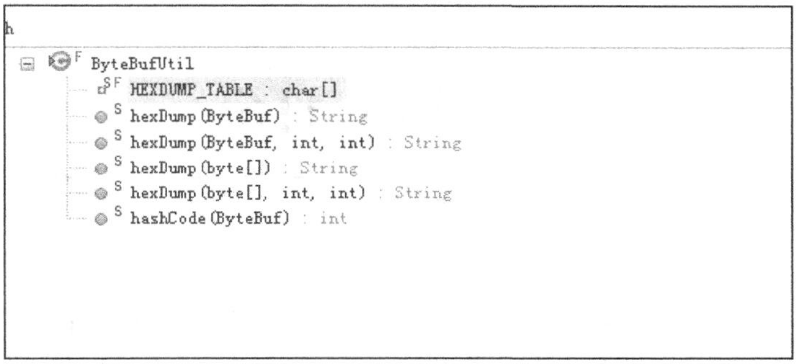

```
ByteBufUtil
    SF HEXDUMP_TABLE : char[]
    S hexDump(ByteBuf) : String
    S hexDump(ByteBuf, int, int) : String
    S hexDump(byte[]) : String
    S hexDump(byte[], int, int) : String
    S hashCode(ByteBuf) : int
```

图 15-71　hexDump 方法

15.4　总结

本章节重点介绍了 ByteBuf 的 API 功能和其源码实现，同时介绍了与 ByteBuf 密切相关的工具类和辅助类。ByteBuf 是 Netty 架构中最重要、最基础的数据结构，熟练地掌握和使用它是学好 Netty 的基本要求，也是成长为高级 Netty 开发人员的必经之路。

由于 ByteBuf 的功能复杂性，它的子类实现非常庞大，在本书中进行穷举是不现实的。读者学习完本章之后，对 Netty ByteBuf 的设计理念和重要类库的实现原理都有了比较深入的了解。以此为基础，再去学习其他相关联的类库会容易很多。

下个章节，我们继续学习 Netty 的另两个重要类库：Channel 和 Unsafe。

第 16 章

Channel 和 Unsafe

提起 Channel，读者朋友们可能并不陌生——JDK 的 NIO 类库的一个重要组成部分，就是 java.nio.SocketChannel 和 java.nio.ServerSocketChannel，它们用于非阻塞的 I/O 操作。

类似于 NIO 的 Channel，Netty 提供了自己的 Channel 和其子类实现，用于异步 I/O 操作和其他相关的操作。

Unsafe 是个内部接口，聚合在 Channel 中协助进行网络读写相关的操作，因为它的设计初衷就是 Channel 的内部辅助类，不应该被 Netty 框架的上层使用者调用，所以被命名为 Unsafe。这里不能仅从字面理解认为它是不安全的操作，而要从整个架构的设计层面体会它的设计初衷和职责。

本章主要内容包括：

◎ Channel 功能说明

◎ Unsafe 功能说明

◎ Channel 的主要实现子类源码分析

◎ Unsafe 的主要实现子类源码分析

16.1 Channel 功能说明

io.netty.channel.Channel 是 Netty 网络操作抽象类，它聚合了一组功能，包括但不限于网路的读、写，客户端发起连接，主动关闭连接，链路关闭，获取通信双方的网络地址等。它也包含了 Netty 框架相关的一些功能，包括获取该 Chanel 的 EventLoop，获取缓冲分配器 ByteBufAllocator 和 pipeline 等。

下面我们先从 Channel 的接口分析，讲解它的主要 API 和功能，然后再一起看下它的子类的相关功能实现，最后再对重要子类和接口进行源码分析。

16.1.1 Channel 的工作原理

Channel 是 Netty 抽象出来的网络 I/O 读写相关的接口，为什么不使用 JDK NIO 原生的 Channel 而要另起炉灶呢，主要原因如下。

（1）JDK 的 SocketChannel 和 ServerSocketChannel 没有统一的 Channel 接口供业务开发者使用，对于用户而言，没有统一的操作视图，使用起来并不方便。

（2）JDK 的 SocketChannel 和 ServerSocketChannel 的主要职责就是网络 I/O 操作，由于它们是 SPI 类接口，由具体的虚拟机厂家来提供，所以通过继承 SPI 功能类来扩展其功能的难度很大；直接实现 ServerSocketChannel 和 SocketChannel 抽象类，其工作量和重新开发一个新的 Channel 功能类是差不多的。

（3）Netty 的 Channel 需要能够跟 Netty 的整体架构融合在一起，例如 I/O 模型、基于 ChannelPipeline 的定制模型，以及基于元数据描述配置化的 TCP 参数等，这些 JDK 的 SocketChannel 和 ServerSocketChannel 都没有提供，需要重新封装。

（4）自定义的 Channel，功能实现更加灵活。

基于上述 4 个原因，Netty 重新设计了 Channel 接口，并且给予了很多不同的实现。它的设计原理比较简单，但是功能却比较繁杂，主要的设计理念如下。

（1）在 Channel 接口层，采用 Facade 模式进行统一封装，将网络 I/O 操作、网络 I/O 相关联的其他操作封装起来，统一对外提供。

（2）Channel 接口的定义尽量大而全，为 SocketChannel 和 ServerSocketChannel 提供统一的视图，由不同子类实现不同的功能，公共功能在抽象父类中实现，最大程度地实现功能和接口的重用。

（3）具体实现采用聚合而非包含的方式，将相关的功能类聚合在 Channel 中，由 Channel 统一负责分配和调度，功能实现更加灵活。

16.1.2 Channel 的功能介绍

Channel 的功能比较繁杂，我们通过分类的方式对它的主要功能进行介绍。

1. 网络 I/O 操作

Channel 网络 I/O 相关的方法定义如图 16-1 所示。

图 16-1　Channel 的 I/O 操作 API 列表

下面我们对这些 API 的功能进行分类说明，读写相关的 API 列表。

（1）Channel read()：从当前的 Channel 中读取数据到第一个 inbound 缓冲区中，如果数据被成功读取，触发 ChannelHandler.channelRead(ChannelHandlerContext, Object)事件。读取操作 API 调用完成之后，紧接着会触发 ChannelHandler.channelReadComplete (Channel HandlerContext)事件，这样业务的 ChannelHandler 可以决定是否需要继续读取数据。如果已经有读操作请求被挂起，则后续的读操作会被忽略。

（2）ChannelFuture write(Object msg)：请求将当前的 msg 通过 ChannelPipeline 写入到目标 Channel 中。注意，write 操作只是将消息存入到消息发送环形数组中，并没有真正被发送，只有调用 flush 操作才会被写入到 Channel 中，发送给对方。

（3）ChannelFuture write(Object msg, ChannelPromise promise)：功能与 write(Object msg)

相同，但是携带了 ChannelPromise 参数负责设置写入操作的结果。

（4）ChannelFuture writeAndFlush(Object msg, ChannelPromise promise)：与方法（3）功能类似，不同之处在于它会将消息写入 Channel 中发送，等价于单独调用 write 和 flush 操作的组合。

（5）ChannelFuture writeAndFlush(Object msg)：功能等同于方法（4），但是没有携带 writeAndFlush(Object msg)参数。

（6）Channel flush()：将之前写入到发送环形数组中的消息全部写入到目标 Chanel 中，发送给通信对方。

（7）ChannelFuture close(ChannelPromise promise)：主动关闭当前连接，通过 ChannelPromise 设置操作结果并进行结果通知，无论操作是否成功，都可以通过 ChannelPromise 获取操作结果。该操作会级联触发 ChannelPipeline 中所有 ChannelHandler 的 ChannelHandler.close(ChannelHandlerContext, ChannelPromise)事件。

（8）ChannelFuture disconnect(ChannelPromise promise)：请求断开与远程通信对端的连接并使用 ChannelPromise 来获取操作结果的通知消息。该方法会级联触发 ChannelHandler.disconnect(ChannelHandlerContext, ChannelPromise)事件。

（9）ChannelFuture connect(SocketAddress remoteAddress)：客户端使用指定的服务端地址 remoteAddress 发起连接请求，如果连接因为应答超时而失败，ChannelFuture 中的操作结果就是 ConnectTimeoutException 异常；如果连接被拒绝，操作结果为 ConnectException。该方法会级联触发 ChannelHandler.connect(ChannelHandlerContext, SocketAddress, SocketAddress, ChannelPromise)事件。

（10）ChannelFuture connect(SocketAddress remoteAddress, SocketAddress localAddress)：与方法（9）功能类似，唯一不同的就是先绑定指定的本地地址 localAddress，然后再连接服务端。

（11）ChannelFuture connect(SocketAddress remoteAddress, ChannelPromise promise)：与方法（9）功能类似，唯一不同的是携带了 ChannelPromise 参数用于写入操作结果。

（12）connect(SocketAddress remoteAddress, SocketAddress localAddress, ChannelPromise promise)：与方法（11）功能类似，唯一不同的就是绑定了本地地址。

（13）ChannelFuture bind(SocketAddress localAddress)：绑定指定的本地 Socket 地址 localAddress，该方法会级联触发 ChannelHandler.bind(ChannelHandlerContext, SocketAddress, ChannelPromise)事件。

（14）ChannelFuture bind(SocketAddress localAddress, ChannelPromise promise)：与方法（13）功能类似，多携带了了一个 ChannelPromise 用于写入操作结果。

（15）ChannelConfig config()：获取当前 Channel 的配置信息，例如 CONNECT_TIMEOUT_MILLIS。

（16）boolean isOpen()：判断当前 Channel 是否已经打开。

（17）boolean isRegistered()：判断当前 Channel 是否已经注册到 EventLoop 上。

（18）boolean isActive()：判断当前 Channel 是否已经处于激活状态。

（19）ChannelMetadata metadata()：获取当前 Channel 的元数据描述信息，包括 TCP 参数配置等。

（20）SocketAddress localAddress()：获取当前 Channel 的本地绑定地址。

（21）SocketAddress remoteAddress()：获取当前 Channel 通信的远程 Socket 地址。

2. 其他常用的 API 功能说明

第一个比较重要的方法是 eventLoop()。Channel 需要注册到 EventLoop 的多路复用器上，用于处理 I/O 事件，通过 eventLoop()方法可以获取到 Channel 注册的 EventLoop。EventLoop 本质上就是处理网络读写事件的 Reactor 线程。在 Netty 中，它不仅仅用来处理网络事件，也可以用来执行定时任务和用户自定义 NioTask 等任务。

第二个比较常用的方法是 metadata()方法。熟悉 TCP 协议的读者可能知道，当创建 Socket 的时候需要指定 TCP 参数，例如接收和发送的 TCP 缓冲区大小、TCP 的超时时间、是否重用地址等。在 Netty 中，每个 Channel 对应一个物理连接，每个连接都有自己的 TCP 参数配置。所以，Channel 会聚合一个 ChannelMetadata 用来对 TCP 参数提供元数据描述信息，通过 metadata()方法就可以获取当前 Channel 的 TCP 参数配置。

第三个方法是 parent()。对于服务端 Channel 而言，它的父 Channel 为空；对于客户端 Channel，它的父 Channel 就是创建它的 ServerSocketChannel。

第四个方法是用户获取 Channel 标识的 id()，它返回 ChannelId 对象，ChannelId 是 Channel 的唯一标识，它的可能生成策略如下。

（1）机器的 MAC 地址（EUI-48 或者 EUI-64）等可以代表全局唯一的信息；

（2）当前的进程 ID；

（3）当前系统时间的毫秒——System.currentTimeMillis();

（4）当前系统时间纳秒数——System.nanoTime();

（5）32 位的随机整型数；

（6）32 位自增的序列数。

16.2 Channel 源码分析

Channel 的实现子类非常多，继承关系复杂，从学习的角度我们抽取最重要的两个——Channel-io.netty.channel.socket.nio.NioServerSocketChannel 和 io.netty.channel.socket.nio.NioSocketChannel 进行重点分析。如果读者对其他的 Channel 实现细节感兴趣，可以按照本书的指导自行阅读。

16.2.1 Channel 的主要继承关系类图

为了便于学习和阅读源码，我们分别看下 NioSocketChannel 和 NioServerSocketChannel 的继承关系类图。

服务端 NioServerSocketChannel 的继承关系类图如图 16-2 所示。

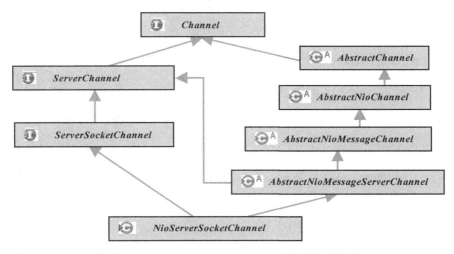

图 16-2 NioServerSocketChannel 继承关系类图

客户端 NioSocketChannel 的继承关系类图如图 16-3 所示。

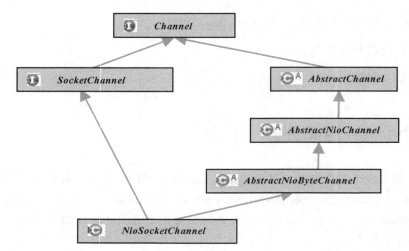

图 16-3　NioSocketChannel 继承关系类图

16.2.2　AbstractChannel 源码分析

1. 成员变量定义

在分析 AbstractChannel 源码之前，我们先看下它的成员变量定义，如图 16-4 所示。

首先定义了两个静态全局异常，如下。

- CLOSED_CHANNEL_EXCEPTION：链路已经关闭已经异常；
- NOT_YET_CONNECTED_EXCEPTION：物理链路尚未建立异常。

声明完上述两个异常之后，通过静态块将它们的堆栈设置为空的 StackTraceElement。

estimatorHandle 用于预测下一个报文的大小，它基于之前数据的采样进行分析预测。

根据之前的 Channel 原理分析，我们知道 AbstractChannel 采用聚合的方式封装各种功能，从成员变量的定义可以看出，它聚合了以下内容。

- parent：代表父类 Channel；
- id：采用默认方式生成的全局唯一 ID；

- unsafe：Unsafe 实例；
- pipeline：当前 Channel 对应的 DefaultChannelPipeline；
- eventLoop：当前 Channel 注册的 EventLoop；

……

```
    private static final InternalLogger logger =
InternalLoggerFactory.getInstance(AbstractChannel.class);

    static final ClosedChannelException CLOSED_CHANNEL_EXCEPTION =
new ClosedChannelException();
    static final NotYetConnectedException
NOT_YET_CONNECTED_EXCEPTION = new NotYetConnectedException();

    static {
CLOSED_CHANNEL_EXCEPTION.setStackTrace(EmptyArrays.EMPTY_STACK_TRACE);

NOT_YET_CONNECTED_EXCEPTION.setStackTrace(EmptyArrays.EMPTY_STACK_TRACE);
    }

    private MessageSizeEstimator.Handle estimatorHandle;

    private final Channel parent;
    private final ChannelId id = DefaultChannelId.newInstance();
    private final Unsafe unsafe;
    private final DefaultChannelPipeline pipeline;
    private final ChannelFuture succeededFuture = new
SucceededChannelFuture(this, null);
    private final VoidChannelPromise voidPromise = new
VoidChannelPromise(this, true);
    private final VoidChannelPromise unsafeVoidPromise = new
VoidChannelPromise(this, false);
    private final CloseFuture closeFuture = new CloseFuture(this);

    private volatile SocketAddress localAddress;
    private volatile SocketAddress remoteAddress;
    private final EventLoop eventLoop;
    private volatile boolean registered;

    /** Cache for the string representation of this channel */
    private boolean strValActive;
    private String strVal;
```

图 16-4　AbstractChannel 成员变量定义

在此不一一枚举。通过变量定义可以看出，AbstractChannel 聚合了所有 Channel 使用到的能力对象，由 AbstractChannel 提供初始化和统一封装，如果功能和子类强相关，则定义成抽象方法由子类具体实现，下面的小节就对它的主要 API 进行源码分析。

2. 核心 API 源码分析

首先看下网络读写操作，前面介绍网络 I/O 操作时讲到，它会触发 ChannelPipeline 中对应的事件方法。Netty 基于事件驱动，我们也可以理解为当 Chnanel 进行 I/O 操作时会产生对应的 I/O 事件，然后驱动事件在 ChannelPipeline 中传播，由对应的 ChannelHandler 对事件进行拦截和处理，不关心的事件可以直接忽略。采用事件驱动的方式可以非常轻松地通过事件定义来划分事件拦截切面，方便业务的定制和功能扩展，相比 AOP，其性能更高，但是功能却基本等价。

网络 I/O 操作直接调用 DefaultChannelPipeline 的相关方法，由 DefaultChannelPipeline 中对应的 ChannelHandler 进行具体的逻辑处理，如图 16-5 所示。

```
@Override
public ChannelFuture connect(SocketAddress remoteAddress,
SocketAddress localAddress) {
    return pipeline.connect(remoteAddress, localAddress);
}

@Override
public ChannelFuture disconnect() {
    return pipeline.disconnect();
}

@Override
public ChannelFuture close() {
    return pipeline.close();
}

@Override
public Channel flush() {
    pipeline.flush();
    return this;
}
```

图 16-5　AbstractChannel 网络 I/O 操作源码实现

AbstractChannel 也提供了一些公共 API 的具体实现，例如 localAddress() 和 remoteAddress() 方法，它的源码实现如图 16-6 所示。

```
@Override
public SocketAddress remoteAddress() {
    SocketAddress remoteAddress = this.remoteAddress;
    if (remoteAddress == null) {
        try {
            this.remoteAddress = remoteAddress = unsafe().remoteAddress();
        } catch (Throwable t) {
            // Sometimes fails on a closed socket in Windows.
            return null;
        }
    }
    return remoteAddress;
}
```

图 16-6　AbstractChannel remoteAddress 方法实现

首先从缓存的成员变量中获取，如果第一次调用为空，需要通过 unsafe 的 remoteAddress 获取，它是个抽象方法，具体由对应的 Channel 子类实现。

16.2.3　AbstractNioChannel 源码分析

1. 成员变量定义

首先，还是从成员变量定义入手，来了解下它的功能实现，成员变量定义如图 16-7 所示。

```
private static final InternalLogger logger =
        InternalLoggerFactory.getInstance(AbstractNioChannel.class);

private final SelectableChannel ch;
protected final int readInterestOp;
private volatile SelectionKey selectionKey;
private volatile boolean inputShutdown;

/**
 * The future of the current connection attempt.  If not null, subsequent
 * connection attempts will fail.
 */
private ChannelPromise connectPromise;
private ScheduledFuture<?> connectTimeoutFuture;
private SocketAddress requestedRemoteAddress;
```

图 16-7　AbstractNioChannel 成员变量定义

由于 NIO Channel、NioSocketChannel 和 NioServerSocketChannel 需要共用，所以定义了一个 java.nio.SocketChannel 和 java.nio.ServerSocketChannel 的公共父类 SelectableChannel，用于设置 SelectableChannel 参数和进行 I/O 操作。

第二个参数是 readInterestOp，它代表了 JDK SelectionKey 的 OP_READ。

随后定义了一个 volatile 修饰的 SelectionKey，该 SelectionKey 是 Channel 注册到 EventLoop 后返回的选择键。由于 Channel 会面临多个业务线程的并发写操作，当 SelectionKey 由 SelectionKey 修改之后，为了能让其他业务线程感知到变化，所以需要使用 volatile 保证修改的可见性，后面的多线程章节会专门对 volatile 的使用进行说明。

最后定义了代表连接操作结果的 ChannelPromise 以及连接超时定时器 ScheduledFuture 和请求的通信地址信息。

2. 核心 API 源码分析

我们一起看下在 AbstractNioChannel 实现的主要 API，首先是 Channel 的注册，如图 16-8 所示。

```java
@Override
protected void doRegister() throws Exception {
    boolean selected = false;
    for (;;) {
        try {
            selectionKey = javaChannel().register(eventLoop().selector, 0, this);
            return;
        } catch (CancelledKeyException e) {
            if (!selected) {
                // Force the Selector to select now as the "canceled" SelectionKey may still be
                // cached and not removed because no Select.select(..) operation was called yet.
                eventLoop().selectNow();
                selected = true;
            } else {
                // We forced a select operation on the selector before but the SelectionKey is still cached
                // for whatever reason. JDK bug ?
                throw e;
            }
        }
    }
}
```

图 16-8　AbstractNioChannel 的注册方法实现

定义一个布尔类型的局部变量 selected 来标识注册操作是否成功，调用 SelectableChannel 的 register 方法，将当前的 Channel 注册到 EventLoop 的多路复用器上，SelectableChannel 的注册方法定义如图 16-9 所示。

```
public abstract SelectionKey register(Selector sel, int ops, Object att)
    throws ClosedChannelException;
```

图 16-9　JDK SelectableChannel 的注册方法定义

注册 Channel 的时候需要指定监听的网络操作位来表示 Channel 对哪几类网络事件感兴趣，具体的定义如下。

◎ public static final int OP_READ = 1 << 0：读操作位；

◎ public static final int OP_WRITE = 1 << 2：写操作位；

◎ public static final int OP_CONNECT = 1 << 3：客户端连接服务端操作位；

◎ public static final int OP_ACCEPT = 1 << 4：服务端接收客户端连接操作位。

AbstractNioChannel 注册的是 0，说明对任何事件都不感兴趣，仅仅完成注册操作。注册的时候可以指定附件，后续 Channel 接收到网络事件通知时可以从 SelectionKey 中重新获取之前的附件进行处理，此处将 AbstractNioChannel 的实现子类自身当作附件注册。如果注册 Channel 成功，则返回 selectionKey，通过 selectionKey 可以从多路复用器中获取 Channel 对象。

如果当前注册返回的 selectionKey 已经被取消，则抛出 CancelledKeyException 异常，捕获该异常进行处理。如果是第一次处理该异常，调用多路复用器的 selectNow()方法将已经取消的 selectionKey 从多路复用器中删除掉。操作成功之后，将 selected 置为 true，说明之前失效的 selectionKey 已经被删除掉。继续发起下一次注册操作，如果成功则退出，如果仍然发生 CancelledKeyException 异常，说明我们无法删除已经被取消的 selectionKey，按照 JDK 的 API 说明，这种意外不应该发生。如果发生这种问题，则说明可能 NIO 的相关类库存在不可恢复的 BUG，直接抛出 CancelledKeyException 异常到上层进行统一处理。

下面继续看另一个比较重要的方法：准备处理读操作之前需要设置网路操作位为读，代码如图 16-10 所示。

```
@Override
protected void doBeginRead() throws Exception {
    if (inputShutdown) {
        return;
    }

    final SelectionKey selectionKey = this.selectionKey;
    if (!selectionKey.isValid()) {
        return;
    }

    final int interestOps = selectionKey.interestOps();
    if ((interestOps & readInterestOp) == 0) {
        selectionKey.interestOps(interestOps | readInterestOp);
    }
}
```

图 16-10　修改网络操作位为读

先判断下 Channel 是否关闭，如果处于关闭中，则直接返回。获取当前的 SelectionKey 进行判断，如果可用，说明 Channel 当前状态正常，则可以进行正常的操作位修改。将 SelectionKey 当前的操作位与读操作位进行按位与操作，如果等于 0，说明目前并没有设置读操作位，通过 interestOps | readInterestOp 设置读操作位，最后调用 selectionKey 的 interestOps 方法重新设置通道的网络操作位，这样就可以监听网络的读事件了。

实际上，对于读操作位的判断和修改与 JDK NIO SelectionKey 的相关方法实现是等价的，如图 16-11 所示。

```
public final boolean isReadable() {
    return (readyOps() & OP_READ) != 0;
}
```

图 16-11　判断当前 SelectionKey 是否可读

16.2.4　AbstractNioByteChannel 源码分析

由于成员变量只有一个 Runnable 类型的 flushTask 来负责继续写半包消息，所以对成员变量不再单独进行介绍。

最主要的方法就是 doWrite（ChannelOutboundBuffer in），下面一起看看它的实现，由于该方法过长，所以我们按照其逻辑进行拆分介绍。如图 16-12 所示。

```
@Override
protected void doWrite(ChannelOutboundBuffer in) throws Exception {
    int writeSpinCount = -1;
    for (;;) {
        Object msg = in.current(true);
        if (msg == null) {
            // Wrote all messages.
            clearOpWrite();
            break;
        }
```

图 16-12　doWrite（ChannelOutboundBuffer in）源码片段 1

从发送消息环形数组 ChannelOutboundBuffer 弹出一条消息，判断该消息是否为空，如果为空，说明消息发送数组中所有待发送的消息都已经发送完成，清除半包标识，然后退出循环。清除半包标识的 clearOpWrite 方法实现如图 16-13 所示。

```
protected final void clearOpWrite() {
    final SelectionKey key = selectionKey();
    final int interestOps = key.interestOps();
    if ((interestOps & SelectionKey.OP_WRITE) != 0) {
        key.interestOps(interestOps & ~SelectionKey.OP_WRITE);
    }
}
```

图 16-13　清除写半包标识

从当前 SelectionKey 中获取网络操作位，然后与 SelectionKey.OP_WRITE 做按位与，如果不等于 0，说明当前的 SelectionKey 是 isWritable 的，需要清除写操作位。清除方法很简单，就是 SelectionKey.OP_WRITE 取非之后与原操作位做按位与操作，清除

SelectionKey 的写操作位。

继续看源码，如果需要发送的消息不为空，则继续处理。如图 16-14 所示。

```
if (msg instanceof ByteBuf) {
    ByteBuf buf = (ByteBuf) msg;
    int readableBytes = buf.readableBytes();
    if (readableBytes == 0) {
        in.remove();
        continue;
    }
    boolean setOpWrite = false;
    boolean done = false;
    long flushedAmount = 0;
    if (writeSpinCount == -1) {
        writeSpinCount = config().getWriteSpinCount();
    }
```

图 16-14　doWrite（ChannelOutboundBuffer in）源码片段 2

首先判断需要发送的消息是否是 ByteBuf 类型，如果是，则进行强制类型转换，将其转换成 ByteBuf 类型，判断当前消息的可读字节数是否为 0，如果为 0，说明该消息不可读，需要丢弃。从环形发送数组中删除该消息，继续循环处理其他的消息。

声明消息发送相关的成员变量，包括：写半包标识、消息是否全部发送标识、发送的总消息字节数。

这些局部变量创建完成之后，对循环发送次数进行判断，如果为-1，则从 Channel 配置对象中获取循环发送次数。循环发送次数是指当一次发送没有完成时（写半包），继续循环发送的次数。设置写半包最大循环次数的原因是当循环发送的时候，I/O 线程会一直尝试进行写操作，此时 I/O 线程无法处理其他的 I/O 操作，例如读新的消息或者执行定时任务和 NioTask 等，如果网络 I/O 阻塞或者对方接收消息太慢，可能会导致线程假死。

继续看循环发送的代码如图 16-15 所示。

```
for (int i = writeSpinCount - 1; i >= 0; i --) {
    int localFlushedAmount = doWriteBytes(buf);
    if (localFlushedAmount == 0) {
        setOpWrite = true;
        break;
    }
    flushedAmount += localFlushedAmount;
    if (!buf.isReadable()) {
        done = true;
        break;
    }
}
```

图 16-15　doWrite（ChannelOutboundBuffer in）源码片段 3

调用 doWriteBytes 进行消息发送，不同的 Channel 子类有不同的实现，因此它是抽象方法。如果本次发送的字节数为 0，说明发送 TCP 缓冲区已满，发生了 ZERO_WINDOW。此时再次发送仍然可能出现写 0 字节，空循环会占用 CPU 的资源，导致 I/O 线程无法处理其他 I/O 操作，所以将写半包标识 setOpWrite 设置为 true，退出循环，释放 I/O 线程。

如果发送的字节数大于 0，则对发送总数进行计数。判断当前消息是否已经发送成功（缓冲区没有可读字节），如果发送成功则设置 done 为 true，退出当前循环。

消息发送操作完成之后调用 ChannelOutboundBuffer 更新发送进度信息，然后对发送结果进行判断。如果发送成功，则将已经发送的消息从发送数组中删除；否则调用 incompleteWrite 方法，设置写半包标识，启动刷新线程继续发送之前没有发送完全的半包消息（写半包）。如图 16-16 所示。

```
in.progress(flushedAmount);
if (done) {
    in.remove();
} else {
    incompleteWrite(setOpWrite);
    break;
}
```

图 16-16　doWrite（ChannelOutboundBuffer in）源码片段 4

处理半包发送任务的方法 incompleteWrite 的实现如图 16-17 所示。

```
protected final void incompleteWrite(boolean setOpWrite) {
    // Did not write completely.
    if (setOpWrite) {
        setOpWrite();
    } else {
        // Schedule flush again later so other tasks can be picked up in the meantime
        Runnable flushTask = this.flushTask;
        if (flushTask == null) {
            flushTask = this.flushTask = new Runnable() {
                @Override
                public void run() {
                    flush();
                }
            };
        }
        eventLoop().execute(flushTask);
    }
}
```

图 16-17　启动写半包任务线程用于后续继续发送半包消息

首先判断是否需要设置写半包标识，如果需要则调用 setOpWrite 设置写半包标识，代码如图 16-18 所示。

```
protected final void setOpWrite() {
    final SelectionKey key = selectionKey();
    final int interestOps = key.interestOps();
    if ((interestOps & SelectionKey.OP_WRITE) == 0) {
        key.interestOps(interestOps | SelectionKey.OP_WRITE);
    }
}
```

图 16-18　设置写半包标识

设置写半包标识就是将 SelectionKey 设置成可写的，通过原操作位与 SelectionKey.OP_WRITE 做按位或操作即可实现。

如果 SelectionKey 的 OP_WRITE 被设置，多路复用器会不断轮询对应的 Channel，用于处理没有发送完成的半包消息，直到清除 SelectionKey 的 OP_WRITE 操作位。因此，设置了 OP_WRITE 操作位后，就不需要启动独立的 Runnable 来负责发送半包消息了。

如果没有设置 OP_WRITE 操作位，需要启动独立的 Runnable，将其加入到 EventLoop 中执行，由 Runnable 负责半包消息的发送。它的实现很简单，就是调用 flush()方法来发送缓冲数组中的消息。

消息发送的另一个分支是文件传输，由于它的实现原理与 ByteBuf 类似，限于篇幅，在此不再详细说明，感兴趣的读者可以自己独立完成分析。

16.2.5　AbstractNioMessageChannel 源码分析

由于 AbstractNioMessageChannel 没有自己的成员变量，所以我们直接对其方法进行说明。

它的主要实现方法只有一个：doWrite(ChannelOutboundBuffer in)，下面首先看下它的源码，如图 16-19 所示。

在循环体内对消息进行发送，从 ChannelOutboundBuffer 中弹出一条消息进行处理，如果消息为空，说明发送缓冲区为空，所有消息都已经被发送完成。清除写半包标识，退出循环。

与 AbstractNioByteChannel 的循环发送类似，利用 writeSpinCount 对单条消息进行发送，调用 doWriteMessage(Object msg, ChannelOutboundBuffer in)判断消息是否发送成功，

如果成功，则将发送标识 done 设置为 true，退出循环；否则继续执行循环，直到执行 writeSpinCount 次。

```
@Override
protected void doWrite(ChannelOutboundBuffer in) throws Exception {
    final SelectionKey key = selectionKey();
    final int interestOps = key.interestOps();
    for (;;) {
        Object msg = in.current();
        if (msg == null) {
            // Wrote all messages.
            if ((interestOps & SelectionKey.OP_WRITE) != 0) {
                key.interestOps(interestOps & ~SelectionKey.OP_WRITE);
            }
            break;
        }
        boolean done = false;
        for (int i = config().getWriteSpinCount() - 1; i >= 0; i --) {
            if (doWriteMessage(msg, in)) {
                done = true;
                break;
            }
        }
        if (done) {
            in.remove();
        } else {
            // Did not write all messages.
            if ((interestOps & SelectionKey.OP_WRITE) == 0) {
                key.interestOps(interestOps | SelectionKey.OP_WRITE);
            }
            break;
        }
    }
}
```

图 16-19　AbstractNioMessageChannel 消息发送

发送操作完成之后，判断发送结果，如果当前的消息被完全发送出去，则将该消息从缓冲数组中删除；否则设置半包标识，注册 SelectionKey.OP_WRITE 到多路复用器上，由多路复用器轮询对应的 Channel 重新发送尚未发送完全的半包消息。

通过代码分析我们发现，AbstractNioMessageChannel 和 AbstractNioByteChannel 的消息发送实现比较相似，不同之处在于：一个发送的是 ByteBuf 或者 FileRegion，它们可以直接被发送；另一个发送的则是 POJO 对象。

16.2.6　AbstractNioMessageServerChannel 源码分析

AbstractNioMessageServerChannel 的实现非常简单，它定义了一个 EventLoopGroup 类

型的 childGroup，用于给新接入的客户端 NioSocketChannel 分配 EventLoop，它的源码实现如图 16-20 所示。

```
public abstract class AbstractNioMessageServerChannel extends
AbstractNioMessageChannel implements ServerChannel {

    private final EventLoopGroup childGroup;

    protected AbstractNioMessageServerChannel(
    Channel parent, EventLoop eventLoop, EventLoopGroup childGroup,
    SelectableChannel ch, int readInterestOp) {
        super(parent, eventLoop, ch, readInterestOp);
        this.childGroup = childGroup;
    }

    @Override
    public EventLoopGroup childEventLoopGroup() {
        return childGroup;
    }
}
```

图 16-20　AbstractNioMessageServerChannel 源码

每当服务端接入一个新的客户端连接 NioSocketChannel 时，都会调用 childEventLoopGroup 方法获取 EventLoopGroup 线程组，用于给 NioSocketChannel 分配 Reactor 线程 EventLoop，相关分配代码如图 16-21 所示。

```
@Override
protected int doReadMessages(List<Object> buf) throws Exception {
    SocketChannel ch = javaChannel().accept();
    try {
        if (ch != null) {
            buf.add(new NioSocketChannel(this,
childEventLoopGroup().next(), ch));
            return 1;
        }
```

图 16-21　通过 childEventLoopGroup 方法进行 I/O 线程分配

16.2.7　NioServerSocketChannel 源码分析

NioServerSocketChannel 的实现比较简单，下面我们重点分析主要 API 的实现，首先看它的成员变量定义和静态方法，如图 16-22 所示。

```
    private static final ChannelMetadata METADATA = new
ChannelMetadata(false);
    private static final InternalLogger logger =
InternalLoggerFactory.getInstance(NioServerSocketChannel.class);

    private static ServerSocketChannel newSocket() {
        try {
            return ServerSocketChannel.open();
        } catch (IOException e) {
            throw new ChannelException(
                "Failed to open a server socket.", e);
        }
    }

    private final ServerSocketChannelConfig config;
```

图 16-22 NioServerSocketChannel 成员变量定义

首先创建了静态的 ChannelMetadata 成员变量，然后定义了 ServerSocketChannelConfig 用于配置 ServerSocketChannel 的 TCP 参数。静态的 newSocket 方法用于通过 ServerSocketChannel 的 open 打开新的 ServerSocketChannel 通道。

接着我们再看下 ServerSocketChannel 相关的接口实现：isActive、remoteAddress、javaChannel 和 doBind，它们的源码如图 16-23 所示。

```
@Override
public boolean isActive() {
    return javaChannel().socket().isBound();
}
@Override
public InetSocketAddress remoteAddress() {
    return null;
}
@Override
protected ServerSocketChannel javaChannel() {
    return (ServerSocketChannel) super.javaChannel();
}
@Override
protected SocketAddress localAddress0() {
    return javaChannel().socket().getLocalSocketAddress();
}
@Override
protected void doBind(SocketAddress localAddress) throws Exception {
    javaChannel().socket().bind(localAddress, config.getBacklog());
}
```

图 16-23 NioServerSocketChannel 本地实现相关方法

通过 java.net.ServerSocket 的 isBound 方法判断服务端监听端口是否处于绑定状态，它的 remoteAddress 为空。javaChannel 的实现是 java.nio.ServerSocketChannel，服务端在进行端口绑定的时候，可以指定 backlog，也就是允许客户端排队的最大长度。相关 API 说明如图 16-24 所示。

```
● void java.net.ServerSocket.bind(SocketAddress endpoint, int backlog) throws IOException
Binds the ServerSocket to a specific address (IP address and port number).
If the address is null, then the system will pick up an ephemeral port and a valid local address to bind the
socket.
The backlog argument is the requested maximum number of pending connections on the socket. Its exact
semantics are implementation specific. In particular, an implementation may impose a maximum length or may
choose to ignore the parameter altogther. The value provided should be greater than 0. If it is less than or
equal to 0, then an implementation specific default will be used.
Parameters:
        endpoint The IP address & port number to bind to.
        backlog requested maximum length of the queue of incoming connections.
Throws:
        IOException - if the bind operation fails, or if the socket is already bound.
        SecurityException - if a SecurityManager is present and its checkListen method doesn't allow the
        operation.
        IllegalArgumentException - if endpoint is a SocketAddress subclass not supported by this socket
```

图 16-24 ServerSocket 的绑定方法

下面继续看服务端 Channel 的 doReadMessages(List<Object> buf)的实现，如图 16-25 所示。

```
@Override
protected int doReadMessages(List<Object> buf) throws Exception {
    SocketChannel ch = javaChannel().accept();
    try {
        if (ch != null) {
            buf.add(new NioSocketChannel(this, childEventLoopGroup().next(), ch));
            return 1;
        }
    } catch (Throwable t) {
        logger.warn("Failed to create a new channel from an accepted socket.", t);
        try {
            ch.close();
        } catch (Throwable t2) {
            logger.warn("Failed to close a socket.", t2);
        }
    }
    return 0;
}
```

图 16-25 NioServerSocketChannel doReadMessages 方法

首先通过 ServerSocketChannel 的 accept 接收新的客户端连接，如果 SocketChannel 不为空，则利用当前的 NioServerSocketChannel、EventLoop 和 SocketChannel 创建新的

NioSocketChannel，并将其加入到 List<Object> buf 中，最后返回 1，表示服务端消息读取成功。

对于 NioServerSocketChannel，它的读取操作就是接收客户端的连接，创建 NioSocketChannel 对象。

最后看下与服务端 Channel 无关的接口定义，由于这些方法是客户端 Channel 相关的，因此，对于服务端 Channel 无须实现。如果这些方法被误调，则返回 UnsupportedOperationException 异常，这些方法的源码如图 16-26 所示。

```
@Override
protected boolean doConnect(
        SocketAddress remoteAddress, SocketAddress localAddress) throws Exception {
    throw new UnsupportedOperationException();
}
@Override
protected void doFinishConnect() throws Exception {
    throw new UnsupportedOperationException();
}
@Override
protected SocketAddress remoteAddress0() {
    return null;
}
@Override
protected void doDisconnect() throws Exception {
    throw new UnsupportedOperationException();
}
@Override
protected boolean doWriteMessage(Object msg, ChannelOutboundBuffer in)
throws Exception {
    throw new UnsupportedOperationException();
}
```

图 16-26　NioServerSocketChannel 不支持的方法列表

16.2.8　NioSocketChannel 源码分析

1. 连接操作

我们重点分析与客户端连接相关的 API 实现，首先看连接方法的实现，如图 16-27 所示。

判断本地 Socket 地址是否为空，如果不为空则调用 java.nio.channels.SocketChannel.socket().bind() 方法绑定本地地址。如果绑定成功，则继续调用 java.nio.channels.SocketChannel.connect(SocketAddress remote) 发起 TCP 连接。对连接结果进行判断，连接结果有以下三种可能。

(1)连接成功,返回 true;

(2)暂时没有连接上,服务端没有返回 ACK 应答,连接结果不确定,返回 false;

(3)连接失败,直接抛出 I/O 异常。

```
@Override
protected boolean doConnect(SocketAddress remoteAddress,
SocketAddress localAddress) throws Exception {
    if (localAddress != null) {
        javaChannel().socket().bind(localAddress);
    }
    boolean success = false;
    try {
        boolean connected = javaChannel().connect(remoteAddress);
        if (!connected) {
            selectionKey().interestOps(SelectionKey.OP_CONNECT);
        }
        success = true;
        return connected;
    } finally {
        if (!success) {
            doClose();
        }
    }
}
```

图 16-27　NioSocketChannel 的连接方法

如果是结果(2),需要将 NioSocketChannel 中的 selectionKey 设置为 OP_CONNECT,监听连接网络操作位。如果抛出了 I/O 异常,说明客户端的 TCP 握手请求直接被 REST 或者被拒绝,此时需要关闭客户端连接,代码如图 16-28 所示。

```
@Override
protected void doClose() throws Exception {
    javaChannel().close();
}
```

图 16-28　连接失败,关闭客户端

2. 写半包

分析完连接操作之后,继续分析写操作,由于它的实现比较复杂,所以仍然需要将其拆分后分段进行分析,代码如图 16-29 所示。

获取待发送的 ByteBuf 个数,如果小于等于 1,则调用父类 AbstractNioByteChannel 的 doWrite 方法,操作完成之后退出。

在批量发送缓冲区的消息之前,先对一系列的局部变量进行赋值,首先,获取需要发送的 ByteBuffer 数组个数 nioBufferCnt,然后,从 ChannelOutboundBuffer 中获取需要发送的总字节数,从 NioSocketChannel 中获取 NIO 的 SocketChannel,将是否发送完成标识设置为 false,将是否有写半包标识设置为 false。如图 16-30 所示。

```
@Override
protected void doWrite(ChannelOutboundBuffer in) throws Exception {
    for (;;) {
        // Do non-gathering write for a single buffer case.
        final int msgCount = in.size();
        if (msgCount <= 1) {
            super.doWrite(in);
            return;
        }

        // Ensure the pending writes are made of ByteBufs only.
        ByteBuffer[] nioBuffers = in.nioBuffers();

        if (nioBuffers == null) {
            super.doWrite(in);
            return;
        }
```

图 16-29　NioSocketChannel 的写方法片段 1

```
int nioBufferCnt = in.nioBufferCount();
long expectedWrittenBytes = in.nioBufferSize();
final SocketChannel ch = javaChannel();
long writtenBytes = 0;
boolean done = false;
boolean setOpWrite = false;
```

图 16-30　NioSocketChannel 的写方法片段 2

继续分析循环发送的代码,代码如图 16-31 所示。

```
for (int i = config().getWriteSpinCount() - 1; i >= 0; i --) {
    final long localWrittenBytes = ch.write(nioBuffers, 0, nioBufferCnt);
```

图 16-31　NioSocketChannel 的写方法片段 3

就像循环读一样,我们需要对一次 Selector 轮询的写操作次数进行上限控制,因为如果 TCP 的发送缓冲区满,TCP 处于 KEEP-ALIVE 状态,消息会无法发送出去,如果不对上限进行控制,就会长时间地处于发送状态,Reactor 线程无法及时读取其他消息和执行排队的 Task。所以,我们必须对循环次数上限做控制。

调用 NIO SocketChannel 的 write 方法,它有三个参数:第一个是需要发送的 ByteBuffer

数组，第二个是数组的偏移量，第三个参数是发送的 ByteBuffer 个数。返回值是写入 SocketChannel 的字节个数。

下面对写入的字节进行判断，如果为 0，说明 TCP 发送缓冲区已满，很有可能无法再写进去，因此从循环中跳出，同时将写半包标识设置为 true，用于向多路复用器注册写操作位，告诉多路复用器有没发完的半包消息，需要轮询出就绪的 SocketChannel 继续发送。代码如图 16-32 所示。

```
if (localWrittenBytes == 0) {
    setOpWrite = true;
    break;
}
```

图 16-32　NioSocketChannel 的写方法片段 4

发送操作完成后进行两个计算：需要发送的字节数要减去已经发送的字节数；发送的字节总数+已经发送的字节数。更新完这两个变量后，判断缓冲区中所有的消息是否已经发送完成。如果是，则把发送完成标识设置为 true 同时退出循环；如果没有发送完成，则继续循环。从循环发送中退出之后，首先对发送完成标识 done 进行判断，如果发送完成，则循环释放已经发送的消息。环形数组的发送缓冲区释放完成后，取消半包标识，告诉多路复用器消息已经全部发送完成。代码如图 16-33 所示。

```
if (done) {
    // Release all buffers
    for (int i = msgCount; i > 0; i --) {
        in.remove();
    }

    // Finish the write loop if no new messages were flushed by in.remove().
    if (in.isEmpty()) {
        clearOpWrite();
        break;
    }
}
```

图 16-33　NioSocketChannel 的写方法片段 5

当缓冲区中的消息没有发送完成，甚至某个 ByteBuffer 只发送了几个字节，出现了所谓的"写半包"时，该怎么办？下面我们继续看看 Netty 是如何处理"写半包"的，如图 16-34 所示。

```
for (int i = msgCount; i > 0; i --) {
    final ByteBuf buf = (ByteBuf) in.current();
    final int readerIndex = buf.readerIndex();
    final int readableBytes = buf.writerIndex() - readerIndex;
    if (readableBytes < writtenBytes) {
        in.progress(readableBytes);
        in.remove();
        writtenBytes -= readableBytes;
    } else if (readableBytes > writtenBytes) {
        buf.readerIndex(readerIndex + (int) writtenBytes);
        in.progress(writtenBytes);
        break;
    } else { // readableBytes == writtenBytes
        in.progress(readableBytes);
        in.remove();
        break;
    }
}
```

图 16-34 NioSocketChannel 的写方法片段 6

首先，循环遍历发送缓冲区，对消息的发送结果进行判断，下面具体展开进行说明。

（1）从 ChannelOutboundBuffer 弹出第一条发送的 ByteBuf，然后获取该 ByteBuf 的读索引和可读字节数。

（2）对可读字节数和发送的总字节数进行比较，如果发送的字节数大于可读的字节数，说明当前的 ByteBuf 已经被完全发送出去，更新 ChannelOutboundBuffer 的发送进度信息，将已经发送的 ByteBuf 删除，释放相关资源。最后，发送的字节数要减去第一条发送的字节数，得到后续消息发送的总字节数，然后继续循环判断第二条消息、第三条消息……

（3）如果可读的消息大于已经发送的总字节数，说明这条消息没有被完整地发送出去，仅仅发送了部分数据报，也就是出现了所谓的"写半包"问题。此时，需要更新可读的索引为当前索引 + 已经发送的总字节数，然后更新 ChannelOutboundBuffer 的发送进度信息，退出循环。

（4）如果可读字节数等于已经发送的总字节数，则说明最后一次发送的消息是个整包消息，没有剩余的半包消息待发送。更新发送进度信息，将最后一条已发送的消息从缓冲区中删除，最后退出循环。

循环发送操作完成之后，更新 SocketChannel 的操作位为 OP_WRITE，由多路复用器在下一次轮询中触发 SocketChannel，继续处理没有发送完成的半包消息。

3. 读写操作

NioSocketChannel 的读写操作实际上是基于 NIO 的 SocketChannel 和 Netty 的 ByteBuf

封装而成，下面我们首先分析从 SocketChannel 中读取数据报，如图 16-35 所示。

```
for (int i = msgCount; i > 0; i --) {
    final ByteBuf buf = (ByteBuf) in.current();
    final int readerIndex = buf.readerIndex();
    final int readableBytes = buf.writerIndex() - readerIndex;

@Override
protected int doReadBytes(ByteBuf byteBuf) throws Exception {
    return byteBuf.writeBytes(javaChannel(), byteBuf.writableBytes());
}
```

图 16-35　NioSocketChannel 读取数据报

它有两个参数，说明如下。

◎ java.nio.channels.SocketChannel：JDK NIO 的 SocketChannel；

◎ length：ByteBuf 的可写最大字节数。

实际上就是从 SocketChannel 中读取 L 个字节到 ByteBuf 中，L 为 ByteBuf 可写的字节数，下面我们看下 ByteBuf writeBytes 方法的实现，如图 16-36 所示。

```
@Override
public int writeBytes(ScatteringByteChannel in, int length) throws IOException {
    ensureWritable(length);
    int writtenBytes = setBytes(writerIndex, in, length);
    if (writtenBytes > 0) {
        writerIndex += writtenBytes;
    }
    return writtenBytes;
}
```

图 16-36　从 SocketChannel 读取数据报到 ByteBuf 中

首先分析 setBytes(int index, ScatteringByteChannel in, int length)在 UnpooledHeapByteBuf 中的实现，如图 16-37 所示。

```
@Override
public int setBytes(int index, ScatteringByteChannel in, int length) throws IOException {
    ensureAccessible();
    try {
        return in.read((ByteBuffer) internalNioBuffer().clear().position(index).limit(index + length));
    } catch (ClosedChannelException e) {
        return -1;
    }
}
```

图 16-37　UnpooledHeapByteBuf 的 set 方法实现

从 SocketChannel 中读取字节数组到缓冲区 java.nio.ByteBuffer 中，它的起始 position 为 writeIndex，limit 为 writeIndex + length，JDK ByteBuffer 的相关 DOC 说明如图 16-38 所示。

```
● int java.nio.channels.ReadableByteChannel.read(ByteBuffer dst) throws IOException
Reads a sequence of bytes from this channel into the given buffer.
An attempt is made to read up to r bytes from the channel, where r is the number of bytes remaining in the
buffer, that is, dst.remaining(), at the moment this method is invoked.
Suppose that a byte sequence of length n is read, where 0 <= n <= r. This byte sequence will be transferred
into the buffer so that the first byte in the sequence is at index p and the last byte is at index p + n - 1,
where p is the buffer's position at the moment this method is invoked. Upon return the buffer's position will be
equal to p + n; its limit will not have changed.
A read operation might not fill the buffer, and in fact it might not read any bytes at all. Whether or not it
does so depends upon the nature and state of the channel. A socket channel in non-blocking mode, for example,
cannot read any more bytes than are immediately available from the socket's input buffer; similarly, a file
channel cannot read any more bytes than remain in the file. It is guaranteed, however, that if a channel is in
blocking mode and there is at least one byte remaining in the buffer then this method will block until at least
one byte is read.
This method may be invoked at any time. If another thread has already initiated a read operation upon this
channel, however, then an invocation of this method will block until the first operation is complete.
```

图 16-38　java.nio.ByteBuffer 的 read 方法

16.3　Unsafe 功能说明

Unsafe 接口实际上是 Channel 接口的辅助接口，它不应该被用户代码直接调用。实际的 I/O 读写操作都是由 Unsafe 接口负责完成的。下面我们一起看下它的 API 定义。

表 16-1　Unsafe API 功能列表

方法名	返回值	功能说明
invoker()	ChannelHandlerInvoker	返回默认使用的 ChannelHandlerInvoker
localAddress()	SocketAddress	返回本地绑定的 Socket 地址
remoteAddress()	SocketAddress	返回通信对端的 Socket 地址
register(ChannelPromise promise)	void	注册 Channel 到多路复用器上，一旦注册操作完成，通知 ChannelFuture
bind(SocketAddress localAddress, ChannelPromise promise)	void	绑定指定的本地地址 localAddress 到当前的 Channel 上，一旦完成，通知 ChannelFuture
connect(SocketAddress remoteAddress, SocketAddress localAddress, ChannelPromise promise)	void	绑定本地的 localAddress 之后，连接服务端，一旦操作完成，通知 ChannelFuture

续表

方法名	返回值	功能说明
disconnect(ChannelPromise promise)	void	断开 Channel 的连接，一旦完成，通知 ChannelFuture
close(ChannelPromise promise)	void	关闭 Channel 的连接，一旦完成，通知 ChannelFuture
closeForcibly()	void	强制立即关闭连接
beginRead()	void	设置网络操作位为读用于读取消息
write(Object msg, ChannelPromise promise)	void	发送消息，一旦完成，通知 ChannelFuture
flush()	void	将发送缓冲数组中的消息写入到 Channel 中
voidPromise()	ChannelPromise	返回一个特殊的可重用和传递的 ChannelPromise，它不用于操作成功或者失败的通知器，仅仅作为一个容器被使用
outboundBuffer()	ChannelOutboundBuffer	返回消息发送缓冲区

16.4 Unsafe 源码分析

实际的网络 I/O 操作基本都是由 Unsafe 功能类负责实现的，下面我们一起看下它的主要功能子类和重要的 API 实现。

16.4.1 Unsafe 继承关系类图

首先看下如图 16-39 所示 Unsafe 接口的类继承关系图。

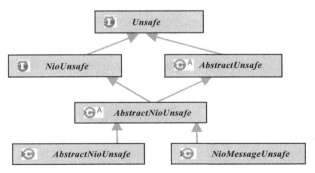

图 16-39　Unsafe 继承关系图

16.4.2 AbstractUnsafe 源码分析

1. register 方法

register 方法主要用于将当前 Unsafe 对应的 Channel 注册到 EventLoop 的多路复用器上，然后调用 DefaultChannelPipeline 的 fireChannelRegistered 方法。如果 Channel 被激活，则调用 DefaultChannelPipeline 的 fireChannelActive 方法。源码如图 16-40 所示。

首先判断当前所在的线程是否是 Channel 对应的 NioEventLoop 线程，如果是同一个线程，则不存在多线程并发操作问题，直接调用 register0 进行注册；如果是由用户线程或者其他线程发起的注册操作，则将注册操作封装成 Runnable，放到 NioEventLoop 任务队列中执行。注意：如果直接执行 register0 方法，会存在多线程并发操作 Channel 的问题。

```
@Override
public final void register(final ChannelPromise promise) {
    if (eventLoop.inEventLoop()) {
        register0(promise);
    } else {
        try {
            eventLoop.execute(new Runnable() {
                @Override
                public void run() {
                    register0(promise);
                }
            });
        } catch (Throwable t) {
            logger.warn(
                "Force-closing a channel whose registration task was not accepted by an event loop: {}",
                AbstractChannel.this, t);
            closeForcibly();
            closeFuture.setClosed();
            promise.setFailure(t);
        }
    }
}
```

图 16-40 AbstractUnsafe 的 register 方法

下面继续看 register0 方法的实现，代码如图 16-41 所示。

首先调用 ensureOpen 方法判断当前 Channel 是否打开，如果没有打开则无法注册，直接返回。校验通过后调用 doRegister 方法，它由 AbstractNioUnsafe 对应的 AbstractNioChannel 实现，代码如图 16-42 所示。

```
private void register0(ChannelPromise promise) {
    try {
        // check if the channel is still open as it could be closed in the mean time when the register
        // call was outside of the eventLoop
        if (!ensureOpen(promise)) {
            return;
        }
        doRegister();
        registered = true;
        promise.setSuccess();
        pipeline.fireChannelRegistered();
        if (isActive()) {
            pipeline.fireChannelActive();
        }
    } catch (Throwable t) {
        // Close the channel directly to avoid FD leak.
        closeForcibly();
        closeFuture.setClosed();
        if (!promise.tryFailure(t)) {
            logger.warn(
                "Tried to fail the registration promise, but it is complete already. " +
                "Swallowing the cause of the registration failure:", t);
        }
    }
}
```

图 16-41　AbstractUnsafe 的 register0 方法

```
@Override
protected void doRegister() throws Exception {
    boolean selected = false;
    for (;;) {
        try {
            selectionKey = javaChannel().register(eventLoop().selector, 0, this);
            return;
        } catch (CancelledKeyException e) {
            if (!selected) {
                // Force the Selector to select now as the "canceled" SelectionKey may still be
                // cached and not removed because no Select.select(..) operation was called yet.
                eventLoop().selectNow();
                selected = true;
            } else {
                // We forced a select operation on the selector before but the SelectionKey is still cached
                // for whatever reason. JDK bug ?
                throw e;
            }
        }
    }
}
```

图 16-42　AbstractNioChannel 的 doRegister 方法

该方法在前面的 AbstractNioChannel 源码分析中已经介绍过，此处不再赘述。如果 doRegister 方法没有抛出异常，则说明 Channel 注册成功。将 ChannelPromise 的结果设置为成功，调用 ChannelPipeline 的 fireChannelRegistered 方法，判断当前的 Channel 是否已经被激活，如果已经被激活，则调用 ChannelPipeline 的 fireChannelActive 方法。

如果注册过程中发生了异常，则强制关闭连接，将异常堆栈信息设置到 ChannelPromise 中。

2. bind 方法

bind 方法主要用于绑定指定的端口，对于服务端，用于绑定监听端口，可以设置 backlog 参数；对于客户端，主要用于指定客户端 Channel 的本地绑定 Socket 地址。代码实现如图 16-43 所示。

```
boolean wasActive = isActive();
    try {
        doBind(localAddress);
    } catch (Throwable t) {
        promise.setFailure(t);
        closeIfClosed();
        return;
    }
    if (!wasActive && isActive()) {
        invokeLater(new Runnable() {
            @Override
            public void run() {
                pipeline.fireChannelActive();
            }
        });
    }
    promise.setSuccess();
```

图 16-43　AbstractUnsafe 的 bind 方法实现

调用 doBind 方法，对于 NioSocketChannel 和 NioServerSocketChannel 有不同的实现，客户端的实现代码如图 16-44 所示。

```
@Override
protected void doBind(SocketAddress localAddress) throws Exception {
    javaChannel().socket().bind(localAddress);
}
```

图 16-44　NioSocketChannel 的 doBind 方法实现

服务端的 doBind 方法实现如图 16-45 所示。

```
@Override
protected void doBind(SocketAddress localAddress) throws Exception {
    javaChannel().socket().bind(localAddress, config.getBacklog());
}
```

图 16-45　NioServerSocketChannel 的 doBind 方法实现

如果绑定本地端口发生异常，则将异常设置到 ChannelPromise 中用于通知 ChannelFuture，随后调用 closeIfClosed 方法来关闭 Channel。

3. disconnect 方法

disconnect 用于客户端或者服务端主动关闭连接,它的代码如图 16-46 所示。

```
@Override
public final void disconnect(final ChannelPromise promise) {
    boolean wasActive = isActive();
    try {
        doDisconnect();
    } catch (Throwable t) {
        promise.setFailure(t);
        closeIfClosed();
        return;
    }
    if (wasActive && !isActive()) {
        invokeLater(new Runnable() {
            @Override
            public void run() {
                pipeline.fireChannelInactive();
            }
        });
    }
    promise.setSuccess();
    closeIfClosed(); // doDisconnect() might have closed the channel
}
```

图 16-46 AbstractUnsafe 的 disconnect 方法实现

4. close 方法

在链路关闭之前需要首先判断是否处于刷新状态,如果处于刷新状态说明还有消息尚未发送出去,需要等到所有消息发送完成再关闭链路,因此,将关闭操作封装成 Runnable 稍后再执行。如图 16-47 所示。

```
@Override
public final void close(final ChannelPromise promise) {
    if (inFlush0) {
        invokeLater(new Runnable() {
            @Override
            public void run() {
                close(promise);
            }
        });
        return;
    }
    if (closeFuture.isDone()) {
        // Closed already.
        promise.setSuccess();
        return;
    }
```

图 16-47 AbstractUnsafe 的 close 方法片段 1

如果链路没有处于刷新状态，需要从 closeFuture 中判断关闭操作是否完成，如果已经完成，不需要重复关闭链路，设置 ChannelPromise 的操作结果为成功并返回。

执行关闭操作，将消息发送缓冲数组设置为空，通知 JVM 进行内存回收。调用抽象方法 doClose 关闭链路。源码如图 16-48 所示。

```
boolean wasActive = isActive();
ChannelOutboundBuffer outboundBuffer = this.outboundBuffer;
this.outboundBuffer = null; // Disallow adding any messages and
flushes to outboundBuffer.
try {
    doClose();
    closeFuture.setClosed();
    promise.setSuccess();
} catch (Throwable t) {
    closeFuture.setClosed();
    promise.setFailure(t);
}
```

图 16-48　AbstractUnsafe 的 close 方法片段 2

如果关闭操作成功，设置 ChannelPromise 结果为成功。如果操作失败，则设置异常对象到 ChannelPromise 中。

调用 ChannelOutboundBuffer 的 close 方法释放缓冲区的消息，随后构造链路关闭通知 Runnable 放到 NioEventLoop 中执行。源码如图 16-49 所示。

```
// Fail all the queued messages
try {
    outboundBuffer.failFlushed(CLOSED_CHANNEL_EXCEPTION);
    outboundBuffer.close(CLOSED_CHANNEL_EXCEPTION);
} finally {
    if (wasActive && !isActive()) {
        invokeLater(new Runnable() {
            @Override
            public void run() {
                pipeline.fireChannelInactive();
            }
        });
    }
    deregister();
}
```

图 16-49　AbstractUnsafe 的 close 方法片段 3

最后，调用 deregister 方法，将 Channel 从多路复用器上取消注册，代码实现如图 16-50 所示。

```
@Override
protected void doDeregister() throws Exception {
    eventLoop().cancel(selectionKey());
}
```

图 16-50　AbstractUnsafe 的 close 方法片段 4

NioEventLoop 的 cancel 方法实际将 selectionKey 对应的 Channel 从多路复用器上去注册，NioEventLoop 的相关代码如图 16-51 所示。

```
void cancel(SelectionKey key) {
  key.cancel();
  cancelledKeys ++;
  if (cancelledKeys >= CLEANUP_INTERVAL) {
    cancelledKeys = 0;
    needsToSelectAgain = true;
  }
}
```

图 16-51　取消 Channel 的注册

5. write 方法

write 方法实际上将消息添加到环形发送数组中，并不是真正的写 Channel，它的代码如图 16-52 所示。

```
@Override
public void write(Object msg, ChannelPromise promise) {
  if (!isActive()) {
    // Mark the write request as failure if the channel is inactive.
    if (isOpen()) {
      promise.tryFailure(NOT_YET_CONNECTED_EXCEPTION);
    } else {
      promise.tryFailure(CLOSED_CHANNEL_EXCEPTION);
    }
    // release message now to prevent resource-leak
    ReferenceCountUtil.release(msg);
  } else {
    outboundBuffer.addMessage(msg, promise);
  }
}
```

图 16-52　写操作

如果 Channel 没有处于激活状态，说明 TCP 链路还没有真正建立成功，当前 Channel 存在以下两种状态。

（1）Channel 打开，但是 TCP 链路尚未建立成功：NOT_YET_CONNECTED_EXCEPTION；

（2）Channel 已经关闭：CLOSED_CHANNEL_EXCEPTION。

对链路状态进行判断，给 ChannelPromise 设置对应的异常，然后调用 ReferenceCountUtil 的 release 方法释放发送的 msg 对象。

如果链路状态正常，则将需要发送的 msg 和 promise 放入发送缓冲区中（环形数组）。

6. flush 方法

flush 方法负责将发送缓冲区中待发送的消息全部写入到 Channel 中，并发送给通信对方。它的代码如图 16-53 所示。

```
@Override
public void flush() {
    ChannelOutboundBuffer outboundBuffer = this.outboundBuffer;
    if (outboundBuffer == null) {
        return;
    }
    outboundBuffer.addFlush();
    flush0();
}
```

图 16-53　刷新操作

首先将发送环形数组的 unflushed 指针修改为 tail，标识本次要发送消息的缓冲区范围。然后调用 flush0 进行发送，由于 flush0 代码非常简单，我们重点分析 doWrite 方法，代码如图 16-54 所示。

```
@Override
protected void doWrite(ChannelOutboundBuffer in) throws Exception {
    for (;;) {
        // Do non-gathering write for a single buffer case.
        final int msgCount = in.size();
        if (msgCount <= 1) {
            super.doWrite(in);
            return;
        }
```

图 16-54　doWrite 方法代码片段 1

首先计算需要发送的消息个数（unflushed - flush），如果只有 1 个消息需要发送，则调用父类的写操作，我们分析 AbstractNioByteChannel 的 doWrite() 方法，代码如图 16-55 所示。

```
@Override
protected void doWrite(ChannelOutboundBuffer in) throws Exception {
    int writeSpinCount = -1;
    for (;;) {
        Object msg = in.current(true);
        if (msg == null) {
            // Wrote all messages.
            clearOpWrite();
            break;
        }
```

图 16-55　doWrite 方法代码片段 2

因为只有一条消息需要发送，所以直接从 ChannelOutboundBuffer 中获取当前需要发送的消息，代码如图 16-56 所示。

```
public Object current(boolean preferDirect) {
    if (isEmpty()) {
        return null;
    } else {
        Object msg = buffer[flushed].msg;
        if (threadLocalDirectBufferSize <= 0 || !preferDirect) {
            return msg;
        }
        if (msg instanceof ByteBuf) {
            ByteBuf buf = (ByteBuf) msg;
            if (buf.isDirect()) {
                return buf;
            } else {
                int readableBytes = buf.readableBytes();
                if (readableBytes == 0) {
                    return buf;
                }
                // Non-direct buffers are copied into JDK's own internal direct buffer on every I/O.
                // We can do a better job by using our pooled allocator. If the current allocator does not
                // pool a direct buffer, we use a ThreadLocal based pool.
                ByteBufAllocator alloc = channel.alloc();
                ByteBuf directBuf;
                if (alloc.isDirectBufferPooled()) {
                    directBuf = alloc.directBuffer(readableBytes);
                } else {
                    directBuf = ThreadLocalPooledByteBuf.newInstance();
                }
```

图 16-56　doWrite 方法代码片段 3

首先，获取需要发送的消息，如果消息为 ByteBuf 且它分配的是 JDK 的非堆内存，则直接返回。对返回的消息进行判断，如果为空，说明该消息已经发送完成并被回收，然后执行清空 OP_WRITE 操作位的 clearOpWrite 方法，代码如图 16-57 所示。

```
protected final void clearOpWrite() {
    final SelectionKey key = selectionKey();
    final int interestOps = key.interestOps();
    if ((interestOps & SelectionKey.OP_WRITE) != 0) {
        key.interestOps(interestOps & ~SelectionKey.OP_WRITE);
    }
}
```

图 16-57　doWrite 方法代码片段 4

继续向下分析，如果需要发送的 ByteBuf 已经没有可写的字节了，则说明已经发送完成，将该消息从环形队列中删除，然后继续循环，代码如图 16-58 所示。

```
if (msg instanceof ByteBuf) {
    ByteBuf buf = (ByteBuf) msg;
    int readableBytes = buf.readableBytes();
    if (readableBytes == 0) {
        in.remove();
        continue;
    }
```

图 16-58　doWrite 方法代码片段 5

下面我们分析下 ChannelOutboundBuffer 的 remove 方法，如图 16-59 所示。

```
public boolean remove() {
    if (isEmpty()) {
        return false;
    }
    Entry e = buffer[flushed];
    Object msg = e.msg;
    if (msg == null) {
        return false;
    }
    ChannelPromise promise = e.promise;
    int size = e.pendingSize;
    e.clear();
    flushed = flushed + 1 & buffer.length - 1;
    safeRelease(msg);
    promise.trySuccess();
    decrementPendingOutboundBytes(size);
    return true;
}
```

图 16-59　doWrite 方法代码片段 6

首先判断环形队列中是否还有需要发送的消息，如果没有，则直接返回。如果非空，则首先获取 Entry，然后对其进行资源释放，同时对需要发送的索引 flushed 进行更新。所有操作执行完之后，调用 decrementPendingOutboundBytes 减去已经发送的字节数，该方法跟 incrementPendingOutboundBytes 类似，会进行发送低水位的判断和事件通知，此处不再赘述。

我们接着继续对消息的发送进行分析，代码如图 16-60 所示。

```
boolean setOpWrite = false;
boolean done = false;
long flushedAmount = 0;
if (writeSpinCount == -1) {
    writeSpinCount = config().getWriteSpinCount();
}
for (int i = writeSpinCount - 1; i >= 0; i --) {
    int localFlushedAmount = doWriteBytes(buf);
    if (localFlushedAmount == 0) {
        setOpWrite = true;           第一步：对写入的字节个数进行判读，如果为 0 说明 TCP 的发送缓
        break;                       冲已满，需要退出并监听写操作
    }
    flushedAmount += localFlushedAmount;
    if (!buf.isReadable()) {
        done = true;
        break;
    }
}
```

图 16-60　doWrite 方法代码片段 7

首先将半包标识设置为 false，从 DefaultSocketChannelConfig 中获取循环发送的次数，进行循环发送，对发送方法 doWriteBytes 展开分析，如图 16-61 所示。

```
@Override
protected int doWriteBytes(ByteBuf buf) throws Exception {
    final int expectedWrittenBytes = buf.readableBytes();
    final int writtenBytes = buf.readBytes(javaChannel(),
expectedWrittenBytes);
    return writtenBytes;
}
```

图 16-61　doWrite 方法代码片段 8

ByteBuf 的 readBytes() 方法的功能是将当前 ByteBuf 中的可写字节数组写入到指定的 Channel 中。方法的第一个参数是 Channel，此处就是 SocketChannel，第二个参数是写入的字节数组长度，它等于 ByteBuf 的可读字节数，返回值是写入的字节个数。由于我们将 SocketChannel 设置为异步非阻塞模式，所以写操作不会阻塞。

从写操作中返回，需要对写入的字节数进行判断，如果为 0，说明 TCP 发送缓冲区已满，不能继续再向里面写入消息，因此，将写半包标识设置为 true，然后退出循环，执行后续排队的其他任务或者读操作，等待下一次 selector 的轮询继续触发写操作。

对写入的字节数进行累加，判断当前的 ByteBuf 中是否还有没有发送的字节，如果没有可发送的字节，则将 done 设置为 true，退出循环。

从循环发送状态退出后，首先根据实际发送的字节数更新发送进度，实际就是发送的字节数和需要发送的字节数的一个比值。执行完进度更新后，判断本轮循环是否将需要发送的消息全部发送完成，如果发送完成则将该消息从循环队列中删除；否则，设置多路复用器的 OP_WRITE 操作位，用于通知 Reactor 线程还有半包消息需要继续发送。

16.4.3　AbstractNioUnsafe 源码分析

AbstractNioUnsafe 是 AbstractUnsafe 类的 NIO 实现，它主要实现了 connect、finishConnect 等方法，下面我们对重点 API 实现进行源码分析。

1. connect 方法

首先获取当前的连接状态进行缓存，然后发起连接操作，代码如图 16-62 所示。

需要指出的是，SocketChannel 执行 connect() 操作有三种可能的结果。

（1）连接成功，返回 true；

（2）暂时没有连接上，服务端没有返回 ACK 应答，连接结果不确定，返回 false；

```
@Override
protected boolean doConnect(SocketAddress remoteAddress,
SocketAddress localAddress) throws Exception {
    if (localAddress != null) {
        javaChannel().socket().bind(localAddress);
    }
    boolean success = false;
    try {
        boolean connected = javaChannel().connect(remoteAddress);
        if (!connected) {
            selectionKey().interestOps(SelectionKey.OP_CONNECT);
        }
        success = true;
        return connected;
    } finally {
        if (!success) {
            doClose();
        }
    }
}
```

如果指定了本地绑定端口，执行绑定操作发起异步 TCP 连接，可能连接成功，也可能暂时没有连接成功。如果没有立即连接成功，则监听连接操作

图 16-62　AbstractNioUnsafe 的 connect 方法代码片段 1

（3）连接失败，直接抛出 I/O 异常。

如果是第（2）种结果，需要将 NioSocketChannel 中的 selectionKey 设置为 OP_CONNECT，监听连接应答消息。

异步连接返回之后，需要判断连接结果，如果连接成功，则触发 ChannelActive 事件，代码如图 16-63 所示。

```
private void fulfillConnectPromise(ChannelPromise promise, boolean wasActive) {
    // trySuccess() will return false if a user cancelled the connection attempt.
    boolean promiseSet = promise.trySuccess();
    // Regardless if the connection attempt was cancelled, channelActive() event should be triggered,
    // because what happened is what happened.
    if (!wasActive && isActive()) {
        pipeline().fireChannelActive();
    }
    // If a user cancelled the connection attempt, close the channel, which is followed by channelInactive().
    if (!promiseSet) {
        close(voidPromise());
    }
}
```

图 16-63　AbstractNioUnsafe 的 connect 方法代码片段 2

这里对 ChannelActive 事件处理不再进行详细说明，它最终会将 NioSocketChannel 中的 selectionKey 设置为 SelectionKey.OP_READ，用于监听网络读操作位。

如果没有立即连接上服务端，则执行如图 16-64 所示分支。

```
// Schedule connect timeout.
    int connectTimeoutMillis = config().getConnectTimeoutMillis();
    if (connectTimeoutMillis > 0) {
        connectTimeoutFuture = eventLoop().schedule(new Runnable() {
            @Override
            public void run() {
                ChannelPromise connectPromise = AbstractNioChannel.this.connectPromise;
                ConnectTimeoutException cause =
                        new ConnectTimeoutException("connection timed out: " + remoteAddress);
                if (connectPromise != null && connectPromise.tryFailure(cause)) {
                    close(voidPromise());
                }
            }
        }, connectTimeoutMillis, TimeUnit.MILLISECONDS);
    }
    promise.addListener(new ChannelFutureListener() {
        @Override
        public void operationComplete(ChannelFuture future) throws Exception {
            if (future.isCancelled()) {
                if (connectTimeoutFuture != null) {
                    connectTimeoutFuture.cancel(false);
                }
                connectPromise = null;
                close(voidPromise());
            }
        }
    });
```

图 16-64　AbstractNioUnsafe 的 connect 方法代码片段 3

上面的操作有两个目的。

（1）根据连接超时时间设置定时任务，超时时间到之后触发校验，如果发现连接并没有完成，则关闭连接句柄，释放资源，设置异常堆栈并发起去注册。

（2）设置连接结果监听器，如果接收到连接完成通知则判断连接是否被取消，如果被取消则关闭连接句柄，释放资源，发起取消注册操作。

2. finishConnect 方法

客户端接收到服务端的 TCP 握手应答消息，通过 SocketChannel 的 finishConnect 方法对连接结果进行判断，代码如图 16-65 所示。

```
@Override
    public void finishConnect() {
        // Note this method is invoked by the event loop only if the connection attempt was
        // neither cancelled nor timed out.
        assert eventLoop().inEventLoop();
        assert connectPromise != null;
        try {
            boolean wasActive = isActive();
            doFinishConnect();
```

图 16-65　AbstractNioUnsafe 的 finishConnect 方法代码片段 1

首先缓存连接状态，当前返回 false，然后执行 doFinishConnect 方法判断连接结果，代码如图 16-66 所示。

```
@Override
    protected void doFinishConnect() throws Exception {
        if (!javaChannel().finishConnect()) {
            throw new Error();
        }
    }
```

图 16-66　AbstractNioUnsafe 的 finishConnect 方法代码片段 2

通过 SocketChannel 的 finishConnect 方法判断连接结果，执行该方法返回三种可能结果。

◎ 连接成功返回 true；

◎ 连接失败返回 false；

◎ 发生链路被关闭、链路中断等异常，连接失败。

只要连接失败，就抛出 Error()，由调用方执行句柄关闭等资源释放操作，如果返回成功，则执行 fulfillConnectPromise 方法，它负责将 SocketChannel 修改为监听读操作位，用来监听网络的读事件，代码如图 16-67 所示。

```
private void fulfillConnectPromise(ChannelPromise promise, boolean wasActive) {
        // trySuccess() will return false if a user cancelled the connection attempt.
        boolean promiseSet = promise.trySuccess();
        // Regardless if the connection attempt was cancelled, channelActive() event should be triggered,
        // because what happened is what happened.
        if (!wasActive && isActive()) {
            pipeline().fireChannelActive();
        }
```

图 16-67　AbstractNioUnsafe 的 fulfillConnectPromise 方法

最后对连接超时进行判断：如果连接超时时仍然没有接收到服务端的 ACK 应答消息，则由定时任务关闭客户端连接，将 SocketChannel 从 Reactor 线程的多路复用器上摘除，释放资源，代码如图 16-68 所示。

```
} finally {
    // Check for null as the connectTimeoutFuture is only created if a
    connectTimeoutMillis > 0 is used
    // See https://github.com/netty/netty/issues/1770
    if (connectTimeoutFuture != null) {
        connectTimeoutFuture.cancel(false);
    }
    connectPromise = null;
}
```

图 16-68　AbstractNioUnsafe 的 finishConnect 方法代码片段 3

16.4.4　NioByteUnsafe 源码分析

我们重点分析它的 read 方法，源码如图 16-69 所示。

```
@Override
public void read() {
    final ChannelConfig config = config();
    final ChannelPipeline pipeline = pipeline();
    final ByteBufAllocator allocator = config.getAllocator();
    final int maxMessagesPerRead = config.getMaxMessagesPerRead();
    RecvByteBufAllocator.Handle allocHandle = this.allocHandle;
    if (allocHandle == null) {
        this.allocHandle = allocHandle =
config.getRecvByteBufAllocator().newHandle();
    }
```

图 16-69　NioByteUnsafe 的 read 方法代码片段 1

首先，获取 NioSocketChannel 的 SocketChannelConfig，它主要用于设置客户端连接的 TCP 参数，接口如图 16-70 所示。

继续看 allocHandle 的初始化。如果是首次调用，从 SocketChannelConfig 的 RecvByteBufAllocator 中创建 Handle。下面我们对 RecvByteBufAllocator 进行简单地代码分析：RecvByteBufAllocator 默认有两种实现，分别是 AdaptiveRecvByteBufAllocator 和 FixedRecvByteBufAllocator。由于 FixedRecvByteBufAllocator 的实现比较简单，我们重点分析 AdaptiveRecvByteBufAllocator 的实现。如图 16-71 所示。

顾名思义，AdaptiveRecvByteBufAllocator 指的是缓冲区大小可以动态调整的 ByteBuf 分配器。它的成员变量定义如图 16-72 所示。

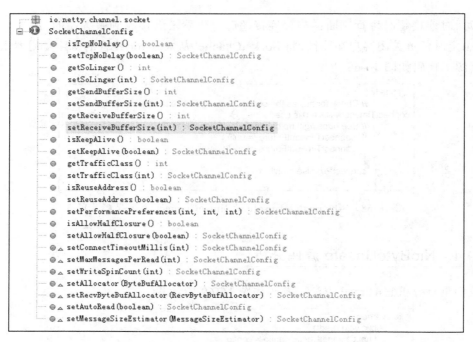

图 16-70 SocketChannelConfig 的 API 列表

图 16-71 RecvByteBufAllocator 接口的继承关系

```
public class AdaptiveRecvByteBufAllocator implements RecvByteBufAllocator {
    static final int DEFAULT_MINIMUM = 64;
    static final int DEFAULT_INITIAL = 1024;

    static final int DEFAULT_MAXIMUM = 65536;
    private static final int INDEX_INCREMENT = 4;
    private static final int INDEX_DECREMENT = 1;
```

图 16-72 RecvByteBufAllocator 的成员变量定义

它分别定义了三个系统默认值：最小缓冲区长度 64 字节、初始容量 1024 字节、最大容量 65536 字节。还定义了两个动态调整容量时的步进参数：扩张的步进索引为 4、收缩的步进索引为 1。

最后，定义了长度的向量表 SIZE_TABLE 并初始化它，初始值如图 16-73 所示。

```
0-->16   1-->32   2-->48   3-->64   4-->80   5-->96   6-->112  7-->128  8-->144
9-->160
10-->176  11-->192  12-->208  13-->224  14-->240  15-->256  16-->272
17-->288  18-->304
19-->320  20-->336  21-->352  22-->368  23-->384  24-->400  25-->416
26-->432  27-->448
28-->464  29-->480  30-->496  31-->512  32-->1024  33-->2048  34-->4096
35-->8192  36-->16384
37-->32768      38-->65536     39-->131072     40-->262144    41-->524288
42-->1048576  43-->2097152  44-->4194304  45-->8388608
46-->16777216    47-->33554432    48-->67108864    49-->134217728
50-->268435456  51-->536870912  52-->1073741824
```

图 16-73　SIZE_TABLE 的扩张

向量数组的每个值都对应一个 Buffer 容量，当容量小于 512 的时候，由于缓冲区已经比较小，需要降低步进值，容量每次下调的幅度要小些；当大于 512 时，说明需要解码的消息码流比较大，这时采用调大步进幅度的方式减少动态扩张的频率，所以它采用 512 的倍数进行扩张。

接下来我们重点分析下 AdaptiveRecvByteBufAllocator 的方法。

方法 1：getSizeTableIndex(final int size)，代码如图 16-74 所示。

```
private static int getSizeTableIndex(final int size) {
    for (int low = 0, high = SIZE_TABLE.length - 1;;) {
        if (high < low) {
            return low;
        }
        if (high == low) {
            return high;
        }

        int mid = low + high >>> 1;
        int a = SIZE_TABLE[mid];
        int b = SIZE_TABLE[mid + 1];
        if (size > b) {
            low = mid + 1;
        } else if (size < a) {
            high = mid - 1;
        } else if (size == a) {
            return mid;
        } else {
            return mid + 1;
        }
    }
}
```

图 16-74　获取向量表索引

根据容量 Size 查找容量向量表对应的索引——这是个典型的二分查找法，由于它的算法非常经典，也比较简单，此处不再赘述。

下面我们分析下它的内部静态类 HandleImpl，首先，还是看下它的成员变量，如图 16-75 所示。

```
private static final class HandleImpl implements Handle {
    private final int minIndex;
    private final int maxIndex;
    private int index;
    private int nextReceiveBufferSize;
    private boolean decreaseNow;
```

图 16-75　HandleImpl 的成员变量

它有 5 个成员变量，分别是：对应向量表的最小索引、最大索引、当前索引、下一次预分配的 Buffer 大小和是否立即执行容量收缩操作。

我们重点分析它的 record(int actualReadBytes)方法：当 NioSocketChannel 执行完读操作后，会计算获得本次轮询读取的总字节数，它就是参数 actualReadBytes，执行 record 方法，根据实际读取的字节数对 ByteBuf 进行动态伸缩和扩张，代码如图 16-76 所示。

```
@Override
public void record(int actualReadBytes) {
    if (actualReadBytes <= SIZE_TABLE[Math.max(0, index - INDEX_DECREMENT - 1)]) {
        if (decreaseNow) {
            index = Math.max(index - INDEX_DECREMENT, minIndex);
            nextReceiveBufferSize = SIZE_TABLE[index];
            decreaseNow = false;
        } else {
            decreaseNow = true;
        }
    } else if (actualReadBytes >= nextReceiveBufferSize) {
        index = Math.min(index + INDEX_INCREMENT, maxIndex);
        nextReceiveBufferSize = SIZE_TABLE[index];
        decreaseNow = false;
    }
}
```

图 16-76　ByteBuf 动态伸缩

首先，对当前索引做步进缩减，然后获取收缩后索引对应的容量，与实际读取的字节数进行比对，如果发现小于收缩后的容量，则重新对当前索引进行赋值，取收缩后的索引和最小索引中的较大者作为最新的索引。然后，为下一次缓冲区容量分配赋值——新的索引对应容量向量表中的容量。相反，如果当前实际读取的字节数大于之前预分配的初始容量，则说明实际分配的容量不足，需要动态扩张。重新计算索引，选取当前索引+扩张步进和最大索引中的较小作为当前索引值，然后对下次缓冲区的容量值进行重新分配，完成缓冲区容量的动态扩张。

通过上述分析我们得知，AdaptiveRecvByteBufAllocator 就是根据本次读取的实际字节数对下次接收缓冲区的容量进行动态调整。

使用动态缓冲区分配器的优点如下。

（1）Netty 作为一个通用的 NIO 框架，并不对用户的应用场景进行假设，可以使用它做流媒体传输，也可以用它做聊天工具。不同的应用场景，传输的码流大小千差万别，无论初始化分配的是 32KB 还是 1MB，都会随着应用场景的变化而变得不适应。因此，Netty 根据上次实际读取的码流大小对下次的接收 Buffer 缓冲区进行预测和调整，能够最大限度地满足不同行业的应用场景。

（2）性能更高，容量过大会导致内存占用开销增加，后续的 Buffer 处理性能会下降；容量过小时需要频繁地内存扩张来接收大的请求消息，同样会导致性能下降。

（3）更节约内存。假如通常情况下请求消息平均值为 1MB 左右，接收缓冲区大小为 1.2MB。突然某个客户发送了一个 10MB 的流媒体附件，接收缓冲区扩张为 10MB 以接纳该附件，如果缓冲区不能收缩，每次缓冲区创建都会分配 10MB 的内存，但是后续所有的消息都是 1MB 左右，这样会导致内存的浪费，如果并发客户端过多，可能会发生内存溢出，最终宕机。

看完了 AdaptiveRecvByteBufAllocator，我们继续分析读操作。

首先通过接收缓冲区分配器的 Handler 计算获得下次预分配的缓冲区容量 byteBufCapacity，如图 16-77 所示。紧接着根据缓冲区容量进行缓冲区分配，Netty 的缓冲区种类很多，此处重点介绍的是消息的读取，因此对缓冲区不展开说明。

```
int byteBufCapacity = allocHandle.guess();
int totalReadAmount = 0;
do {
    byteBuf = allocator.ioBuffer(byteBufCapacity);
```

图 16-77　NioByteUnsafe 的 read 方法代码片段 2

接收缓冲区 ByteBuf 分配完成后，进行消息的异步读取，代码如图 16-78 所示。

```
int localReadAmount = doReadBytes(byteBuf);
```

图 16-78　NioByteUnsafe 的 read 方法代码片段 3

它是个抽象方法，具体实现在 NioSocketChannel 中，代码如图 16-79 所示。

```
@Override
protected int doReadBytes(ByteBuf byteBuf) throws Exception {
    return byteBuf.writeBytes(javaChannel(), byteBuf.writableBytes());
}
```

图 16-79　NioByteUnsafe 的 read 方法代码片段 4

其中 javaChannel() 返回的是 SocketChannel，代码如图 16-80 所示。

```
@Override
protected SocketChannel javaChannel() {
    return (SocketChannel) super.javaChannel();
}
```

图 16-80　NioByteUnsafe 的 read 方法代码片段 5

byteBuf.writableBytes() 返回本次可读的最大长度，我们继续展开看最终是如何从 Channel 中读取码流的，代码如图 16-81 所示。

```
@Override
public int writeBytes(ScatteringByteChannel in, int length) throws IOException
{
    ensureWritable(length);
    int writtenBytes = setBytes(writerIndex, in, length);
    if (writtenBytes > 0) {
        writerIndex += writtenBytes;
    }
    return writtenBytes;
}
```

图 16-81　NioByteUnsafe 的 read 方法代码片段 6

对 setBytes 方法展开分析如图 16-82 所示。

```
@Override
public int setBytes(int index, ScatteringByteChannel in, int length) throws
IOException {
    ensureAccessible();
    try {
        return in.read((ByteBuffer)
internalNioBuffer().clear().position(index).limit(index + length));
    } catch (ClosedChannelException e) {
        return -1;
    }
}
```

图 16-82　NioByteUnsafe 的 read 方法代码片段 7

由于 SocketChannel 的 read 方法参数是 Java NIO 的 ByteBuffer，所以，需要先将 Netty 的 ByteBuf 转换成 JDK 的 ByteBuffer，随后调用 ByteBuffer 的 clear 方法对指针进行重置用于新消息的读取，随后将 position 指针指到初始读 index，读取的上限设置为 index +读取的长度。最后调用 read 方法将 SocketChannel 中就绪的码流读取到 ByteBuffer 中，完成消息的读取，返回读取的字节数。

完成消息的异步读取后，需要对本次读取的字节数进行判断，有以下三种可能：

（1）返回 0，表示没有就绪的消息可读；

（2）返回值大于 0，读到了消息；

（3）返回值-1，表示发生了 I/O 异常，读取失败。

下面我们继续看 Netty 的后续处理，首先对读取的字节数进行判断，如果等于或者小于 0，表示没有就绪的消息可读或者发生了 I/O 异常，此时需要释放接收缓冲区；如果读取的字节数小于 0，则需要将 close 状态位置位，用于关闭连接，释放句柄资源。置位完成之后，退出循环。源码如图 16-83 所示。

```
if (localReadAmount <= 0) {
    // not was read release the buffer
    byteBuf.release();
    close = localReadAmount < 0;
    break;
}
```

图 16-83　NioByteUnsafe 的 read 方法代码片段 8

完成一次异步读之后，就会触发一次 ChannelRead 事件，这里要特别提醒大家的是：完成一次读操作，并不意味着读到了一条完整的消息，因为 TCP 底层存在组包和粘包，所以，一次读操作可能包含多条消息，也可能是一条不完整的消息。因此不要把它跟读取的消息个数等同起来。在没有做任何半包处理的情况下，以 ChannelRead 的触发次数做计数器来进行性能分析和统计，是完全错误的。当然，如果你使用了半包解码器或者处理了半包，就能够实现一次 ChannelRead 对应一条完整的消息。

触发和完成 ChannelRead 事件调用之后，将接收缓冲区释放，代码如图 16-84 所示。

```
pipeline.fireChannelRead(byteBuf);
byteBuf = null;
```

图 16-84　NioByteUnsafe 的 read 方法代码片段 9

因为一次读操作未必能够完成 TCP 缓冲区的全部读取工作，所以，读操作在循环体中进行,每次读取操作完成之后,会对读取的字节数进行累加,代码如图 16-85 所示。

```
if (totalReadAmount >= Integer.MAX_VALUE - localReadAmount) {
    // Avoid overflow.
    totalReadAmount = Integer.MAX_VALUE;
    break;
}
totalReadAmount += localReadAmount;
```

图 16-85　NioByteUnsafe 的 read 方法代码片段 10

在累加之前,需要对长度上限做保护,如果累计读取的字节数已经发生溢出,则将读取到的字节数设置为整型的最大值,然后退出循环。原因是本次循环已经读取过多的字节,需要退出,否则会影响后面排队的 Task 任务和写操作的执行。如果没有溢出,则执行累加操作。代码如图 16-86 所示。

```
if (localReadAmount < writable) {
    // Read less than what the buffer can hold,
    // which might mean we drained the recv buffer completely.
    break;
}
```

图 16-86　NioByteUnsafe 的 read 方法代码片段 11

最后,对本次读取的字节数进行判断,如果小于缓冲区可写的容量,说明 TCP 缓冲区已经没有就绪的字节可读,读取操作已经完成,需要退出循环。如果仍然有未读的消息,则继续执行读操作。连续的读操作会阻塞排在后面的任务队列中待执行的 Task,以及写操作,所以,要对连续读操作做上限控制,默认值为 16 次,无论 TCP 缓冲区有多少码流需要读取,只要连续 16 次没有读完,都需要强制退出,等待下次 selector 轮询周期再执行。如图 16-87 所示。

```
} while (++ messages < maxMessagesPerRead);
```

图 16-87　NioByteUnsafe 的 read 方法代码片段 12

完成多路复用器本轮读操作之后,触发 ChannelReadComplete 事件,随后调用接收缓冲区容量分配器的 Hanlder 的记录方法,将本次读取的总字节数传入到 record() 方法中进行缓冲区的动态分配,为下一次读取选取更加合适的缓冲区容量,代码如图 16-88 所示。

```
pipeline.fireChannelReadComplete();
allocHandle.record(totalReadAmount);
```

图 16-88　NioByteUnsafe 的 read 方法代码片段 13

上面我们提到,如果读到的返回值为-1,表明发生了 I/O 异常,需要关闭连接,释放资源，代码如图 16-89 所示。

```
if (close) {
        closeOnRead(pipeline);
        close = false;
}
```

图 16-89　NioByteUnsafe 的 read 方法代码片段 14

至此，请求消息的异步读取源码我们已经分析完成。

16.5　总结

本章介绍了 Netty 最重要的接口之一——Channel 的设计原理和功能列表，并对其主要实现子类 NioSocketChannel 和 NioServerSocketChannel 的源码进行了分析，涉及到了"半包读"和"半包写"的相关知识。

由于 Channel 的很多 I/O 操作都是通过其内部聚合的 Unsafe 接口及其子类实现的，如果不清楚 Unsafe 相关子类的代码实现，也就无法真正了解清楚 Channel 的实现。因此本章节对 Unsafe 的相关实现也进行了源码分析。

事实上，Channel 的实现子类还有很多，包括用于处理 UDP 的 DatagramChannel、用于本地测试的 EmbeddedChannel 等。限于篇幅，本书无法对这些子类的功能和源码进行一一枚举。感兴趣的读者可以通过阅读 API 文档、学习 Demo 和源码分析相结合的方式掌握这些类库的使用。

第 17 章

ChannelPipeline 和 ChannelHandler

Netty 的 ChannelPipeline 和 ChannelHandler 机制类似于 Servlet 和 Filter 过滤器，这类拦截器实际上是职责链模式的一种变形，主要是为了方便事件的拦截和用户业务逻辑的定制。

Servlet Filter 是 JEE Web 应用程序级的 Java 代码组件，它能够以声明的方式插入到 HTTP 请求响应的处理过程中，用于拦截请求和响应，以便能够查看、提取或以某种方式操作正在客户端和服务器之间交换的数据。拦截器封装了业务定制逻辑，能够实现对 Web 应用程序的预处理和事后处理。

过滤器提供了一种面向对象的模块化机制，用来将公共任务封装到可插入的组件中。这些组件通过 Web 部署配置文件（web.xml）进行声明，可以方便地添加和删除过滤器，无须改动任何应用程序代码或 JSP 页面，由 Servlet 进行动态调用。通过在请求/响应链中使用过滤器，可以对应用程序（而不是以任何方式替代）的 Servlet 或 JSP 页面提供的核心处理进行补充，而不破坏 Servlet 或 JSP 页面的功能。由于是纯 Java 实现，所以 Servlet 过滤器具有跨平台的可重用性，使得它们很容易被部署到任何符合 Servlet 规范的 JEE 环境中。

Netty 的 Channel 过滤器实现原理与 Servlet Filter 机制一致，它将 Channel 的数据管道抽象为 ChannelPipeline，消息在 ChannelPipeline 中流动和传递。ChannelPipeline 持有 I/O 事件拦截器 ChannelHandler 的链表，由 ChannelHandler 对 I/O 事件进行拦截和处理，可以方便地通过新增和删除 ChannelHandler 来实现不同的业务逻辑定制，不需要对已有的 ChannelHandler 进行修改，能够实现对修改封闭和对扩展的支持。

下面我们对 ChannelPipeline 和 ChannelHandler，以及与之相关的 ChannelHandlerContext 进行详细介绍和源码分析。

本章主要内容包括：

◎ ChannelPipeline 功能说明

◎ ChannelPipeline 源码分析

◎ ChannelHandler 功能说明

◎ ChannelHandler 源码分析

17.1 ChannelPipeline 功能说明

ChannelPipeline 是 ChannelHandler 的容器，它负责 ChannelHandler 的管理和事件拦截与调度。

17.1.1 ChannelPipeline 的事件处理

图 17-1 展示了一个消息被 ChannelPipeline 的 ChannelHandler 链拦截和处理的全过程，消息的读取和发送处理全流程描述如下。

（1）底层的 SocketChannel read()方法读取 ByteBuf，触发 ChannelRead 事件，由 I/O 线程 NioEventLoop 调用 ChannelPipeline 的 fireChannelRead(Object msg)方法，将消息（ByteBuf）传输到 ChannelPipeline 中。

（2）消息依次被 HeadHandler、ChannelHandler1、ChannelHandler2……TailHandler 拦截和处理，在这个过程中，任何 ChannelHandler 都可以中断当前的流程，结束消息的传递。

（3）调用 ChannelHandlerContext 的 write 方法发送消息，消息从 TailHandler 开始，途

经 ChannelHandlerN……ChannelHandler1、HeadHandler，最终被添加到消息发送缓冲区中等待刷新和发送，在此过程中也可以中断消息的传递，例如当编码失败时，就需要中断流程，构造异常的 Future 返回。

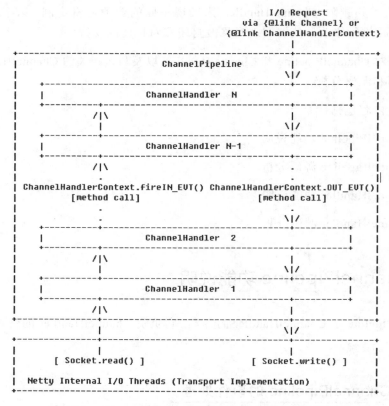

图 17-1　ChannelPipeline 对事件流的拦截和处理流程

Netty 中的事件分为 inbound 事件和 outbound 事件。inbound 事件通常由 I/O 线程触发，例如 TCP 链路建立事件、链路关闭事件、读事件、异常通知事件等，它对应图 17-1 的左半部分。

触发 inbound 事件的方法如下。

（1）ChannelHandlerContext.fireChannelRegistered()：Channel 注册事件；

（2）ChannelHandlerContext.fireChannelActive()：TCP 链路建立成功，Channel 激活事件；

（3）ChannelHandlerContext.fireChannelRead(Object)：读事件；

（4）ChannelHandlerContext.fireChannelReadComplete()：读操作完成通知事件；

（5）ChannelHandlerContext.fireExceptionCaught(Throwable)：异常通知事件；

（6）ChannelHandlerContext.fireUserEventTriggered(Object)：用户自定义事件；

（7）ChannelHandlerContext.fireChannelWritabilityChanged()：Channel 的可写状态变化通知事件；

（8）ChannelHandlerContext.fireChannelInactive()：TCP 连接关闭，链路不可用通知事件。

Outbound 事件通常是由用户主动发起的网络 I/O 操作，例如用户发起的连接操作、绑定操作、消息发送等操作，它对应图 17-1 的右半部分。

触发 outbound 事件的方法如下：

（1）ChannelHandlerContext.bind(SocketAddress, ChannelPromise)：绑定本地地址事件；

（2）ChannelHandlerContext.connect(SocketAddress, SocketAddress, ChannelPromise)：连接服务端事件；

（3）ChannelHandlerContext.write(Object, ChannelPromise)：发送事件；

（4）ChannelHandlerContext.flush()：刷新事件；

（5）ChannelHandlerContext.read()：读事件；

（6）ChannelHandlerContext.disconnect(ChannelPromise)：断开连接事件；

（7）ChannelHandlerContext.close(ChannelPromise)：关闭当前 Channel 事件。

17.1.2　自定义拦截器

ChannelPipeline 通过 ChannelHandler 接口来实现事件的拦截和处理，由于 ChannelHandler 中的事件种类繁多，不同的 ChannelHandler 可能只需要关心其中的某一个或者几个事件，所以，通常 ChannelHandler 只需要继承 ChannelHandlerAdapter 类覆盖自己关心的方法即可。

例如，下面的例子展示了拦截 Channel Active 事件，打印 TCP 链路建立成功日志，代码如下：

```
public class MyInboundHandler extends ChannelHandlerAdapter {
    @Override
    public void channelActive(ChannelHandlerContext ctx) {
        System.out.println("TCP connected!");
        ctx.fireChannelActive();
    }
}
```

下面的例子展示了如何在链路关闭的时候释放资源,示例代码如下。

```
public class MyOutboundHandler extends ChannelHandlerAdapter {
    @Override
    public void close(ChannelHandlerContext ctx,ChannelPromise promise) {
        System.out.println("TCP closing ..");
        Object.release();
        ctx.close(promise);
    }
}
```

17.1.3 构建 pipeline

事实上,用户不需要自己创建 pipeline,因为使用 ServerBootstrap 或者 Bootstrap 启动服务端或者客户端时,Netty 会为每个 Channel 连接创建一个独立的 pipeline。对于使用者而言,只需要将自定义的拦截器加入到 pipeline 中即可。相关的代码如下。

```
pipeline = ch.pipeline();
pipeline.addLast("decoder", new MyProtocolDecoder());
pipeline.addLast("encoder", new MyProtocolEncoder());
```

对于类似编解码这样的 ChannelHandler,它存在先后顺序,例如 MessageToMessageDecoder,在它之前往往需要有 ByteToMessageDecoder 将 ByteBuf 解码为对象,然后对对象做二次解码得到最终的 POJO 对象。Pipeline 支持指定位置添加或者删除拦截器,相关接口定义如图 17-2 所示。

```
addBefore(String, String, ChannelHandler) : ChannelPipeline
addBefore(EventExecutorGroup, String, String, ChannelHandler) : ChannelPipeline
addBefore(ChannelHandlerInvoker, String, String, ChannelHandler) : ChannelPipeline
addAfter(String, String, ChannelHandler) : ChannelPipeline
addAfter(EventExecutorGroup, String, String, ChannelHandler) : ChannelPipeline
addAfter(ChannelHandlerInvoker, String, String, ChannelHandler) : ChannelPipeline
```

图 17-2 按顺序添加 ChannelHandler

17.1.4　ChannelPipeline 的主要特性

ChannelPipeline 支持运行态动态的添加或者删除 ChannelHandler，在某些场景下这个特性非常实用。例如当业务高峰期需要对系统做拥塞保护时，就可以根据当前的系统时间进行判断，如果处于业务高峰期，则动态地将系统拥塞保护 ChannelHandler 添加到当前的 ChannelPipeline 中，当高峰期过去之后，就可以动态删除拥塞保护 ChannelHandler 了。

ChannelPipeline 是线程安全的，这意味着 N 个业务线程可以并发地操作 ChannelPipeline 而不存在多线程并发问题。但是，ChannelHandler 却不是线程安全的，这意味着尽管 ChannelPipeline 是线程安全的，但是用户仍然需要自己保证 ChannelHandler 的线程安全。

17.2　ChannelPipeline 源码分析

ChannelPipeline 的代码相对比较简单，它实际上是一个 ChannelHandler 的容器，内部维护了一个 ChannelHandler 的链表和迭代器，可以方便地实现 ChannelHandler 查找、添加、替换和删除。

17.2.1　ChannelPipeline 的类继承关系图

ChannelPipeline 的类继承关系比较简单，如图 17-3 所示。

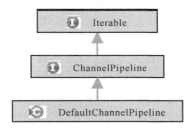

图 17-3　ChannelPipeline 类继承关系图

17.2.2　ChannelPipeline 对 ChannelHandler 的管理

ChannelPipeline 是 ChannelHandler 的管理容器，负责 ChannelHandler 的查询、添加、替换和删除。由于它与 Map 等容器的实现非常类似，所以我们只简单抽取新增接口进行源码分析，其他方法读者可以自行阅读和分析。在 ChannelPipeline 中添加 ChannelHandler 方法如图 17-4 所示。

```
@Override
public ChannelPipeline addBefore(String baseName, String name,
ChannelHandler handler) {
    return addBefore((ChannelHandlerInvoker) null, baseName, name, handler);
}
```

图 17-4　ChannelPipeline 添加 ChannelHandler 方法

直接调用 addBefore(ChannelHandlerInvoker invoker, String baseName, final String name, ChannelHandler handler)方法，代码如图 17-5 所示。

```
@Override
public ChannelPipeline addBefore(
        ChannelHandlerInvoker invoker, String baseName, final String name,
ChannelHandler handler) {
    synchronized (this) {
        DefaultChannelHandlerContext ctx = getContextOrDie(baseName);

        checkDuplicateName(name);

        DefaultChannelHandlerContext newCtx =
                new DefaultChannelHandlerContext(this, invoker, name, handler);

        addBefore0(name, ctx, newCtx);
    }
    return this;
}
```

图 17-5　ChannelPipeline 的 addBefore 方法

由于 ChannelPipeline 支持运行期动态修改，因此存在两种潜在的多线程并发访问场景。

◎ I/O 线程和用户业务线程的并发访问；

◎ 用户多个线程之间的并发访问。

为了保证 ChannelPipeline 的线程安全性，需要通过线程安全容器或者锁来保证并发访问的安全，此处 Netty 直接使用了 synchronized 关键字，保证同步块内的所有操作的原子性。首先根据 baseName 获取它对应的 DefaultChannelHandlerContext，ChannelPipeline 维护了 ChannelHandler 名和 ChannelHandlerContext 实例的映射关系，代码如图 17-6 所示。

```
@Override
public ChannelHandlerContext context(String name) {
    if (name == null) {
        throw new NullPointerException("name");
    }

    synchronized (this) {
        return name2ctx.get(name);
    }
}
```

图 17-6　ChannelPipeline 的 context 方法

对新增的 ChannelHandler 名进行重复性校验，如果已经有同名的 ChannelHandler 存在，则不允许覆盖，抛出 IllegalArgumentException("Duplicate handler name: " + name)异常。校验通过之后，使用新增的 ChannelHandler 等参数构造一个新的 DefaultChannelHandlerContext 实例，代码如图 17-7 所示。

```
DefaultChannelHandlerContext(
        DefaultChannelPipeline pipeline, ChannelHandlerInvoker invoker, String
name, ChannelHandler handler) {
    if (name == null) {
        throw new NullPointerException("name");
    }
    if (handler == null) {
        throw new NullPointerException("handler");
    }

    channel = pipeline.channel;
    this.pipeline = pipeline;
    this.name = name;
    this.handler = handler;

    skipFlags = skipFlags(handler);

    if (invoker == null) {
        this.invoker = channel.unsafe().invoker();
    } else {
        this.invoker = invoker;
    }
}
```

图 17-7　构造新的 DefaultChannelHandlerContext 实例

将新创建的 DefaultChannelHandlerContext 添加到当前的 pipeline 中，代码如图 17-8 所示。

```
private void addBefore0(final String name, DefaultChannelHandlerContext ctx,
DefaultChannelHandlerContext newCtx) {
    checkMultiplicity(newCtx);

    newCtx.prev = ctx.prev;
    newCtx.next = ctx;
    ctx.prev.next = newCtx;
    ctx.prev = newCtx;

    name2ctx.put(name, newCtx);

    callHandlerAdded(newCtx);
}
```

图 17-8　添加 DefaultChannelHandlerContext 到 pipeline

首先需要对添加的 ChannelHandlerContext 做重复性校验，校验代码如图 17-9 所示。

如果 ChannelHandlerContext 不是可以在多个 ChannelPipeline 中共享的，且已经被添加到 ChannelPipeline 中，则抛出 ChannelPipelineException 异常。Handler 指针修改如图 17-10 所示。

```
private static void checkMultiplicity(ChannelHandlerContext ctx) {
    ChannelHandler handler = ctx.handler();
    if (handler instanceof ChannelHandlerAdapter) {
        ChannelHandlerAdapter h = (ChannelHandlerAdapter) handler;
        if (!h.isSharable() && h.added) {
            throw new ChannelPipelineException(
                h.getClass().getName() +
                " is not a @Sharable handler, so can't be added or removed multiple times.");
        }
        h.added = true;
    }
}
```

图 17-9　ChannelHandlerContext 重复性校验

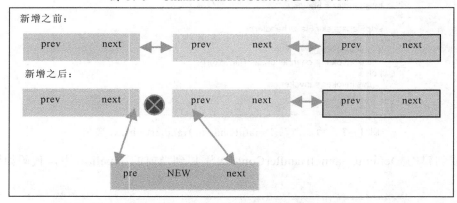

图 17-10　ChannelHandlerContext 位置指针迁移图

加入成功之后，缓存 ChannelHandlerContext，发送新增 ChannelHandlerContext 通知消息。

17.2.3　ChannelPipeline 的 inbound 事件

当发生某个 I/O 事件的时候，例如链路建立、链路关闭、读取操作完成等，都会产生一个事件，事件在 pipeline 中得到传播和处理，它是事件处理的总入口。由于网络 I/O 相关的事件有限，因此 Netty 对这些事件进行了统一抽象，Netty 自身和用户的 ChannelHandler 会对感兴趣的事件进行拦截和处理。

第 17 章 ChannelPipeline 和 ChannelHandler

pipeline 中以 fireXXX 命名的方法都是从 I/O 线程流向用户业务 Handler 的 inbound 事件，它们的实现因功能而异，但是处理步骤类似，总结如下。

（1）调用 HeadHandler 对应的 fireXXX 方法；

（2）执行事件相关的逻辑操作。

以 fireChannelActive 方法为例，调用 head.fireChannelActive() 之后，判断当前的 Channel 配置是否自动读取，如果为真则调用 Channel 的 read 方法，代码如图 17-11 所示。

```
@Override
public ChannelPipeline fireChannelActive() {
  head.fireChannelActive();

  if (channel.config().isAutoRead()) {
    channel.read();
  }

  return this;
}
```

图 17-11　fireChannelActive 方法

17.2.4　ChannelPipeline 的 outbound 事件

由用户线程或者代码发起的 I/O 操作被称为 outbound 事件，事实上 inbound 和 outbound 是 Netty 自身根据事件在 pipeline 中的流向抽象出来的术语，在其他 NIO 框架中并没有这个概念。

inbound 事件相关联的操作如图 17-12 所示。

```
● fireChannelWritabilityChanged() : ChannelPipeline
● bind(SocketAddress) : ChannelFuture
● connect(SocketAddress) : ChannelFuture
● connect(SocketAddress, SocketAddress) : ChannelFuture
● disconnect() : ChannelFuture
● close() : ChannelFuture
● flush() : ChannelPipeline
● bind(SocketAddress, ChannelPromise) : ChannelFuture
● connect(SocketAddress, ChannelPromise) : ChannelFuture
● connect(SocketAddress, SocketAddress, ChannelPromise) : ChannelFuture
● disconnect(ChannelPromise) : ChannelFuture
● close(ChannelPromise) : ChannelFuture
● read() : ChannelPipeline
● write(Object) : ChannelFuture
● write(Object, ChannelPromise) : ChannelFuture
● writeAndFlush(Object, ChannelPromise) : ChannelFuture
● writeAndFlush(Object) : ChannelFuture
```

图 17-12　inbound 事件相关方法

Pipeline 本身并不直接进行 I/O 操作，在前面对 Channel 和 Unsafe 的介绍中我们知道最终都是由 Unsafe 和 Channel 来实现真正的 I/O 操作的。Pipeline 负责将 I/O 事件通过 TailHandler 进行调度和传播，最终调用 Unsafe 的 I/O 方法进行 I/O 操作，相关代码实现如图 17-13 所示。

```
@Override
public ChannelFuture connect(SocketAddress remoteAddress,
SocketAddress localAddress) {
    return tail.connect(remoteAddress, localAddress);
}
```

图 17-13　pipeline 的客户端连接操作

它直接调用 TailHandler 的 connect 方法，最终会调用到 HeadHandler 的 connect 方法，代码如图 17-14 所示。

```
@Override
public void connect(
        ChannelHandlerContext ctx,
        SocketAddress remoteAddress, SocketAddress localAddress,
        ChannelPromise promise) throws Exception {
    unsafe.connect(remoteAddress, localAddress, promise);
}
```

图 17-14　HeadHandler 的 connect 方法

最终由 HeadHandler 调用 Unsafe 的 connect 方法发起真正的连接，pipeline 仅仅负责事件的调度。

17.3　ChannelHandler 功能说明

ChannelHandler 类似于 Servlet 的 Filter 过滤器，负责对 I/O 事件或者 I/O 操作进行拦截和处理，它可以选择性地拦截和处理自己感兴趣的事件，也可以透传和终止事件的传递。

基于 ChannelHandler 接口，用户可以方便地进行业务逻辑定制，例如打印日志、统一封装异常信息、性能统计和消息编解码等。

ChannelHandler 支持注解，目前支持的注解有两种。

- Sharable：多个 ChannelPipeline 共用同一个 ChannelHandler；
- Skip：被 Skip 注解的方法不会被调用，直接被忽略。

17.3.1　ChannelHandlerAdapter 功能说明

对于大多数的 ChannelHandler 会选择性地拦截和处理某个或者某些事件,其他的事件会忽略,由下一个 ChannelHandler 进行拦截和处理。这就会导致一个问题:用户 ChannelHandler 必须要实现 ChannelHandler 的所有接口,包括它不关心的那些事件处理接口,这会导致用户代码的冗余和臃肿,代码的可维护性也会变差。

为了解决这个问题,Netty 提供了 ChannelHandlerAdapter 基类,它的所有接口实现都是事件透传,如果用户 ChannelHandler 关心某个事件,只需要覆盖 ChannelHandlerAdapter 对应的方法即可,对于不关心的,可以直接继承使用父类的方法,这样子类的代码就会非常简洁和清晰。前面几章样例代码中,我们的 ChannelHandler 都是直接继承自 ChannelHandler Adapter,开发起来非常简单和高效。

ChannelHandlerAdapter 相关的代码实现如图 17-15 所示。

```
@Skip
@Override
public void read(ChannelHandlerContext ctx) throws Exception {
    ctx.read();
}
@Skip
@Override
public void write(ChannelHandlerContext ctx, Object msg, ChannelPromise promise) throws Exception {
    ctx.write(msg, promise);
}
```

图 17-15　ChannelHandlerAdapter 源码

从图 17-15 的源码中我们发现这些透传方法被@Skip 注解了,这些方法在执行的过程中会被忽略,直接跳到下一个 ChannelHandler 中执行对应的方法。

17.3.2　ByteToMessageDecoder 功能说明

利用 NIO 进行网络编程时,往往需要将读取到的字节数组或者字节缓冲区解码为业务可以使用的 POJO 对象。为了方便业务将 ByteBuf 解码成业务 POJO 对象,Netty 提供了 ByteToMessageDecoder 抽象工具解码类。

用户的解码器继承 ByteToMessageDecoder,只需要实现 void decode(ChannelHandler Context ctx, ByteBuf in, List<Object> out)抽象方法即可完成 ByteBuf 到 POJO 对象的解码。

由于 ByteToMessageDecoder 并没有考虑 TCP 粘包和组包等场景，读半包需要用户解码器自己负责处理。正因为如此，对于大多数场景不会直接继承 ByteToMessageDecoder，而是继承另外一些更高级的解码器来屏蔽半包的处理，下面的小节我们会对它们进行一一介绍。

17.3.3 MessageToMessageDecoder 功能说明

MessageToMessageDecoder 实际上是 Netty 的二次解码器，它的职责是将一个对象二次解码为其他对象。

为什么称它为二次解码器呢？我们知道，从 SocketChannel 读取到的 TCP 数据报是 ByteBuffer，实际就是字节数组，我们首先需要将 ByteBuffer 缓冲区中的数据报读取出来，并将其解码为 Java 对象；然后对 Java 对象根据某些规则做二次解码，将其解码为另一个 POJO 对象。因为 MessageToMessageDecoder 在 ByteToMessageDecoder 之后，所以称之为二次解码器。

二次解码器在实际的商业项目中非常有用，以 HTTP+XML 协议栈为例，第一次解码往往是将字节数组解码成 HttpRequest 对象，然后对 HttpRequest 消息中的消息体字符串进行二次解码，将 XML 格式的字符串解码为 POJO 对象，这就用到了二次解码器。类似这样的场景还有很多，不再一一枚举。

事实上，做一个超级复杂的解码器将多个解码器组合成一个大而全的 MessageToMessageDecoder 解码器似乎也能解决多次解码的问题，但是采用这种方式的代码可维护性会非常差。例如，如果我们打算在 HTTP+XML 协议栈中增加一个打印码流的功能，即首次解码获取 HttpRequest 对象之后打印 XML 格式的码流。如果采用多个解码器组合，在中间插入一个打印消息体的 Handler 即可，不需要修改原有的代码；如果做一个大而全的解码器，就需要在解码的方法中增加打印码流的代码，可扩展性和可维护性都会变差。

用户的解码器只需要实现 void decode(ChannelHandlerContext ctx, I msg, List<Object> out)抽象方法即可，由于它是将一个 POJO 解码为另一个 POJO，所以一般不会涉及到半包的处理，相对于 ByteToMessageDecoder 更加简单些。

17.3.4 LengthFieldBasedFrameDecoder 功能说明

在编解码章节我们讲过 TCP 的粘包导致解码的时候需要考虑如何处理半包的问题，

第 17 章 ChannelPipeline 和 ChannelHandler

前面介绍了 Netty 提供的半包解码器 LineBasedFrameDecoder 和 DelimiterBased FrameDecoder，现在我们继续学习第三种最通用的半包解码器——LengthFieldBasedFrameDecoder。

如何区分一个整包消息，通常有如下 4 种做法。

- ◎ 固定长度，例如每 120 个字节代表一个整包消息，不足的前面补零。解码器在处理这类定常消息的时候比较简单，每次读到指定长度的字节后再进行解码。
- ◎ 通过回车换行符区分消息，例如 FTP 协议。这类区分消息的方式多用于文本协议。
- ◎ 通过分隔符区分整包消息。
- ◎ 通过指定长度来标识整包消息。

如果消息是通过长度进行区分的，LengthFieldBasedFrameDecoder 都可以自动处理粘包和半包问题，只需要传入正确的参数，即可轻松搞定"读半包"问题。

下面我们看看如何通过参数组合的不同来实现不同的"半包"读取策略。第一种常用的方式是消息的第一个字段是长度字段，后面是消息体，消息头中只包含一个长度字段。它的消息结构定义如图 17-16 所示。

```
+--------+----------------+
| Length | Actual Content |
| 0x000C | "HELLO, WORLD" |
+--------+----------------+
```

图 17-16 解码前的字节缓冲区（14 字节）

使用以下参数组合进行解码。

- ◎ lengthFieldOffset = 0；
- ◎ lengthFieldLength = 2；
- ◎ lengthAdjustment = 0；
- ◎ initialBytesToStrip = 0。

解码后的字节缓冲区内容如图 17-17 所示。

图 17-17 包含消息长度字段（14 字节）

因为通过 ByteBuf.readableBytes()方法我们可以获取当前消息的长度，所以解码后的字节缓冲区可以不携带长度字段，由于长度字段在起始位置并且长度为 2，所以将 initialBytesToStrip 设置为 2，参数组合修改为：

◎ lengthFieldOffset = 0；

◎ lengthFieldLength = 2；

◎ lengthAdjustment = 0；

◎ initialBytesToStrip = 2。

解码后的字节缓冲区内容如图 17-18 所示。

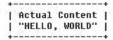

图 17-18　仅包含消息体（12 字节）

从图 17-18 的解码结果看，解码后的字节缓冲区丢弃了长度字段，仅仅包含消息体，不过通过 ByteBuf.readableBytes()方法仍然能够获取到长度字段的值。

在大多数的应用场景中，长度仅用来标识消息体的长度，这类协议通常由消息长度字段+消息体组成，如图 17-18 所示的例子。但是，对于一些协议，长度还包含了消息头的长度。在这种应用场景中，往往需要使用 lengthAdjustment 进行修正，修正后的参数组合方式如下。由于整个消息的长度往往都大于消息体的长度，所以，lengthAdjustment 为负数，图 17-19 展示了通过指定 lengthAdjustment 字段来包含消息头的长度。

◎ lengthFieldOffset = 0；

◎ lengthFieldLength = 2；

◎ lengthAdjustment = -2；

◎ initialBytesToStrip = 0。

```
解码前 (14 bytes)                      解码后 (14 bytes)
+--------+----------------+      +--------+----------------+
| Length | Actual Content |----->| Length | Actual Content |
| 0x000E | "HELLO, WORLD" |      | 0x000E | "HELLO, WORLD" |
+--------+----------------+      +--------+----------------+
```

图 17-19　包含消息头长度的解码

第 17 章　ChannelPipeline 和 ChannelHandler

由于协议种类繁多，并不是所有的协议都将长度字段放在消息头的首位，当标识消息长度的字段位于消息头的中间或者尾部时，需要使用 lengthFieldOffset 字段进行标识，下面的参数组合给出了如何解决消息长度字段不在首位的问题。

- ◎ lengthFieldOffset = 2；
- ◎ lengthFieldLength = 3；
- ◎ lengthAdjustment = 0；
- ◎ initialBytesToStrip = 0。

```
解码前 (17 bytes)                              解码后 (17 bytes)
+----------+----------+----------------+     +----------+----------+----------------+
| Header 1 |  Length  | Actual Content |---->| Header 1 |  Length  | Actual Content |
|  0xCAFE  | 0x00000C | "HELLO, WORLD" |     |  0xCAFE  | 0x00000C | "HELLO, WORLD" |
+----------+----------+----------------+     +----------+----------+----------------+
```

图 17-20　通过定义长度偏移量解决长度字段不在首位的问题

由于消息头 1 的长度为 2，所以长度字段的偏移量为 2；消息长度字段 Length 为 3，所以 lengthFieldLength 值为 3。由于长度字段仅仅标识消息体的长度，所以 lengthAdjustment 和 initialBytesToStrip 都为 0。

最后一种场景是长度字段夹在两个消息头之间或者长度字段位于消息头的中间，前后都有其他消息头字段，在这种场景下如果想忽略长度字段以及其前面的其他消息头字段，则可以通过 initialBytesToStrip 参数来跳过要忽略的字节长度，它的组合效果如下。

- ◎ lengthFieldOffset = 1；
- ◎ lengthFieldLength = 2；
- ◎ lengthAdjustment = 1；
- ◎ initialBytesToStrip = 3。

```
解码前(16 bytes)                              解码后 (13 bytes)
+------+--------+------+----------------+     +------+----------------+
| HDR1 | Length | HDR2 | Actual Content |---->| HDR2 | Actual Content |
| 0xCA | 0x000C | 0xFE | "HELLO, WORLD" |     | 0xFE | "HELLO, WORLD" |
+------+--------+------+----------------+     +------+----------------+
```

图 17-21　initialBytesToStrip 参数的使用

首先，由于 HDR1 的长度为 1，所以长度字段的偏移量 lengthFieldOffset 为 1；长度字段为 2 个字节，所以 lengthFieldLength 为 2。由于长度字段是消息体的长度，解码后如果携带消息头中的字段，则需要使用 lengthAdjustment 进行调整，此处它的值为 1，表示的是 HDR2 的长度，最后由于解码后的缓冲区要忽略长度字段和 HDR1 部分，所以 lengthAdjustment 为 3。解码后的结果为 13 个字节，HDR1 和 Length 字段被忽略。

事实上，通过 4 个参数的不同组合，可以达到不同的解码效果，用户在使用过程中可以根据业务的实际情况进行灵活调整。

由于 TCP 存在粘包和组包问题，所以通常情况下必须自己处理半包消息。利用 LengthFieldBasedFrameDecoder 解码器可以自动解决半包问题，它通常的用法如下。

```
pipeline.addLast("frameDecoder", new LineBasedFrameDecoder(80));
pipeline.addLast("stringDecoder", new StringDecoder(CharsetUtil.UTF_8));
```

在 pipeline 中增加 LineBasedFrameDecoder 解码器，指定正确的参数组合，它可以将 Netty 的 ByteBuf 解码成单个的整包消息，后面的业务解码器拿到的就是个完整的数据报，正常进行解码即可，不再需要额外考虑"读半包"问题，方便了业务消息的解码。

17.3.5　MessageToByteEncoder 功能说明

MessageToByteEncoder 负责将 POJO 对象编码成 ByteBuf，用户的编码器继承 MessageToByteEncoder，实现 void encode(ChannelHandlerContext ctx, I msg, ByteBuf out) 接口接口，示例代码如下。

```
public class IntegerEncoder extends MessageToByteEncoder<Integer> {
    @Override
    public void encode(ChannelHandlerContext ctx, Integer msg, ByteBuf out)
      throws Exception {
        out.writeInt(msg);
    }
}
```

17.3.6　MessageToMessageEncoder 功能说明

将一个 POJO 对象编码成另一个对象，以 HTTP+XML 协议为例，它的一种实现方式是：先将 POJO 对象编码成 XML 字符串，再将字符串编码为 HTTP 请求或者应答消息。对于复杂协议，往往需要经历多次编码，为了便于功能扩展，可以通过多个编码器组合来实现相关功能。

用户的解码器继承 MessageToMessageEncoder 解码器，实现 void encode(ChannelHandlerContext ctx, I msg, List<Object> out)方法即可。注意，它与 MessageToByteEncoder 的区别是输出是对象列表而不是 ByteBuf，示例代码如下。

```
public class IntegerToStringEncoder extends MessageToMessageEncoder <Integer>
{
    @Override
    public void encode(ChannelHandlerContext ctx, Integer message,
      List<Object> out)
          throws Exception
    {
        out.add(message.toString());
    }
}
```

17.3.7　LengthFieldPrepender 功能说明

如果协议中的第一个字段为长度字段，Netty 提供了 LengthFieldPrepender 编码器，它可以计算当前待发送消息的二进制字节长度，将该长度添加到 ByteBuf 的缓冲区头中，如图 17-22 所示。

```
编码前(12 bytes)              编码后(14 bytes)
+----------------+            +--------+----------------+
| "HELLO, WORLD" |  ----->    + 0x000C | "HELLO, WORLD" |
+----------------+            +--------+----------------+
```

图 17-22　LengthFieldPrepender 编码器

通过 LengthFieldPrepender 可以将待发送消息的长度写入到 ByteBuf 的前 2 个字节，编码后的消息组成为长度字段+原消息的方式。

通过设置 LengthFieldPrepender 为 true，消息长度将包含长度本身占用的字节数，打开 LengthFieldPrepender 后，图 17-22 示例中的编码结果如图 17-23 所示。

```
编码前(12 bytes)              编码后(14 bytes)
+----------------+            +--------+----------------+
| "HELLO, WORLD" |  ----->    + 0x000E | "HELLO, WORLD" |
+----------------+            +--------+----------------+
```

图 17-23　打开 LengthFieldPrepender 开关后的编码结果

17.4 ChannelHandler 源码分析

17.4.1 ChannelHandler 的类继承关系图

相对于 ByteBuf 和 Channel，ChannelHandler 的类继承关系稍微简单些，但是它的子类非常多。由于 ChannelHandler 是 Netty 框架和用户代码的主要扩展和定制点，所以它的子类种类繁多、功能各异，系统 ChannelHandler 主要分类如下。

◎ ChannelPipeline 的系统 ChannelHandler，用于 I/O 操作和对事件进行预处理，对于用户不可见，这类 ChannelHandler 主要包括 HeadHandler 和 TailHandler；

◎ 编解码 ChannelHandler，包括 ByteToMessageCodec、MessageToMessageDecoder 等，这些编解码类本身又包含多种子类，如图 17-24 所示。

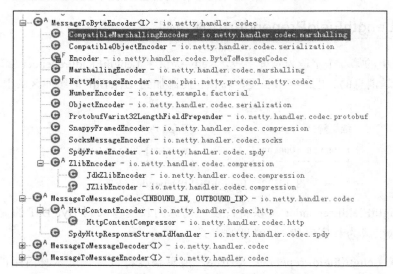

图 17-24　编解码 ChannelHandler

◎ 其他系统功能性 ChannelHandler，包括流量整型 Handler、读写超时 Handler、日志 Handler 等。

本章节仅给出讲解到的编解码类，其他不再一一枚举。

第 17 章　ChannelPipeline 和 ChannelHandler

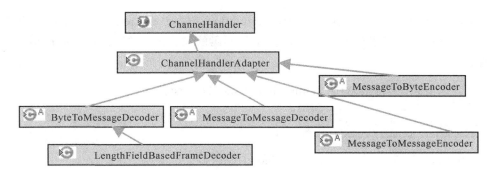

图 17-24　编解码 ChannelHandler 子类继承关系图

17.4.2　ByteToMessageDecoder 源码分析

顾名思义，ByteToMessageDecoder 解码器用于将 ByteBuf 解码成 POJO 对象，下面一起看它的实现。

首先看 channelRead 方法的源码，如图 17-25 所示。

```
@Override
public void channelRead(ChannelHandlerContext ctx, Object msg) throws Exception {
    if (msg instanceof ByteBuf) {
        RecyclableArrayList out = RecyclableArrayList.newInstance();
        try {
            ByteBuf data = (ByteBuf) msg;
            first = cumulation == null;
            if (first) {
                cumulation = data;
            } else {
                if (cumulation.writerIndex() > cumulation.maxCapacity() - data.readableBytes()) {
                    expandCumulation(ctx, data.readableBytes());
                }
                cumulation.writeBytes(data);
                data.release();
            }
            callDecode(ctx, cumulation, out);
```

图 17-25　ByteToMessageDecoder 的 channelRead 方法

首先判断需要解码的 msg 对象是否是 ByteBuf，如果是 ByteBuf 才需要进行解码，否则直接透传。

通过 cumulation 是否为空判断解码器是否缓存了没有解码完成的半包消息，如果为空，说明是首次解码或者最近一次已经处理完了半包消息，没有缓存的半包消息需要处理，直接将需要解码的 ByteBuf 赋值给 cumulation；如果 cumulation 缓存有上次没有解码完成的 ByteBuf，则进行复制操作，将需要解码的 ByteBuf 复制到 cumulation 中，它的原理如下。

半包解码前：（半包消息 1= cumulation.readableBytes()）

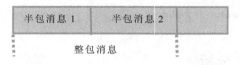

半包解码后：（半包消息 2= msg.readableBytes()）

在复制之前需要对 cumulation 的可写缓冲区进行判断，如果不足则需要动态扩展，扩展的代码如图 17-26 所示。

```
private void expandCumulation(ChannelHandlerContext ctx, int readable) {
    ByteBuf oldCumulation = cumulation;
    cumulation = ctx.alloc().buffer(oldCumulation.readableBytes() + readable);
    cumulation.writeBytes(oldCumulation);
    oldCumulation.release();
}
```

图 17-26　ByteToMessageDecoder 的 expandCumulation 方法

扩展的代码很简单，利用字节缓冲区分配器重新分配一个新的 ByteBuf，将老的 cumulation 复制到新的 ByteBuf 中，释放 cumulation。需要注意的是，此处内存扩展没有采用倍增或者步进的方式，分配的缓冲区恰恰够用，此处的算法可以优化下，以防止连续半包导致的频繁缓冲区扩张和内存复制。

复制操作完成之后释放需要解码的 ByteBuf 对象，调用 callDecode 方法进行解码，代码如图 17-27 所示。

对 ByteBuf 进行循环解码，循环的条件是解码缓冲区对象中有可读的字节，调用抽象 decode 方法，由用户的子类解码器进行解码，方法定义如图 17-28 所示。

解码后需要对当前的 pipeline 状态和解码结果进行判断，代码如图 17-29 所示。

第 17 章　ChannelPipeline 和 ChannelHandler

```
protected void callDecode(ChannelHandlerContext ctx, ByteBuf in,
List<Object> out) {
    try {
        while (in.isReadable()) {
            int outSize = out.size();
            int oldInputLength = in.readableBytes();
            decode(ctx, in, out);
            if (ctx.isRemoved()) {
                break;
            }
```

图 17-27　ByteToMessageDecoder 的 callDecode 方法

```
protected abstract void decode(ChannelHandlerContext ctx, ByteBuf in,
List<Object> out) throws Exception;
```

图 17-28　ByteToMessageDecoder 的 decode 抽象方法

```
if (outSize == out.size()) {
    if (oldInputLength == in.readableBytes()) {
        break;
    } else {
        continue;
    }
}
if (oldInputLength == in.readableBytes()) {
    throw new DecoderException(
        StringUtil.simpleClassName(getClass()) +
        ".decode() did not read anything but decoded a message.");
}
if (isSingleDecode()) {
    break;
}
```

图 17-29　ByteToMessageDecoder 的 callDecode 方法

如果当前的 ChannelHandlerContext 已经被移除，则不能继续进行解码，直接退出循环；如果输出的 out 列表长度没变化，说明解码没有成功，需要针对以下不同场景进行判断。

（1）如果用户解码器没有消费 ByteBuf，则说明是个半包消息，需要由 I/O 线程继续读取后续的数据报，在这种场景下要退出循环。

（2）如果用户解码器消费了 ByteBuf，说明可以解码可以继续进行。

从图 17-29 所示代码可以看出，业务解码器需要遵守 Netty 的某些契约，解码器才能正常工作，否则可能会导致功能错误，最重要的契约就是：如果业务解码器认为当前的字节缓冲区无法完成业务层的解码，需要将 readIndex 复位，告诉 Netty 解码条件不满足应当退出解码，继续读取数据报。

（3）如果用户解码器没有消费 ByteBuf，但是却解码出了一个或者多个对象，这种行为被认为是非法的，需要抛出 DecoderException 异常。

（4）最后通过 isSingleDecode 进行判断，如果是单条消息解码器，第一次解码完成之后就退出循环。

17.4.3　MessageToMessageDecoder 源码分析

MessageToMessageDecoder 负责将一个 POJO 对象解码成另一个 POJO 对象，下面一起看下它的源码实现。

首先看 channelRead 方法的源码，如图 17-30 所示。

先通过 RecyclableArrayList 创建一个新的可循环利用的 RecyclableArrayList，然后对解码的消息类型进行判断，通过类型参数校验器看是否是可接收的类型，如果是则校验通过，参数类型校验的代码如图 17-31 所示。

校验通过之后，直接调用 decode 抽象方法，由具体实现子类进行消息解码，解码抽象方法定义如图 17-32 所示。

解码完成之后，调用 ReferenceCountUtil 的 release 方法来释放被解码的 msg 对象。

如果需要解码的对象不是当前解码器可以接收和处理的类型，则将它加入到 RecyclableArrayList 中不进行解码。

```
@Override
public void channelRead(ChannelHandlerContext ctx, Object msg) throws Exception {
    RecyclableArrayList out = RecyclableArrayList.newInstance();
    try {
        if (acceptInboundMessage(msg)) {
            @SuppressWarnings("unchecked")
            I cast = (I) msg;
            try {
                decode(ctx, cast, out);
            } finally {
                ReferenceCountUtil.release(cast);
            }
        } else {
            out.add(msg);
        }
    } catch (DecoderException e) {
        throw e;
    } catch (Exception e) {
        throw new DecoderException(e);
    } finally {
        int size = out.size();
        for (int i = 0; i < size; i ++) {
            ctx.fireChannelRead(out.get(i));
        }
        out.recycle();
    }
}
```

图 17-30　MessageToMessageDecoder 的 channelRead 方法

```
public boolean acceptInboundMessage(Object msg) throws Exception {
    return matcher.match(msg);
}
```

图 17-31　MessageToMessageDecoder 的参数校验

```
protected abstract void decode(ChannelHandlerContext ctx, I msg, List<Object> out) throws Exception;
```

图 17-32　MessageToMessageDecoder 的抽象 decode 方法定义

最后，对 RecyclableArrayList 进行遍历，循环调用 ChannelHandlerContext 的 fireChannelRead 方法，通知后续的 ChannelHandler 继续进行处理。循环通知完成之后，通过 recycle 方法释放 RecyclableArrayList 对象。

17.4.4　LengthFieldBasedFrameDecoder 源码分析

本节我们一起来学习最通用和重要的解码器——基于消息长度的半包解码器，首先看它的入口方法，源码如图 17-33 所示。

```
@Override
protected final void decode(ChannelHandlerContext ctx, ByteBuf in, List<Object> out) throws Exception {
    Object decoded = decode(ctx, in);
    if (decoded != null) {
        out.add(decoded);
    }
}
```

图 17-33　LengthFieldBasedFrameDecoder 的 decode 方法

调用内部的 decode(ChannelHandlerContext ctx, ByteBuf in)方法，如果解码成功，将其加入到输出的 List<Object> out 列表中。

下面继续看 decode(ChannelHandlerContext ctx, ByteBuf in)的实现，如图 17-34 所示。

```
protected Object decode(ChannelHandlerContext ctx, ByteBuf in) throws Exception {
    if (discardingTooLongFrame) {
        long bytesToDiscard = this.bytesToDiscard;
        int localBytesToDiscard = (int) Math.min(bytesToDiscard, in.readableBytes());
        in.skipBytes(localBytesToDiscard);
        bytesToDiscard -= localBytesToDiscard;
        this.bytesToDiscard = bytesToDiscard;

        failIfNecessary(false);
    }
```

图 17-34　LengthFieldBasedFrameDecoder 的 decode 方法片段 1

判断 discardingTooLongFrame 标识，看是否需要丢弃当前可读的字节缓冲区，如果为真，则执行丢弃操作，具体如下。

判断需要丢弃的字节长度，由于丢弃的字节数不能大于当前缓冲区可读的字节数，所以需要通过 Math.min(bytesToDiscard, in.readableBytes())函数进行选择，取 bytesToDiscard 和缓冲区可读字节数之中的最小值。计算获取需要丢弃的字节数之后，调用 ByteBuf 的 skipBytes 方法跳过需要忽略的字节长度，然后 bytesToDiscard 减去已经忽略的字节长度。最后判断是否已经达到需要忽略的字节数，达到的话对 discardingTooLongFrame 等进行置位，代码如图 17-35 所示。

```
private void failIfNecessary(boolean firstDetectionOfTooLongFrame) {
    if (bytesToDiscard == 0) {
        // Reset to the initial state and tell the handlers that
        // the frame was too large.
        long tooLongFrameLength = this.tooLongFrameLength;
        this.tooLongFrameLength = 0;
        discardingTooLongFrame = false;
        if (!failFast ||
            failFast && firstDetectionOfTooLongFrame) {
            fail(tooLongFrameLength);
        }
    } else {
        // Keep discarding and notify handlers if necessary.
        if (failFast && firstDetectionOfTooLongFrame) {
            fail(tooLongFrameLength);
        }
    }
}
```

图 17-35　LengthFieldBasedFrameDecoder 的 failIfNecessary 方法

对当前缓冲区的可读字节数和长度偏移量进行对比，如果小于长度偏移量，则说明当前缓冲区的数据报不够，需要返回空，由 I/O 线程继续读取后续的数据报。如图 17-36 所示。

```
if (in.readableBytes() < lengthFieldEndOffset) {
    return null;
}

int actualLengthFieldOffset = in.readerIndex() + lengthFieldOffset;
long frameLength = getUnadjustedFrameLength(in,
actualLengthFieldOffset, lengthFieldLength, byteOrder);
```

图 17-36　LengthFieldBasedFrameDecoder 的 decode 方法片段 2

通过读索引和 lengthFieldOffset 计算获取实际的长度字段索引，然后通过索引值获取消息报文的长度字段，代码如图 17-37 所示。

根据长度字段自身的字节长度进行判断，共有以下 6 种可能的取值。

◎ 长度所占字节为 1，通过 ByteBuf 的 getUnsignedByte 方法获取长度值；

◎ 长度所占字节为 2，通过 ByteBuf 的 getUnsignedShort 方法获取长度值；

```
protected long getUnadjustedFrameLength(ByteBuf buf, int offset, int length,
ByteOrder order) {
    buf = buf.order(order);
    long frameLength;
    switch (length) {
    case 1:
        frameLength = buf.getUnsignedByte(offset);
        break;
    case 2:
        frameLength = buf.getUnsignedShort(offset);
        break;
    case 3:
        frameLength = buf.getUnsignedMedium(offset);
        break;
    case 4:
        frameLength = buf.getUnsignedInt(offset);
        break;
    case 8:
        frameLength = buf.getLong(offset);
        break;
    default:
        throw new DecoderException(
            "unsupported lengthFieldLength: " + lengthFieldLength + " (expected: 1, 2, 3, 4, or 8)");
    }
    return frameLength;
}
```

图 17-37　LengthFieldBasedFrameDecoder 的 getUnadjustedFrameLength 方法

◎ 长度所占字节为 3，通过 ByteBuf 的 getUnsignedMedium 方法获取长度值；

◎ 长度所占字节为 4，通过 ByteBuf 的 getUnsignedInt 方法获取长度值；

◎ 长度所占字节为 8，通过 ByteBuf 的 getLong 方法获取长度值；

◎ 其他长度不支持，抛出 DecoderException 异常。

获取长度之后，就需要对长度进行合法性判断，同时根据其他解码参数进行长度调整，代码如图 17-38 所示。

```
if (frameLength < 0) {
    in.skipBytes(lengthFieldEndOffset);
    throw new CorruptedFrameException(
        "negative pre-adjustment length field: " + frameLength);
}
frameLength += lengthAdjustment + lengthFieldEndOffset;
if (frameLength < lengthFieldEndOffset) {
    in.skipBytes(lengthFieldEndOffset);
    throw new CorruptedFrameException(
        "Adjusted frame length (" + frameLength + ") is less " +
        "than lengthFieldEndOffset: " + lengthFieldEndOffset);
}
```

图 17-38 LengthFieldBasedFrameDecoder 的 decode 方法片段 3

如果长度小于 0，说明报文非法，跳过 lengthFieldEndOffset 个字节，抛出 CorruptedFrameException 异常。

根据 lengthFieldEndOffset 和 lengthAdjustment 字段进行长度修正，如果修正后的报文长度小于 lengthFieldEndOffset，则说明是非法数据报，需要抛出 CorruptedFrameException 异常。

如果修正后的报文长度大于 ByteBuf 的最大容量，说明接收到的消息长度大于系统允许的最大长度上限，需要设置 discardingTooLongFrame，计算需要丢弃的字节数，根据情况选择是否需要抛出解码异常。

丢弃的策略如下：frameLength 减去 ByteBuf 的可读字节数就是需要丢弃的字节长度，如果需要丢弃的字节数 discard 小于缓冲区可读的字节数，则直接丢弃整包消息。如果需要丢弃的字节数大于当前的可读字节数，说明即便将当前所有可读的字节数全部丢弃，也无法完成任务，则设置 discardingTooLongFrame 标识为 true，下次解码的时候继续丢弃。丢弃操作完成之后，调用 failIfNecessary 方法根据实际情况抛出异常。如图 17-39 所示。

```
int frameLengthInt = (int) frameLength;
if (in.readableBytes() < frameLengthInt) {
    return null;
}
if (initialBytesToStrip > frameLengthInt) {
    in.skipBytes(frameLengthInt);
    throw new CorruptedFrameException(
        "Adjusted frame length (" + frameLength + ") is less " +
        "than initialBytesToStrip: " + initialBytesToStrip);
}
in.skipBytes(initialBytesToStrip);

// extract frame
int readerIndex = in.readerIndex();

int actualFrameLength = frameLengthInt - initialBytesToStrip;

ByteBuf frame = extractFrame(ctx, in, readerIndex, actualFrameLength);

in.readerIndex(readerIndex + actualFrameLength);

return frame;
```

图 17-39 LengthFieldBasedFrameDecoder 的 decode 方法片段 4

如果当前的可读字节数小于 frameLength，说明是个半包消息，需要返回空，由 I/O 线程继续读取后续的数据报，等待下次解码。

对需要忽略的消息头字段进行判断，如果大于消息长度 frameLength，说明码流非法，需要忽略当前的数据报，抛出 CorruptedFrameException 异常。通过 ByteBuf 的 skipBytes 方法忽略消息头中不需要的字段，得到整包 ByteBuf。

通过 extractFrame 方法获取解码后的整包消息缓冲区，代码如图 17-40 所示。

```
protected ByteBuf extractFrame(ChannelHandlerContext ctx, ByteBuf buffer, int index, int length) {
    ByteBuf frame = ctx.alloc().buffer(length);
    frame.writeBytes(buffer, index, length);
    return frame;
}
```

图 17-40　LengthFieldBasedFrameDecoder 的 extractFrame 方法

根据消息的实际长度分配一个新的 ByteBuf 对象，将需要解码的 ByteBuf 可写缓冲区复制到新创建的 ByteBuf 中并返回，返回之后更新原解码缓冲区 ByteBuf 为原读索引+消息报文的实际长度（actualFrameLength）。

至此，基于长度的半包解码器介绍完毕，对于使用者而言，实际不需要对 LengthFieldBasedFrameDecoder 进行定制。只需要了解每个参数的用法，再结合用户的业务场景进行参数设置，即可实现半包消息的自动解码，后面的业务解码器得到的是个完整的整包消息，不用再额外考虑如何处理半包。这极大地降低了开发难度，提升了开发效率。

17.4.5　MessageToByteEncoder 源码分析

MessageToByteEncoder 负责将用户的 POJO 对象编码成 ByteBuf，以便通过网络进行传输。下面一起看它的源码实现，如图 17-41 所示。

```
@Override
public void write(ChannelHandlerContext ctx, Object msg, ChannelPromise promise) throws Exception {
    ByteBuf buf = null;
    try {
        if (acceptOutboundMessage(msg)) {
            @SuppressWarnings("unchecked")
            I cast = (I) msg;
            if (preferDirect) {
                buf = ctx.alloc().ioBuffer();
            } else {
                buf = ctx.alloc().heapBuffer();
            }
            try {
                encode(ctx, cast, buf);
            } finally {
                ReferenceCountUtil.release(cast);
            }
```

图 17-41　MessageToByteEncoder 的 write 方法片段 1

首先判断当前编码器是否支持需要发送的消息,如果不支持则直接透传;如果支持则判断缓冲区的类型,对于直接内存分配 ioBuffer(堆外内存),对于堆内存通过 heapBuffer 方法分配。

编码使用的缓冲区分配完成之后,调用 encode 抽象方法进行编码,方法定义如图 17-42 所示。

```
protected abstract void encode(ChannelHandlerContext ctx, I msg, ByteBuf out)
throws Exception;
```

图 17-42　MessageToByteEncoder 的抽象 encode 方法

编码完成之后,调用 ReferenceCountUtil 的 release 方法释放编码对象 msg。对编码后的 ByteBuf 进行以下判断。

◎ 如果缓冲区包含可发送的字节,则调用 ChannelHandlerContext 的 write 方法发送 ByteBuf;

◎ 如果缓冲区没有包含可写的字节,则需要释放编码后的 ByteBuf,写入一个空的 ByteBuf 到 ChannelHandlerContext 中。

发送操作完成之后,在方法退出之前释放编码缓冲区 ByteBuf 对象。

17.4.6　MessageToMessageEncoder 源码分析

MessageToMessageEncoder 负责将一个 POJO 对象编码成另一个 POJO 对象,例如将 XML Document 对象编码成 XML 格式的字符串。下面一起看它的源码实现,如图 17-43 所示。

```
public void write(ChannelHandlerContext ctx, Object msg, ChannelPromise
promise) throws Exception {
    RecyclableArrayList out = null;
    try {
        if (acceptOutboundMessage(msg)) {
            out = RecyclableArrayList.newInstance();
            @SuppressWarnings("unchecked")
            I cast = (I) msg;
            try {
                encode(ctx, cast, out);
            } finally {
                ReferenceCountUtil.release(cast);
            }
            if (out.isEmpty()) {
                out.recycle();
                out = null;
                throw new EncoderException(
                    StringUtil.simpleClassName(this) + " must produce at least one message.");
            } else {
                ctx.write(msg, promise);
            }
```

图 17-43　MessageToMessageEncoder 的抽象 write 方法

第 17 章 ChannelPipeline 和 ChannelHandler

与之前的编码器类似，创建 RecyclableArrayList 对象，判断当前需要编码的对象是否是编码器可处理的类型，如果不是，则忽略，执行下一个 ChannelHandler 的 write 方法。

具体的编码方法实现由用户子类编码器负责完成，如果编码后的 RecyclableArrayList 为空，说明编码没有成功，释放 RecyclableArrayList 引用。

如果编码成功，则通过遍历 RecyclableArrayList，循环发送编码后的 POJO 对象，代码如图 17-44 所示。

```
finally {
    if (out != null) {
        final int sizeMinusOne = out.size() - 1;
        if (sizeMinusOne >= 0) {
            for (int i = 0; i < sizeMinusOne; i ++) {
                ctx.write(out.get(i));
            }
            ctx.write(out.get(sizeMinusOne), promise);
        }
        out.recycle();
    }
}
```

图 17-44　循环发送编码后的 POJO 对象

17.4.7　LengthFieldPrepender 源码分析

LengthFieldPrepender 负责在待发送的 ByteBuf 消息头中增加一个长度字段来标识消息的长度，它简化了用户的编码器开发，使用户不需要额外去设置这个长度字段。下面我们来看下它的实现，如图 17-45 所示。

首先对长度字段进行设置，如果需要包含消息长度自身，则在原来长度的基础之上再加上 lengthFieldLength 的长度。

如果调整后的消息长度小于 0，则抛出参数非法异常。对消息长度自身所占的字节数进行判断，以便采用正确的方法将长度字段写入到 ByteBuf 中，共有以下 6 种可能。

- ◎ 长度字段所占字节为 1：如果使用 1 个 Byte 字节代表消息长度，则最大长度需要小于 256 个字节。对长度进行校验，如果校验失败，则抛出参数非法异常；若校验通过，则创建新的 ByteBuf 并通过 writeByte 将长度值写入到 ByteBuf 中。
- ◎ 长度字段所占字节为 2：如果使用 2 个 Byte 字节代表消息长度，则最大长度需要小于 65536 个字节，对长度进行校验，如果校验失败，则抛出参数非法异常；若校验通过，则创建新的 ByteBuf 并通过 writeShort 将长度值写入到 ByteBuf 中。

```
protected void encode(ChannelHandlerContext ctx, ByteBuf msg, List<Object>
out) throws Exception {
    int length = msg.readableBytes() + lengthAdjustment;
    if (lengthIncludesLengthFieldLength) {
        length += lengthFieldLength;
    }
    if (length < 0) {
        throw new IllegalArgumentException(
            "Adjusted frame length (" + length + ") is less than zero");
    }
    switch (lengthFieldLength) {
    case 1:
        if (length >= 256) {
            throw new IllegalArgumentException(
                "length does not fit into a byte: " + length);
        }
        out.add(ctx.alloc().buffer(1).writeByte((byte) length));
        break;
```

图 17-45　LengthFieldPrepender 的 encode 方法片段 1

◎ 长度字段所占字节为 3：如果使用 3 个 Byte 字节代表消息长度，则最大长度需要小于 16777216 个字节，对长度进行校验，如果校验失败，则抛出参数非法异常；若校验通过，则创建新的 ByteBuf 并通过 writeMedium 将长度值写入到 ByteBuf 中。

◎ 长度字段所占字节为 4：创建新的 ByteBuf，并通过 writeInt 将长度值写入到 ByteBuf 中。

◎ 长度字段所占字节为 8：创建新的 ByteBuf，并通过 writeLong 将长度值写入到 ByteBuf 中。

◎ 其他长度值：直接抛出 Error。

最后将原需要发送的 ByteBuf 复制到 List<Object> out 中，完成编码。

17.5　总结

本章介绍了 ChannelPipeline 和 ChannelHandler 的功能及原理，并给出了使用建议，指出了需要注意的细节。

最后，对 ChannelPipeline 和 ChannelHandler 的主要功能子类进行了源码分析。通过学习源码，相信读者不仅仅能学到 Netty 的一些高级用法，而且能够举一反三，通过按需扩展和功能定制来更好的满足业务的差异化需求。

第 18 章

EventLoop 和 EventLoopGroup

从本章开始我们将学习 Netty 的线程模型。Netty 框架的主要线程就是 I/O 线程，线程模型设计的好坏，决定了系统的吞吐量、并发性和安全性等架构质量属性。

Netty 的线程模型被精心地设计，既提升了框架的并发性能，又能在很大程度避免锁，局部实现了无锁化设计。从本章开始，我们将介绍 Netty 的线程模型，同时对它的 NIO 线程 NioEventLoop 进行详尽地源码分析，让读者能够学习到更多 I/O 相关的多线程设计原理和实现。

本章主要内容包括：

◎ Netty 的线程模型

◎ NioEventLoop 源码分析

18.1 Netty 的线程模型

当我们讨论 Netty 线程模型的时候，一般首先会想到的是经典的 Reactor 线程模型，尽管不同的 NIO 框架对于 Reactor 模式的实现存在差异，但本质上还是遵循了 Reactor 的基础线程模型。

下面让我们一起回顾经典的 Reactor 线程模型。

18.1.1 Reactor 单线程模型

Reactor 单线程模型，是指所有的 I/O 操作都在同一个 NIO 线程上面完成。NIO 线程的职责如下。

◎ 作为 NIO 服务端，接收客户端的 TCP 连接；

◎ 作为 NIO 客户端，向服务端发起 TCP 连接；

◎ 读取通信对端的请求或者应答消息；

◎ 向通信对端发送消息请求或者应答消息。

Reactor 单线程模型如图 18-1 所示。

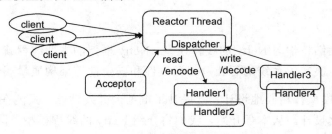

图 18-1　Reactor 单线程模型

由于 Reactor 模式使用的是异步非阻塞 I/O，所有的 I/O 操作都不会导致阻塞，理论上一个线程可以独立处理所有 I/O 相关的操作。从架构层面看，一个 NIO 线程确实可以完成其承担的职责。例如，通过 Acceptor 类接收客户端的 TCP 连接请求消息，当链路建立成功之后，通过 Dispatch 将对应的 ByteBuffer 派发到指定的 Handler 上，进行消息解码。用户线程消息编码后通过 NIO 线程将消息发送给客户端。

在一些小容量应用场景下，可以使用单线程模型。但是这对于高负载、大并发的应用场景却不合适，主要原因如下。

◎ 一个 NIO 线程同时处理成百上千的链路，性能上无法支撑，即便 NIO 线程的 CPU

负荷达到 100%，也无法满足海量消息的编码、解码、读取和发送。
- 当 NIO 线程负载过重之后，处理速度将变慢，这会导致大量客户端连接超时，超时之后往往会进行重发，这更加重了 NIO 线程的负载，最终会导致大量消息积压和处理超时，成为系统的性能瓶颈。
- 可靠性问题：一旦 NIO 线程意外跑飞，或者进入死循环，会导致整个系统通信模块不可用，不能接收和处理外部消息，造成节点故障。

为了解决这些问题，演进出了 Reactor 多线程模型。下面我们一起学习下 Reactor 多线程模型。

18.1.2 Reactor 多线程模型

Rector 多线程模型与单线程模型最大的区别就是有一组 NIO 线程来处理 I/O 操作，它的原理如图 18-2 所示。

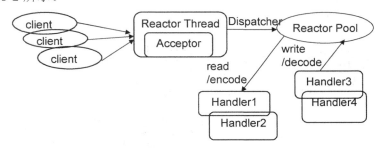

图 18-2 Reactor 多线程模型

Reactor 多线程模型的特点如下。

- 有专门一个 NIO 线程——Acceptor 线程用于监听服务端，接收客户端的 TCP 连接请求。
- 网络 I/O 操作——读、写等由一个 NIO 线程池负责，线程池可以采用标准的 JDK 线程池实现，它包含一个任务队列和 N 个可用的线程，由这些 NIO 线程负责消息的读取、解码、编码和发送。
- 一个 NIO 线程可以同时处理 N 条链路，但是一个链路只对应一个 NIO 线程，防止发生并发操作问题。

在绝大多数场景下，Reactor 多线程模型可以满足性能需求。但是，在个别特殊场景中，一个 NIO 线程负责监听和处理所有的客户端连接可能会存在性能问题。例如并发百万客户端连接，或者服务端需要对客户端握手进行安全认证，但是认证本身非常损耗性能。在这类场景下，单独一个 Acceptor 线程可能会存在性能不足的问题，为了解决性能问题，产生了第三种 Reactor 线程模型——主从 Reactor 多线程模型。

18.1.3 主从 Reactor 多线程模型

主从 Reactor 线程模型的特点是：服务端用于接收客户端连接的不再是一个单独的 NIO 线程，而是一个独立的 NIO 线程池。Acceptor 接收到客户端 TCP 连接请求并处理完成后（可能包含接入认证等），将新创建的 SocketChannel 注册到 I/O 线程池（sub reactor 线程池）的某个 I/O 线程上，由它负责 SocketChannel 的读写和编解码工作。Acceptor 线程池仅仅用于客户端的登录、握手和安全认证，一旦链路建立成功，就将链路注册到后端 subReactor 线程池的 I/O 线程上，由 I/O 线程负责后续的 I/O 操作。

它的线程模型如图 18-3 所示。

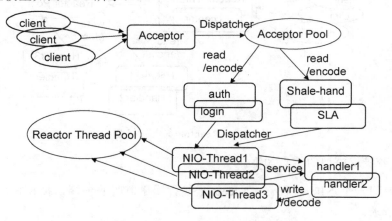

图 18-3 主从 Reactor 多线程模型

利用主从 NIO 线程模型，可以解决一个服务端监听线程无法有效处理所有客户端连接的性能不足问题。因此，在 Netty 的官方 Demo 中，推荐使用该线程模型。

18.1.4 Netty 的线程模型

Netty 的线程模型并不是一成不变的，它实际取决于用户的启动参数配置。通过设置不同的启动参数，Netty 可以同时支持 Reactor 单线程模型、多线程模型和主从 Reactor 多线层模型。

下面让我们通过一张原理图（图 18-4）来快速了解 Netty 的线程模型。

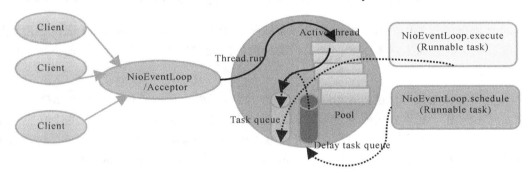

图 18-4　Netty 的线程模型

可以通过如图 18-5 所示的 Netty 服务端启动代码来了解它的线程模型。

```
// Configure the server.
    EventLoopGroup bossGroup = new NioEventLoopGroup();
    EventLoopGroup workerGroup = new NioEventLoopGroup();
    try {
        ServerBootstrap b = new ServerBootstrap();
        b.group(bossGroup, workerGroup)
         .channel(NioServerSocketChannel.class)
         .option(ChannelOption.SO_BACKLOG, 100)
         .handler(new LoggingHandler(LogLevel.INFO))
         .childHandler(new ChannelInitializer<SocketChannel>() {
```

图 18-5　Netty 服务端启动

服务端启动的时候，创建了两个 NioEventLoopGroup，它们实际是两个独立的 Reactor 线程池。一个用于接收客户端的 TCP 连接，另一个用于处理 I/O 相关的读写操作，或者执行系统 Task、定时任务 Task 等。

Netty 用于接收客户端请求的线程池职责如下。

（1）接收客户端 TCP 连接，初始化 Channel 参数；

（2）将链路状态变更事件通知给 ChannelPipeline。

Netty 处理 I/O 操作的 Reactor 线程池职责如下。

（1）异步读取通信对端的数据报，发送读事件到 ChannelPipeline；

（2）异步发送消息到通信对端，调用 ChannelPipeline 的消息发送接口；

（3）执行系统调用 Task；

（4）执行定时任务 Task，例如链路空闲状态监测定时任务。

通过调整线程池的线程个数、是否共享线程池等方式，Netty 的 Reactor 线程模型可以在单线程、多线程和主从多线程间切换，这种灵活的配置方式可以最大程度地满足不同用户的个性化定制。

为了尽可能地提升性能，Netty 在很多地方进行了无锁化的设计，例如在 I/O 线程内部进行串行操作，避免多线程竞争导致的性能下降问题。表面上看，串行化设计似乎 CPU 利用率不高，并发程度不够。但是，通过调整 NIO 线程池的线程参数，可以同时启动多个串行化的线程并行运行，这种局部无锁化的串行线程设计相比一个队列—多个工作线程的模型性能更优。

它的设计原理如图 18-6 所示。

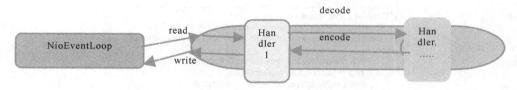

图 18-6　Netty Reactor 线程模型

Netty 的 NioEventLoop 读取到消息之后，直接调用 ChannelPipeline 的 fireChannelRead(Object msg)。只要用户不主动切换线程，一直都是由 NioEventLoop 调用用户的 Handler，期间不进行线程切换。这种串行化处理方式避免了多线程操作导致的锁的竞争，从性能角度看是最优的。

18.1.5　最佳实践

Netty 的多线程编程最佳实践如下。

（1）创建两个 NioEventLoopGroup，用于逻辑隔离 NIO Acceptor 和 NIO I/O 线程。

（2）尽量不要在 ChannelHandler 中启动用户线程（解码后用于将 POJO 消息派发到后端业务线程的除外）。

（3）解码要放在 NIO 线程调用的解码 Handler 中进行，不要切换到用户线程中完成消息的解码。

（4）如果业务逻辑操作非常简单，没有复杂的业务逻辑计算，没有可能会导致线程被阻塞的磁盘操作、数据库操作、网络操作等，可以直接在 NIO 线程上完成业务逻辑编排，不需要切换到用户线程。

（5）如果业务逻辑处理复杂，不要在 NIO 线程上完成，建议将解码后的 POJO 消息封装成 Task，派发到业务线程池中由业务线程执行，以保证 NIO 线程尽快被释放，处理其他的 I/O 操作。

推荐的线程数量计算公式有以下两种。

◎ 公式一：线程数量=（线程总时间/瓶颈资源时间）× 瓶颈资源的线程并行数。
◎ 公式二：QPS=1000/线程总时间×线程数。

由于用户场景的不同，对于一些复杂的系统，实际上很难计算出最优线程配置，只能是根据测试数据和用户场景，结合公式给出一个相对合理的范围，然后对范围内的数据进行性能测试，选择相对最优值。

18.2 NioEventLoop 源码分析

18.2.1 NioEventLoop 设计原理

Netty 的 NioEventLoop 并不是一个纯粹的 I/O 线程，它除了负责 I/O 的读写之外，还兼顾处理以下两类任务。

◎ 系统 Task：通过调用 NioEventLoop 的 execute(Runnable task)方法实现，Netty 有很多系统 Task，创建它们的主要原因是：当 I/O 线程和用户线程同时操作网络资源时，为了防止并发操作导致的锁竞争，将用户线程的操作封装成 Task 放入消息队列中，由 I/O 线程负责执行，这样就实现了局部无锁化。

◎ 定时任务：通过调用 NioEventLoop 的 schedule(Runnable command, long delay,

TimeUnit unit)方法实现。

正是因为 NioEventLoop 具备多种职责，所以它的实现比较特殊，它并不是个简单的 Runnable。我们来看下它的继承关系，如图 18-7 所示。

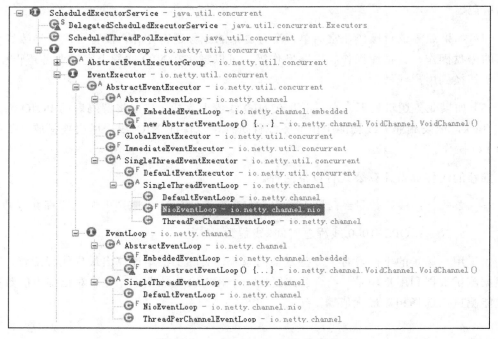

图 18-7　NioEventLoop 继承关系

它实现了 EventLoop 接口、EventExecutorGroup 接口和 ScheduledExecutorService 接口，正是因为这种设计，导致 NioEventLoop 和其父类功能实现非常复杂。下面我们就重点分析它的源码实现，理解它的设计原理。

18.2.2　NioEventLoop 继承关系类图

从下个小节开始，我们将对 NioEventLoop 的源码进行分析。通过源码分析，希望读者能够理解 Netty 的 Reactor 线程设计原理，掌握其精髓。NioEventLoop 继承关系图如图 18-8 所示。

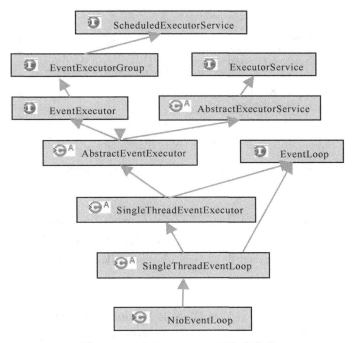

图 18-8　NioEventLoop 继承关系图

18.2.3　NioEventLoop

作为 NIO 框架的 Reactor 线程，NioEventLoop 需要处理网络 I/O 读写事件，因此它必须聚合一个多路复用器对象。下面我们看它的 Selector 定义，如图 18-9 所示。

```
/**
 * The NIO {@link Selector}.
 */
Selector selector;
private SelectedSelectionKeySet selectedKeys;

private final SelectorProvider provider;
```

图 18-9　NioEventLoop 的 Selector 定义

Selector 的初始化非常简单，直接调用 Selector.open()方法就能创建并打开一个新的 Selector。Netty 对 Selector 的 selectedKeys 进行了优化，用户可以通过 io.netty.noKeySet Optimization 开关决定是否启用该优化项。默认不打开 selectedKeys 优化功能。

Selector 的初始化代码如图 18-10 所示。

```
private Selector openSelector() {
  final Selector selector;
  try {
    selector = provider.openSelector();
  } catch (IOException e) {
    throw new ChannelException("failed to open a new selector", e);
  }
  if (DISABLE_KEYSET_OPTIMIZATION) {
    return selector;
  }
```

图 18-10　Selector 的初始化

如果没有开启 selectedKeys 优化开关，通过 provider.openSelector()创建并打开多路复用器之后就立即返回。

如果开启了优化开关，需要通过反射的方式从 Selector 实例中获取 selectedKeys 和 publicSelectedKeys，将上述两个成员变量设置为可写，通过反射的方式使用 Netty 构造的 selectedKeys 包装类 selectedKeySet 将原 JDK 的 selectedKeys 替换掉。

分析完 Selector 的初始化，下面重点看下 run 方法的实现，如图 18-11 所示。

所有的逻辑操作都在 for 循环体内进行，只有当 NioEventLoop 接收到退出指令的时候，才退出循环，否则一直执行下去，这也是通用的 NIO 线程实现方式。

首先需要将 wakenUp 还原为 false，并将之前的 wake up 状态保存到 oldWakenUp 变量中。通过 hasTasks()方法判断当前的消息队列中是否有消息尚未处理，如果有则调用 selectNow()方法立即进行一次 select 操作，看是否有准备就绪的 Channel 需要处理。它的代码实现如图 18-12 所示。

```
@Override
protected void run() {
  for (;;) {
    oldWakenUp = wakenUp.getAndSet(false);
    try {
      if (hasTasks()) {
        selectNow();
      } else {
        select();
        if (wakenUp.get()) {
          selector.wakeup();
        }
```

```
void selectNow() throws IOException {
  try {
    selector.selectNow();
  } finally {
    // restore wakup state if needed
    if (wakenUp.get()) {
      selector.wakeup();
    }
  }
}
```

图 18-11　NioEventLoop 的 run 方法　　　图 18-12　NioEventLoop 的 selectNow()方法

Selector 的 selectNow()方法会立即触发 Selector 的选择操作，如果有准备就绪的 Channel，则返回就绪 Channel 的集合，否则返回 0。选择完成之后，再次判断用户是否调用了 Selector 的 wakeup 方法，如果调用，则执行 selector.wakeup()操作。

下面我们返回到 run 方法，继续分析代码。如果消息队列中没有消息需要处理，则执行 select()方法，由 Selector 多路复用器轮询，看是否有准备就绪的 Channel。它的实现如图 18-13 所示。

取当前系统的纳秒时间，调用 delayNanos()方法计算获得 NioEventLoop 中定时任务的触发时间。

```
private void select() throws IOException {
    Selector selector = this.selector;
    try {
        int selectCnt = 0;
        long currentTimeNanos = System.nanoTime();
        long selectDeadLineNanos = currentTimeNanos + delayNanos(currentTimeNanos);
        for (;;) {
            long timeoutMillis = (selectDeadLineNanos - currentTimeNanos + 500000L) / 1000000L;
            if (timeoutMillis <= 0) {
                if (selectCnt == 0) {
                    selector.selectNow();
                    selectCnt = 1;
                }
                break;
            }
```

图 18-13　NioEventLoop 的 select()方法

计算下一个将要触发的定时任务的剩余超时时间，将它转换成毫秒，为超时时间增加 0.5 毫秒的调整值。对剩余的超时时间进行判断，如果需要立即执行或者已经超时，则调用 selector.selectNow()进行轮询操作，将 selectCnt 设置为 1，并退出当前循环。

将定时任务剩余的超时时间作为参数进行 select 操作，每完成一次 select 操作，对 select 计数器 selectCnt 加 1。

Select 操作完成之后，需要对结果进行判断，如果存在下列任意一种情况，则退出当前循环。

◎　有 Channel 处于就绪状态，selectedKeys 不为 0，说明有读写事件需要处理；
◎　oldWakenUp 为 true；
◎　系统或者用户调用了 wakeup 操作，唤醒当前的多路复用器；
◎　消息队列中有新的任务需要处理。

如果本次 Selector 的轮询结果为空，也没有 wakeup 操作或是新的消息需要处理，则说明是个空轮询，有可能触发了 JDK 的 epoll bug，它会导致 Selector 的空轮询，使 I/O 线程一直处于 100%状态。截止到当前最新的 JDK7 版本，该 bug 仍然没有被完全修复。所以 Netty 需要对该 bug 进行规避和修正。

Bug-id =6403933 的 Selector 堆栈如图 18-14 所示。

该 Bug 的修复策略如下。

（1）对 Selector 的 select 操作周期进行统计；

（2）每完成一次空的 select 操作进行一次计数；

（3）在某个周期（例如 100ms）内如果连续发生 N 次空轮询，说明触发了 JDK NIO 的 epoll()死循环 bug。

```
java.lang.Thread.State: RUNNABLE
    at sun.nio.ch.EPollArrayWrapper.epollWait(Native Method)
    at sun.nio.ch.EPollArrayWrapper.poll(EPollArrayWrapper.java:210)
    at sun.nio.ch.EPollSelectorImpl.doSelect(EPollSelectorImpl.java:65)
    at sun.nio.ch.SelectorImpl.lockAndDoSelect(SelectorImpl.java:69)
    - locked <0x0000000750928190> (a sun.nio.ch.Util$2)
    - locked <0x00000007509281a8> (a java.util.Collections$UnmodifiableSet)
    - locked <0x0000000750946098> (a sun.nio.ch.EPollSelectorImpl)
    at sun.nio.ch.SelectorImpl.select(SelectorImpl.java:80)
```

图 18-14　JDK Selector CPU 100% bug

监测到 Selector 处于死循环后，需要通过重建 Selector 的方式让系统恢复正常，代码如图 18-15 所示。

```
public void rebuildSelector() {
    if (!inEventLoop()) {
        execute(new Runnable() {
            @Override
            public void run() {
                rebuildSelector();
            }
        });
        return;
    }
    final Selector oldSelector = selector;
    final Selector newSelector;
    if (oldSelector == null) {
        return;
    }
    try {
        newSelector = openSelector();
    } catch (Exception e) {
        logger.warn("Failed to create a new Selector.", e);
        return;
    }
```

图 18-15　重建 Selector

首先通过 inEventLoop() 方法判断是否是其他线程发起的 rebuildSelector，如果由其他线程发起，为了避免多线程并发操作 Selector 和其他资源，需要将 rebuildSelector 封装成 Task，放到 NioEventLoop 的消息队列中，由 NioEventLoop 线程负责调用，这样就避免了多线程并发操作导致的线程安全问题。

调用 openSelector 方法创建并打开新的 Selector，通过循环，将原 Selector 上注册的 SocketChannel 从旧的 Selector 上去注册，重新注册到新的 Selector 上，并将老的 Selector 关闭。

相关代码如图 18-16 所示。

```
int nChannels = 0;
for (;;) {
    try {
        for (SelectionKey key: oldSelector.keys()) {
            Object a = key.attachment();
            try {
                if (key.channel().keyFor(newSelector) != null) {
                    continue;
                }

                int interestOps = key.interestOps();
                key.cancel();
                key.channel().register(newSelector, interestOps, a);
                nChannels ++;
            } catch (Exception e) {
                logger.warn("Failed to re-register a Channel to the new Selector.", e);
                if (a instanceof AbstractNioChannel) {
                    AbstractNioChannel ch = (AbstractNioChannel) a;
                    ch.unsafe().close(ch.unsafe().voidPromise());
                } else {
                    @SuppressWarnings("unchecked")
                    NioTask<SelectableChannel> task = (NioTask<SelectableChannel>) a;
                    invokeChannelUnregistered(task, key, e);
                }
            }
        }
    } catch (ConcurrentModificationException e) {
        // Probably due to concurrent modification of the key set.
        continue;
    }
    break;
}
```

图 18-16　将 SocketChannel 重新注册到新建的 Selector 上

通过销毁旧的、有问题的多路复用器，使用新建的 Selector，就可以解决空轮询 Selector 导致的 I/O 线程 CPU 占用 100% 的问题。

如果轮询到了处于就绪状态的 SocketChannel，则需要处理网络 I/O 事件，相关代码如

图 18-17 所示。

```
final long ioStartTime = System.nanoTime();
    needsToSelectAgain = false;
    if (selectedKeys != null) {
        processSelectedKeysOptimized(selectedKeys.flip());
    } else {
        processSelectedKeysPlain(selector.selectedKeys());
    }
    final long ioTime = System.nanoTime() - ioStartTime;
```

图 18-17　进行 I/O 操作

由于默认未开启 selectedKeys 优化功能，所以会进入 processSelectedKeysPlain 分支执行。下面继续分析 processSelectedKeysPlain 的代码实现，如图 18-18 所示。

```
private void processSelectedKeysPlain(Set<SelectionKey> selectedKeys) {
    if (selectedKeys.isEmpty()) {
        return;
    }

    Iterator<SelectionKey> i = selectedKeys.iterator();
    for (;;) {
        final SelectionKey k = i.next();
        final Object a = k.attachment();
        i.remove();
```

图 18-18　遍历 SelectionKey 进行网络读写

对 SelectionKey 进行保护性判断，如果为空则返回。获取 SelectionKey 的迭代器进行循环操作，通过迭代器获取 SelectionKey 和 SocketChannel 的附件对象，将已选择的选择键从迭代器中删除，防止下次被重复选择和处理，代码如图 18-19 所示。

```
if (a instanceof AbstractNioChannel) {
    processSelectedKey(k, (AbstractNioChannel) a);
} else {
    @SuppressWarnings("unchecked")
    NioTask<SelectableChannel> task = (NioTask<SelectableChannel>) a;
    processSelectedKey(k, task);
}
```

图 18-19　判断 SocketChannel 的附件类型

对 SocketChannel 的附件类型进行判读，如果是 AbstractNioChannel 类型，说明它是 NioServerSocketChannel 或者 NioSocketChannel，需要进行 I/O 读写相关的操作；如果它是 NioTask，则对其进行类型转换，调用 processSelectedKey 进行处理。由于 Netty 自身没实现 NioTask 接口，所以通常情况下系统不会执行该分支，除非用户自行注册该 Task 到多路复用器。

下面我们继续分析 I/O 事件的处理，代码如图 18-20 所示。

第 18 章　EventLoop 和 EventLoopGroup

```
private static void processSelectedKey(SelectionKey k, AbstractNioChannel ch) {
    final NioUnsafe unsafe = ch.unsafe();
    if (!k.isValid()) {
        // close the channel if the key is not valid anymore
        unsafe.close(unsafe.voidPromise());
        return;
    }
```

图 18-20　对选择键的状态进行判断

首先从 NioServerSocketChannel 或者 NioSocketChannel 中获取其内部类 Unsafe，判断当前选择键是否可用，如果不可用，则调用 Unsafe 的 close 方法，释放连接资源。

如果选择键可用，则继续对网络操作位进行判断，代码如图 18-21 所示。

```
int readyOps = k.readyOps();
    // Also check for readOps of 0 to workaround possible JDK bug which
may otherwise lead
    // to a spin loop
    if ((readyOps & (SelectionKey.OP_READ |
SelectionKey.OP_ACCEPT)) != 0 || readyOps == 0) {
        unsafe.read();
        if (!ch.isOpen()) {
            // Connection already closed - no need to handle write.
            return;
        }
    }
```

图 18-21　处理读事件

如果是读或者连接操作，则调用 Unsafe 的 read 方法。此处 Unsafe 的实现是个多态，对于 NioServerSocketChannel，它的读操作就是接收客户端的 TCP 连接，相关代码如图 18-22 所示。

```
protected int doReadMessages(List<Object> buf) throws Exception {
    SocketChannel ch = javaChannel().accept();
    try {
        if (ch != null) {
            buf.add(new NioSocketChannel(this, childEventLoopGroup().next(), ch));
            return 1;
        }
    } catch (Throwable t) {
        logger.warn("Failed to create a new channel from an accepted socket.", t);
        try {
            ch.close();
        } catch (Throwable t2) {
            logger.warn("Failed to close a socket.", t2);
        }
    }
    return 0;
}
```

图 18-22　NioServerSocketChannel 的读操作

对于 NioSocketChannel，它的读操作就是从 SocketChannel 中读取 ByteBuffer，相关代码如图 18-23 所示。

```
@Override
protected int doReadBytes(ByteBuf byteBuf) throws Exception {
    return byteBuf.writeBytes(javaChannel(), byteBuf.writableBytes());
}
```

图 18-23　NioSocketChannel 的读操作

如果网络操作位为写，则说明有半包消息尚未发送完成，需要继续调用 flush 方法进行发送，相关的代码如图 18-24 所示。

```
if ((readyOps & SelectionKey.OP_WRITE) != 0) {
    // Call forceFlush which will also take care of clear the OP_WRITE
    once there is nothing left to write
        ch.unsafe().forceFlush();
}
```

图 18-24　调用 flush 方法写半包

如果网络操作位为连接状态，则需要对连接结果进行判读，代码如图 18-25 所示。

```
if ((readyOps & SelectionKey.OP_CONNECT) != 0) {
    // remove OP_CONNECT as otherwise Selector.select(..) will always
    return without blocking
        int ops = k.interestOps();
        ops &= ~SelectionKey.OP_CONNECT;
        k.interestOps(ops);

        unsafe.finishConnect();
}
```

图 18-25　处理连接事件

需要注意的是，在进行 finishConnect 判断之前，需要将网络操作位进行修改，注销掉 SelectionKey.OP_CONNECT。

处理完 I/O 事件之后，NioEventLoop 需要执行非 I/O 操作的系统 Task 和定时任务，代码如图 18-26 所示。

```
final long ioTime = System.nanoTime() - ioStartTime;
final int ioRatio = this.ioRatio;
runAllTasks(ioTime * (100 - ioRatio) / ioRatio);
```

图 18-26　执行非 I/O 任务

由于 NioEventLoop 需要同时处理 I/O 事件和非 I/O 任务，为了保证两者都能得到足够的 CPU 时间被执行，Netty 提供了 I/O 比例供用户定制。如果 I/O 操作多于定时任务和 Task，则可以将 I/O 比例调大，反之则调小，默认值为 50%。

Task 的执行时间根据本次 I/O 操作的执行时间计算得来。下面我们具体看 runAllTasks 方法的实现，如图 18-27 所示。

```
protected boolean runAllTasks(long timeoutNanos) {
    fetchFromDelayedQueue();
    Runnable task = pollTask();
    if (task == null) {
        return false;
    }
}
```

图 18-27　执行非 I/O 任务

首先从定时任务消息队列中弹出消息进行处理，如果消息队列为空，则退出循环。根据当前的时间戳进行判断，如果该定时任务已经或者正处于超时状态，则将其加入到执行 Task Queue 中，同时从延时队列中删除。定时任务如果没有超时，说明本轮循环不需要处理，直接退出即可，代码实现如图 18-28 所示。

```
private void fetchFromDelayedQueue() {
    long nanoTime = 0L;
    for (;;) {
        ScheduledFutureTask<?> delayedTask = delayedTaskQueue.peek();
        if (delayedTask == null) {
            break;
        }
        if (nanoTime == 0L) {
            nanoTime = ScheduledFutureTask.nanoTime();
        }
        if (delayedTask.deadlineNanos() <= nanoTime) {
            delayedTaskQueue.remove();
            taskQueue.add(delayedTask);
        } else {
            break;
        }
    }
}
```

图 18-28　从延时队列中移除消息到 Task Queue

执行 Task Queue 中原有的任务和从延时队列中复制的已经超时或者正处于超时状态的定时任务，代码如图 18-29 所示。

由于获取系统纳秒时间是个耗时的操作，每次循环都获取当前系统纳秒时间进行超时判断会降低性能。为了提升性能，每执行 60 次循环判断一次，如果当前系统时间已经到了分配给非 I/O 操作的超时时间，则退出循环。这是为了防止由于非 I/O 任务过多导致 I/O 操作被长时间阻塞。

```
for (;;) {
    try {
        task.run();
    } catch (Throwable t) {
        logger.warn("A task raised an exception.", t);
    }
    runTasks ++;
    if ((runTasks & 0x3F) == 0) {
        lastExecutionTime = ScheduledFutureTask.nanoTime();
        if (lastExecutionTime >= deadline) {
            break;
        }
    }
}
```

图 18-29 执行普通 Task 和定时任务

最后，判断系统是否进入优雅停机状态，如果处于关闭状态，则需要调用 closeAll 方法，释放资源，并让 NioEventLoop 线程退出循环，结束运行。资源关闭的代码实现如图 18-30 所示。

```
private void closeAll() {
    selectAgain();
    Set<SelectionKey> keys = selector.keys();
    Collection<AbstractNioChannel> channels = new ArrayList<AbstractNioChannel>(keys.size());
    for (SelectionKey k: keys) {
        Object a = k.attachment();
        if (a instanceof AbstractNioChannel) {
            channels.add((AbstractNioChannel) a);
        } else {
            k.cancel();
            NioTask<SelectableChannel> task = (NioTask<SelectableChannel>) a;
            invokeChannelUnregistered(task, k, null);
        }
    }
    for (AbstractNioChannel ch: channels) {
        ch.unsafe().close(ch.unsafe().voidPromise());
    }
}
```

图 18-30 NioEventLoop 线程退出，资源释放

遍历获取所有的 Channel，调用它的 Unsafe.close() 方法关闭所有链路，释放线程池、ChannelPipeline 和 ChannelHandler 等资源。

18.3 总结

本章详细介绍了 Netty 的线程模型以及 NioEventLoop 的实现，通过最佳实践和源码分

析，让读者加深对 Netty 框架线程模型的理解，能够在未来的工作中可以恰到好处地使用它。

对于任何架构，线程模型设计的好坏都直接影响软件的性能和并发处理能力。幸运的是，Netty 的线程模型被精心地设计和实现。相信通过对 Netty 线程模型的学习，广大读者朋友可以举一反三，将 Reactor 线程模型的精髓应用到日常的工作中。

第 19 章

Future 和 Promise

本章我们介绍 Netty 的 Future 和 Promise。从名字可以看出，Future 用于获取异步操作的结果，而 Promise 则比较抽象，无法直接猜测出其功能。本章将介绍 Netty Future 和 Promise 的功能，分析它们的源码，帮助读者掌握其实现原理。

本章主要内容包括：

- ◎ Future 功能
- ◎ Future 源码分析
- ◎ Promise 功能
- ◎ Promise 源码分析

19.1 Future 功能

Future 最早来源于 JDK 的 java.util.concurrent.Future，它用于代表异步操作的结果。相关 API 如图 19-1 所示。

可以通过 get 方法获取操作结果，如果操作尚未完成，则会同步阻塞当前调用的线程；如果不允许阻塞太长时间或者无限期阻塞，可以通过带超时时间的 get 方法获取结果；如果到达超时时间操作仍然没有完成，则抛出 TimeoutException。

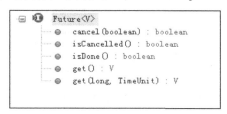

图 19-1　JDK Future 的 API 列表

通过 isDone()方法可以判断当前的异步操作是否完成，如果完成，无论成功与否，都返回 true，否则返回 false。

通过 cancel 可以尝试取消异步操作，它的结果是未知的，如果操作已经完成，或者发生其他未知的原因拒绝取消，取消操作将会失败。

ChannelFuture 功能介绍

由于 Netty 的 Future 都是与异步 I/O 操作相关的，因此，命名为 ChannelFuture，代表它与 Channel 操作相关。

它的 API 接口列表如表 19-1 所示。

表 19-1　ChannelFuture 接口列表

返 回 值	方法名称
ChannelFuture	addListener(GenericFutureListener<? extends Future<? super java.lang.Void>> listener) Adds the specified listener to this future
ChannelFuture	addListeners(GenericFutureListener<? extends Future<? super java.lang.Void>>... listeners) Adds the specified listeners to this future
ChannelFuture	await() Waits for this future to be completed
ChannelFuture	awaitUninterruptibly() Waits for this future to be completed without interruption
Channel	channel() Returns a channel where the I/O operation associated with this future takes place
ChannelFuture	removeListener(GenericFutureListener<? extends Future<? super java.lang.Void>> listener) Removes the specified listener from this future

续表

返 回 值	方法名称
ChannelFuture	removeListeners(GenericFutureListener<? extends Future<? super java.lang.Void>>... listeners) Removes the specified listeners from this future
ChannelFuture	sync() Waits for this future until it is done, and rethrows the cause of the failure if this future failed
ChannelFuture	syncUninterruptibly() Waits for this future until it is done, and rethrows the cause of the failure if this future failed

在 Netty 中，所有的 I/O 操作都是异步的，这意味着任何 I/O 调用都会立即返回，而不是像传统 BIO 那样同步等待操作完成。异步操作会带来一个问题：调用者如何获取异步操作的结果？ChannelFuture 就是为了解决这个问题而专门设计的。下面我们一起看它的原理。

ChannelFuture 有两种状态：uncompleted 和 completed。当开始一个 I/O 操作时，一个新的 ChannelFuture 被创建，此时它处于 uncompleted 状态——非失败、非成功、非取消，因为 I/O 操作此时还没有完成。一旦 I/O 操作完成，ChannelFuture 将会被设置成 completed，它的结果有如下三种可能。

◎ 操作成功；

◎ 操作失败；

◎ 操作被取消。

ChannelFuture 的状态迁移图如图 19-2 所示。

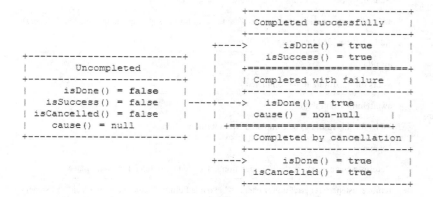

图 19-2　ChannelFuture 状态迁移图

ChannelFuture 提供了一系列新的 API，用于获取操作结果、添加事件监听器、取消 I/O 操作、同步等待等。

我们重点介绍添加监听器的接口。管理监听器相关的接口定义如图 19-3 所示。

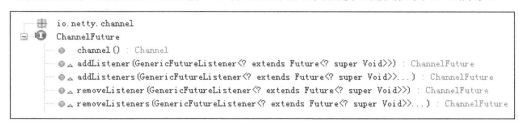

图 19-3　ChannelFuture 管理监听器

Netty 强烈建议直接通过添加监听器的方式获取 I/O 操作结果，或者进行后续的相关操作。

ChannelFuture 可以同时增加一个或者多个 GenericFutureListener，也可以用 remove 方法删除 GenericFutureListener。

GenericFutureListener 的接口定义如图 19-4 所示。

```
void operationComplete(F future) throws Exception;
```

图 19-4　GenericFutureListener 接口定义

当 I/O 操作完成之后，I/O 线程会回调 ChannelFuture 中 GenericFutureListener 的 operationComplete 方法，并把 ChannelFuture 对象当作方法的入参。如果用户需要做上下文相关的操作，需要将上下文信息保存到对应的 ChannelFuture 中。

推荐通过 GenericFutureListener 代替 ChannelFuture 的 get 等方法的原因是：当我们进行异步 I/O 操作时，完成的时间是无法预测的，如果不设置超时时间，它会导致调用线程长时间被阻塞，甚至挂死。而设置超时时间，时间又无法精确预测。利用异步通知机制回调 GenericFutureListener 是最佳的解决方案，它的性能最优。

需要注意的是：不要在 ChannelHandler 中调用 ChannelFuture 的 await()方法，这会导致死锁。原因是发起 I/O 操作之后，由 I/O 线程负责异步通知发起 I/O 操作的用户线程，如果 I/O 线程和用户线程是同一个线程，就会导致 I/O 线程等待自己通知操作完成，这就导致了死锁，这跟经典的两个线程互等待死锁不同，属于自己把自己挂死。

相关代码示例如图 19-5 所示。

```
// BAD - NEVER DO THIS
@Override
public void channelRead(ChannelHandlerContext ctx, GoodByeMessage msg) {
    ChannelFuture future = ctx.channel().close();
    future.awaitUninterruptibly();
    // Perform post-closure operation
    // ...
}

// GOOD
@Override
public void channelRead(ChannelHandlerContext ctx, GoodByeMessage msg) {
    ChannelFuture future = ctx.channel().close();
    future.addListener(new ChannelFutureListener() {
        public void operationComplete(ChannelFuture future) {
            // Perform post-closure operation
            // ...
        }
    });
}
```

图 19-5　ChannelFuture 的正反用法示例

异步 I/O 操作有两类超时：一个是 TCP 层面的 I/O 超时，另一个是业务逻辑层面的操作超时。两者没有必然的联系，但是通常情况下业务逻辑超时时间应该大于 I/O 超时时间，它们两者是包含的关系。

相关代码举例如图 19-6 所示。

```
// GOOD
Bootstrap b = ...;
// Configure the connect timeout option.
b.option(ChannelOption.CONNECT_TIMEOUT_MILLIS, 10000);
ChannelFuture f = b.connect(...);
f.awaitUninterruptibly();

// Now we are sure the future is completed.
assert f.isDone();

if (f.isCancelled()) {
    // Connection attempt cancelled by user
} else if (!f.isSuccess()) {
    f.cause().printStackTrace();
} else {
    // Connection established successfully
}
```

图 19-6　I/O 超时时间配置

ChannelFuture 超时时间配置如图 19-7 所示。

```
Bootstrap b = ...;
ChannelFuture f = b.connect(...);
f.awaitUninterruptibly(10, TimeUnit.SECONDS);
if (f.isCancelled()) {
    // Connection attempt cancelled by user
} else if (!f.isSuccess()) {
    // You might get a NullPointerException here because the future
    // might not be completed yet.
    f.cause().printStackTrace();
} else {
    // Connection established successfully
}
```

图 19-7　ChannelFuture 超时时间配置

需要指出的是：ChannelFuture 超时并不代表 I/O 超时，这意味着 ChannelFuture 超时后，如果没有关闭连接资源，随后连接依旧可能会成功，这会导致严重的问题。所以通常情况下，必须要考虑究竟是设置 I/O 超时还是 ChannelFuture 超时。

19.2　ChannelFuture 源码分析

ChannelFuture 的接口继承关系如图 19-8 所示。

```
io.netty.channel
接口 ChannelFuture

所有超级接口：
    java.util.concurrent.Future<java.lang.Void>

所有已知子接口：
    ChannelProgressiveFuture, ChannelProgressivePromise, ChannelPromise

所有已知实现类：
    DefaultChannelProgressivePromise, DefaultChannelPromise
```

图 19-8　ChannelFuture 接口继承关系图

AbstractFuture

AbstractFuture 实现 Future 接口，它不允许 I/O 操作被取消。下面我们重点看它的代码实现。

获取异步操作结果的代码如图 19-9 所示。

```
public V get() throws InterruptedException, ExecutionException {
    await();
    Throwable cause = cause();
    if (cause == null) {
        return getNow();
    }
    throw new ExecutionException(cause);
}
```

图 19-9　同步获取 I/O 操作结果

首先，调用 await() 方法进行无限期阻塞，当 I/O 操作完成后会被 notify()。程序继续向下执行，检查 I/O 操作是否发生了异常，如果没有异常，则通过 getNow() 方法获取结果并返回。否则，将异常堆栈进行包装，抛出 ExecutionException。

接着我们看支持超时的获取操作结果方法，如图 19-10 所示。

```
@Override
public V get(long timeout, TimeUnit unit) throws InterruptedException,
ExecutionException, TimeoutException {
    if (await(timeout, unit)) {
        Throwable cause = cause();
        if (cause == null) {
            return getNow();
        }
        throw new ExecutionException(cause);
    }
    throw new TimeoutException();
}
```

图 19-10　支持获取超时的方法

支持超时很简单，调用 await(long timeout, TimeUnit unit) 方法即可。如果超时，则抛出 TimeoutException。如果没有超时，则依次判断是否发生了 I/O 异常等情况，操作与无参数的 get 方法相同。

其他 ChannelFuture 的实现子类，由于功能比较简单，读者阅读起来也没太大难度，所以这里不再花费时间进行详细解读，感兴趣的读者可以独立阅读和分析。AbstractFuture 的继承关系如图 19-11 所示。

```
Type hierarchy of 'io.netty.util.concurrent.AbstractFuture':
Object - java.lang
    AbstractFuture<V> - io.netty.util.concurrent
        CompleteFuture<V> - io.netty.util.concurrent
            CompleteChannelFuture - io.netty.channel
                FailedChannelFuture - io.netty.channel
                SucceededChannelFuture - io.netty.channel
            FailedFuture<V> - io.netty.util.concurrent
            SucceededFuture<V> - io.netty.util.concurrent
        DefaultPromise<V> - io.netty.util.concurrent
            DefaultChannelGroupFuture - io.netty.channel.group
            DefaultChannelPromise - io.netty.channel
                CloseFuture - io.netty.channel.AbstractChannel
            DefaultProgressivePromise<V> - io.netty.util.concurrent
                DefaultChannelProgressivePromise - io.netty.channel
                ImmediateProgressivePromise<V> - io.netty.util.concurrent.ImmediateEventExecutor
            ImmediatePromise<V> - io.netty.util.concurrent.ImmediateEventExecutor
            LazyChannelPromise - io.netty.handler.ssl.SslHandler
            PromiseTask<V> - io.netty.util.concurrent
                ScheduledFutureTask<V> - io.netty.util.concurrent
            VoidChannelPromise - io.netty.channel
```

图 19-11　AbstractFuture 的继承关系图

19.3　Promise 功能介绍

Promise 是可写的 Future，Future 自身并没有写操作相关的接口，Netty 通过 Promise 对 Future 进行扩展，用于设置 I/O 操作的结果。Future 相关的接口定义如图 19-12 所示。

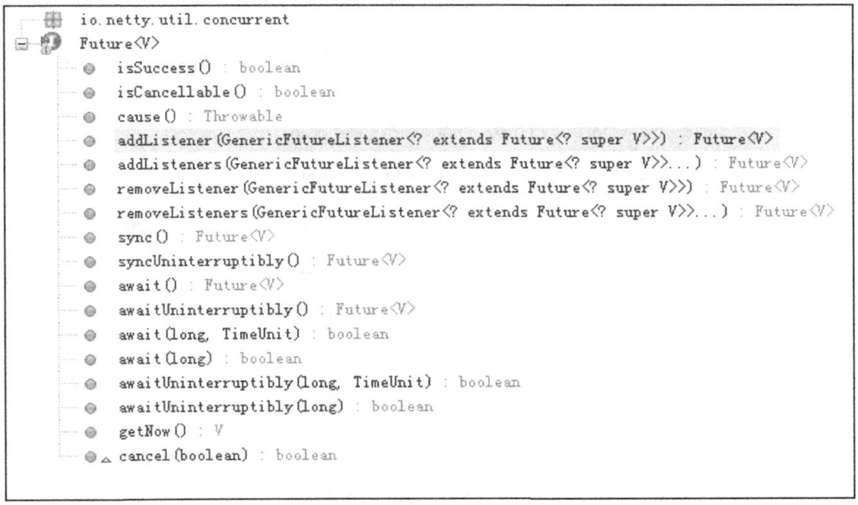

```
io.netty.util.concurrent
Future<V>
    isSuccess() : boolean
    isCancellable() : boolean
    cause() : Throwable
    addListener(GenericFutureListener<? extends Future<? super V>>) : Future<V>
    addListeners(GenericFutureListener<? extends Future<? super V>>...) : Future<V>
    removeListener(GenericFutureListener<? extends Future<? super V>>) : Future<V>
    removeListeners(GenericFutureListener<? extends Future<? super V>>...) : Future<V>
    sync() : Future<V>
    syncUninterruptibly() : Future<V>
    await() : Future<V>
    awaitUninterruptibly() : Future<V>
    await(long, TimeUnit) : boolean
    await(long) : boolean
    awaitUninterruptibly(long, TimeUnit) : boolean
    awaitUninterruptibly(long) : boolean
    getNow() : V
    cancel(boolean) : boolean
```

图 19-12　Netty 的 Future 接口定义

Promise 相关的写操作接口定义如图 19-13 所示。

Promise<V>	setFailure(java.lang.Throwable cause) Marks this future as a failure and notifies all listeners.
Promise<V>	setSuccess(V result) Marks this future as a success and notifies all listeners.
boolean	setUncancellable() Make this future impossible to cancel.
Promise<V>	sync() Waits for this future until it is done, and rethrows the cause of the failure if this future failed
Promise<V>	syncUninterruptibly() Waits for this future until it is done, and rethrows the cause of the failure if this future failed
boolean	tryFailure(java.lang.Throwable cause) Marks this future as a failure and notifies all listeners.
boolean	trySuccess(V result) Marks this future as a success and notifies all listeners.

图 19-13　Promise 写操作相关的接口定义

Netty 发起 I/O 操作的时候，会创建一个新的 Promise 对象，例如调用 ChannelHandlerContext 的 write(Object object)方法时，会创建一个新的 ChannelPromise，相关代码如图 19-14 所示。

```
public ChannelPromise newPromise() {
    return new DefaultChannelPromise(channel(), executor());
}
```

图 19-14　I/O 操作时创建一个新的 Promise

当 I/O 操作发生异常或者完成时，设置 Promise 的结果，代码如图 19-15 所示。

```
@Override
public void write(Object msg, ChannelPromise promise) {
    if (!isActive()) {
        // Mark the write request as failure if the channel is inactive.
        if (isOpen()) {
            promise.tryFailure(NOT_YET_CONNECTED_EXCEPTION);
        } else {
            promise.tryFailure(CLOSED_CHANNEL_EXCEPTION);
        }
        // release message now to prevent resource-leak
        ReferenceCountUtil.release(msg);
    } else {
        outboundBuffer.addMessage(msg, promise);
    }

}
```

图 19-15　I/O 操作异常时调用 tryFailure 方法设置结果

19.4 Promise 源码分析

19.4.1 Promise 继承关系图

由于 I/O 操作种类非常多，因此对应的 Promise 子类也非常繁多，它的继承关系如图 19-16 所示。

图 19-16 Promise 继承关系图

尽管 Promise 的子类种类繁多，但是它的功能相对比较清晰，代码也较为简单，因此我们只分析一个它的实现子类的源码，如果读者对其他子类感兴趣，可以自行学习。

19.4.2 DefaultPromise

下面看比较重要的 setSuccess 方法的实现，如图 19-17 所示。

首先调用 setSuccess0 方法并对其操作结果进行判断，如果操作成功，则调用

notifyListeners 方法通知 listener。

```
@Override
public Promise<V> setSuccess(V result) {
    if (setSuccess0(result)) {
        notifyListeners();
        return this;
    }
    throw new IllegalStateException("complete already: " + this);
}
```

图 19-17　DefaultPromise 的 setSuccess 方法

setSuccess0 方法的实现如图 19-18 所示。

```
private boolean setSuccess0(V result) {
    if (isDone()) {
        return false;
    }

    synchronized (this) {
        if (isDone()) {
            return false;
        }
        if (result == null) {
            this.result = SUCCESS;
        } else {
            this.result = result;
        }
        if (hasWaiters()) {
            notifyAll();
        }
    }
    return true;
}
```

图 19-18　DefaultPromise 的 setSuccess0 私有方法

首先判断当前 Promise 的操作结果是否已经被设置，如果已经被设置，则不允许重复设置，返回设置失败。

由于可能存在 I/O 线程和用户线程同时操作 Promise，所以设置操作结果的时候需要加锁保护，防止并发操作。

对操作结果是否被设置进行二次判断（为了提升并发性能的二次判断），如果已经被设置，则返回操作失败。

对操作结果 result 进行判断，如果为空，说明仅仅需要 notify 在等待的业务线程，不包含具体的业务逻辑对象。因此，将 result 设置为系统默认的 SUCCESS。如果操作结果非空，将结果设置为 result。

如果有正在等待异步 I/O 操作完成的用户线程或者其他系统线程，则调用 notifyAll 方法唤醒所有正在等待的线程。注意，notifyAll 和 wait 方法都必须在同步块内使用。

分析完 setSuccess0 方法，我们继续看 await 方法的实现，如图 19-19 所示。

```
public Promise<V> await() throws InterruptedException {
    if (isDone()) {
        return this;
    }
    if (Thread.interrupted()) {
        throw new InterruptedException(toString());
    }
    synchronized (this) {
        while (!isDone()) {
            checkDeadLock();
            incWaiters();
            try {
                wait();
            } finally {
                decWaiters();
            }
        }
    }
    return this;
}
```

图 19-19　DefaultPromise 的 await 方法

如果当前的 Promise 已经被设置，则直接返回。如果线程已经被中断，则抛出中断异常。通过同步关键字锁定当前 Promise 对象，使用循环判断对 isDone 结果进行判断，进行循环判断的原因是防止线程被意外唤醒导致的功能异常。如果对循环判断的实现原理感兴趣，读者可以查看《Effective Java 中文版第 2 版》第 243 页对 wait 和 notify 用法的讲解。

由于在 I/O 线程中调用 Promise 的 await 或者 sync 方法会导致死锁，所以在循环体中需要对死锁进行保护性校验，防止 I/O 线程被挂死，最后调用 java.lang.Object.wait()方法进行无限期等待，直到 I/O 线程调用 setSuccess 方法、trySuccess 方法、setFailure 或者 tryFailure 方法。

19.5　总结

本章重点介绍了 Future 和 Promise，由于 Netty 中的 I/O 操作种类繁多，所以 Future 和 Promise 的子类也非常繁多。尽管这些子类的功能各异，但本质上都是异步 I/O 操作结果的通知回调类。Future-Listener 机制在 JDK 中的应用已经非常广泛，所以本章并没有对这些子类的实现做过多的源码分析，希望读者在本章源码分析的基础上自行学习其他相关子类的实现。

无论 Future 还是 Promise，都强烈建议读者通过增加监听器 Listener 的方式接收异步 I/O 操作结果的通知，而不是调用 wait 或者 sync 阻塞用户线程。

架构和行业应用篇
Netty 高级特性

第 20 章　Netty 架构剖析

第 21 章　Java 多线程编程在 Netty 中的应用

第 22 章　高性能之道

第 23 章　可靠性

第 24 章　安全性

第 25 章　Netty 未来展望

第 20 章 Netty 架构剖析

本章将重点分析 Netty 的逻辑架构,通过对其关键架构质量属性的分析,让读者朋友能够更加深入地了解 Netty 的设计精髓。

希望读者在今后的架构设计中能够从 Netty 架构中汲取营养,设计出高性能、高可靠性和可扩展的产品。

本章主要内容包括:

◎ Netty 逻辑架构分析

◎ 关键架构质量属性

20.1 Netty 逻辑架构

Netty 采用了典型的三层网络架构进行设计和开发,逻辑架构如图 20-1 所示。

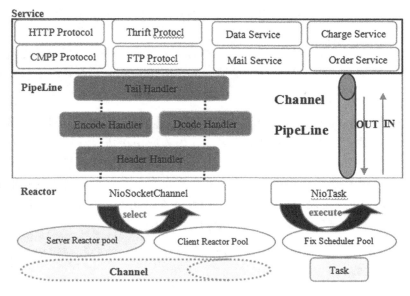

图 20-1　Netty 逻辑架构图

20.1.1　Reactor 通信调度层

它由一系列辅助类完成，包括 Reactor 线程 NioEventLoop 及其父类，NioSocketChannel/NioServerSocketChannel 及其父类，ByteBuffer 以及由其衍生出来的各种 Buffer，Unsafe 以及其衍生出的各种内部类等。该层的主要职责就是监听网络的读写和连接操作，负责将网络层的数据读取到内存缓冲区中，然后触发各种网络事件，例如连接创建、连接激活、读事件、写事件等，将这些事件触发到 PipeLine 中，由 PipeLine 管理的职责链来进行后续的处理。

20.1.2　职责链 ChannelPipeline

它负责事件在职责链中的有序传播，同时负责动态地编排职责链。职责链可以选择监听和处理自己关心的事件，它可以拦截处理和向后/向前传播事件。不同应用的 Handler 节点的功能也不同，通常情况下，往往会开发编解码 Hanlder 用于消息的编解码，它可以将外部的协议消息转换成内部的 POJO 对象，这样上层业务则只需要关心处理业务逻辑即可，不需要感知底层的协议差异和线程模型差异，实现了架构层面的分层隔离。

20.1.3 业务逻辑编排层（Service ChannelHandler）

业务逻辑编排层通常有两类：一类是纯粹的业务逻辑编排，还有一类是其他的应用层协议插件，用于特定协议相关的会话和链路管理。例如 CMPP 协议，用于管理和中国移动短信系统的对接。

架构的不同层面，需要关心和处理的对象都不同，通常情况下，对于业务开发者，只需要关心职责链的拦截和业务 Handler 的编排。因为应用层协议栈往往是开发一次，到处运行，所以实际上对于业务开发者来说，只需要关心服务层的业务逻辑开发即可。各种应用协议以插件的形式提供，只有协议开发人员需要关注协议插件，对于其他业务开发人员来说，只需关心业务逻辑定制。这种分层的架构设计理念实现了 NIO 框架各层之间的解耦，便于上层业务协议栈的开发和业务逻辑的定制。

正是由于 Netty 的分层架构设计非常合理，基于 Netty 的各种应用服务器和协议栈开发才能够如雨后春笋般得到快速发展。

20.2 关键架构质量属性

20.2.1 高性能

影响最终产品的性能因素非常多，其中软件因素如下。

◎ 架构不合理导致的性能问题。

◎ 编码实现不合理导致的性能问题，例如锁的不恰当使用导致性能瓶颈。

硬件因素如下。

◎ 服务器硬件配置太低导致的性能问题。

◎ 带宽、磁盘的 IOPS 等限制导致的 I/O 操作性能差。

◎ 测试环境被共用导致被测试的软件产品受到影响。

尽管影响产品性能的因素非常多，但是架构的性能模型合理与否对性能的影响非常大。如果一个产品的架构设计得不好，无论开发如何努力，都很难开发出一个高性能、高可用的软件产品。

"性能是设计出来的，而不是测试出来的"。下面我们看 Netty 的架构设计是如何实现高性能的。

（1）采用异步非阻塞的 I/O 类库，基于 Reactor 模式实现，解决了传统同步阻塞 I/O 模式下一个服务端无法平滑地处理线性增长的客户端的问题。

（2）TCP 接收和发送缓冲区使用直接内存代替堆内存，避免了内存复制，提升了 I/O 读取和写入的性能。

（3）支持通过内存池的方式循环利用 ByteBuf，避免了频繁创建和销毁 ByteBuf 带来的性能损耗。

（4）可配置的 I/O 线程数、TCP 参数等，为不同的用户场景提供定制化的调优参数，满足不同的性能场景。

（5）采用环形数组缓冲区实现无锁化并发编程，代替传统的线程安全容器或者锁。

（6）合理地使用线程安全容器、原子类等，提升系统的并发处理能力。

（7）关键资源的处理使用单线程串行化的方式，避免多线程并发访问带来的锁竞争和额外的 CPU 资源消耗问题。

（8）通过引用计数器及时地申请释放不再被引用的对象，细粒度的内存管理降低了 GC 的频率，减少了频繁 GC 带来的时延增大和 CPU 损耗。

无论是 Netty 的官方性能测试数据，还是携带业务实际场景的性能测试，Netty 在各个 NIO 框架中综合性能是最高的。下面，我们来看 Netty 官方的性能测试数据，如图 20-2、20-3、20-4 和 20-5 所示。

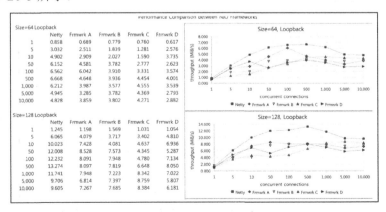

图 20-2　64 和 128 字节测试消息（本机网卡回环）

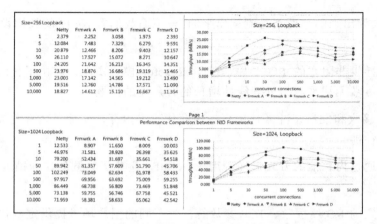

图 20-3 256 和 1K 字节测试消息(本机网卡回环)

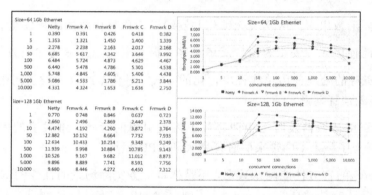

图 20-4 64 和 128 字节测试消息(跨主机通信)

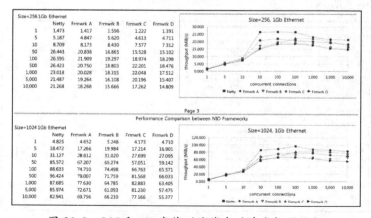

图 20-5 256 和 1K 字节测试消息(跨主机通信)

20.2.2 可靠性

作为一个高性能的异步通信框架，架构的可靠性是大家选择的一个重要依据。下面我们探讨 Netty 架构的可靠性设计。

1. 链路有效性检测

由于长连接不需要每次发送消息都创建链路，也不需要在消息交互完成时关闭链路，因此相对于短连接性能更高。对于长连接，一旦链路建立成功便一直维系双方之间的链路，直到系统退出。

为了保证长连接的链路有效性，往往需要通过心跳机制周期性地进行链路检测。使用周期性心跳的原因是：在系统空闲时，例如凌晨，往往没有业务消息。如果此时链路被防火墙 Hang 住，或者遭遇网络闪断、网络单通等，通信双方无法识别出这类链路异常。等到第二天业务高峰期到来时，瞬间的海量业务冲击会导致消息积压无法发送给对方，由于链路的重建需要时间，这期间业务会大量失败（集群或者分布式组网情况会好一些）。为了解决这个问题，需要周期性的心跳对链路进行有效性检测，一旦发生问题，可以及时关闭链路，重建 TCP 连接。

当有业务消息时，无须心跳检测，可以由业务消息进行链路可用性检测。所以心跳消息往往是在链路空闲时发送的。

为了支持心跳，Netty 提供了如下两种链路空闲检测机制。

- 读空闲超时机制：当连续周期 T 没有消息可读时，触发超时 Handler，用户可以基于读空闲超时发送心跳消息，进行链路检测；如果连续 N 个周期仍然没有读取到心跳消息，可以主动关闭链路。

- 写空闲超时机制：当连续周期 T 没有消息要发送时，触发超时 Handler，用户可以基于写空闲超时发送心跳消息，进行链路检测；如果连续 N 个周期仍然没有接收到对方的心跳消息，可以主动关闭链路。

为了满足不同用户场景的心跳定制，Netty 提供了空闲状态检测事件通知机制，用户可以订阅空闲超时事件、写空闲超时事件、读或者写超时事件，在接收到对应的空闲事件之后，灵活地进行定制。

2. 内存保护机制

Netty 提供多种机制对内存进行保护，包括以下几个方面。

◎ 通过对象引用计数器对 Netty 的 ByteBuf 等内置对象进行细粒度的内存申请和释放，对非法的对象引用进行检测和保护。

◎ 通过内存池来重用 ByteBuf，节省内存。

◎ 可设置的内存容量上限，包括 ByteBuf、线程池线程数等。

AbstractReferenceCountedByteBuf 的内存管理方法实现如图 20-6、20-7 所示。

```java
@Override
public ByteBuf retain() {
    for (;;) {
        int refCnt = this.refCnt;
        if (refCnt == 0) {
            throw new IllegalReferenceCountException(0, 1);
        }
        if (refCnt == Integer.MAX_VALUE) {
            throw new IllegalReferenceCountException(Integer.MAX_VALUE, 1);
        }
        if (refCntUpdater.compareAndSet(this, refCnt, refCnt + 1)) {
            break;
        }
    }
    return this;
}
```

图 20-6　对象引用

```java
@Override
public final boolean release() {
    for (;;) {
        int refCnt = this.refCnt;
        if (refCnt == 0) {
            throw new IllegalReferenceCountException(0, -1);
        }

        if (refCntUpdater.compareAndSet(this, refCnt, refCnt - 1)) {
            if (refCnt == 1) {
                deallocate();
                return true;
            }
            return false;
        }
    }
}
```

图 20-7　对象引用释放

ByteBuf 的解码保护，防止非法码流导致内存溢出，代码如图 20-8 所示。

```
 * @param maxFrameLength
 *        the maximum length of the frame.  If the length of the frame is
 *        greater than this value, {@link TooLongFrameException} will be
 *        thrown.
 * @param lengthFieldOffset
 *        the offset of the length field
 * @param lengthFieldLength
 *        the length of the length field
 */
public LengthFieldBasedFrameDecoder(
        int maxFrameLength,
        int lengthFieldOffset, int lengthFieldLength) {
    this(maxFrameLength, lengthFieldOffset, lengthFieldLength, 0, 0);
}
```

图 20-8　解码器单条消息最大长度上限保护

如果长度解码器没有单个消息最大报文长度限制，当解码错位或者读取到畸形码流时，长度值可能是个超大整数值，例如 4294967296，这很容易导致内存溢出。如果有上限保护，例如单条消息最大不允许超过 10MB，当读取到非法消息长度 4294967296 后，直接抛出解码异常，这样就避免了大内存的分配。

3. 优雅停机

相比于 Netty 的早期版本，Netty 5.0 版本的优雅退出功能做得更加完善。优雅停机功能指的是当系统退出时，JVM 通过注册的 Shutdown Hook 拦截到退出信号量，然后执行退出操作，释放相关模块的资源占用，将缓冲区的消息处理完成或者清空，将待刷新的数据持久化到磁盘或者数据库中，等到资源回收和缓冲区消息处理完成之后，再退出。

优雅停机往往需要设置个最大超时时间 T，如果达到 T 后系统仍然没有退出，则通过 Kill - 9 pid 强杀当前的进程。

Netty 所有涉及到资源回收和释放的地方都增加了优雅退出的方法，它们的相关接口如表 20-1 所示。

表 20-1　Netty 重要资源的优雅退出方法

EventExecutorGroup.shutdownGracefully()	NIO 线程优雅退出
EventExecutorGroup.shutdownGracefully(long quietPeriod, long timeout, TimeUnit unit)	NIO 线程优雅退出，支持设置超时时间
Channel.close()	Channel 的关闭
Unsafe.close(ChannelPromise promise)	Unsafe 的关闭操作，可以设置可写的 Future
Unsafe.closeForcibly()	Unsafe 的强制关闭操作

续表

ChannelPipeline.close()	ChannelPipeline 的关闭
ChannelPipeline.close(ChannelPromise promise)	ChannelPipeline 的关闭，可以设置可写的 Future
ChannelHandler.close(ChannelHandlerContext ctx, ChannelPromise promise)	ChannelHandler 的关闭

20.2.3 可定制性

Netty 的可定制性主要体现在以下几点。

◎ 责任链模式：ChannelPipeline 基于责任链模式开发，便于业务逻辑的拦截、定制和扩展。

◎ 基于接口的开发：关键的类库都提供了接口或者抽象类，如果 Netty 自身的实现无法满足用户的需求，可以由用户自定义实现相关接口。

◎ 提供了大量工厂类，通过重载这些工厂类可以按需创建出用户实现的对象。

◎ 提供了大量的系统参数供用户按需设置，增强系统的场景定制性。

20.2.4 可扩展性

基于 Netty 的基础 NIO 框架，可以方便地进行应用层协议定制，例如 HTTP 协议栈、Thrift 协议栈、FTP 协议栈等。这些扩展不需要修改 Netty 的源码，直接基于 Netty 的二进制类库即可实现协议的扩展和定制。

目前，业界存在大量的基于 Netty 框架开发的协议，例如基于 Netty 的 HTTP 协议、Dubbo 协议、RocketMQ 内部私有协议等。

20.3 总结

本章首先对 Netty 的逻辑架构进行了分层和介绍，让读者能够从架构的层面了解 Netty，随后对 Netty 的关键架构质量属性进行了详细分析和介绍。通过对 Netty 架构的剖析和讲解，希望读者能够掌握如何设计高性能、高可靠性、可定制性和可扩展的软件架构。

第 21 章

Java 多线程编程在 Netty 中的应用

作为异步事件驱动、高性能的 NIO 框架，Netty 代码中大量运用了 Java 多线程编程技巧。并发编程处理的恰当与否，将直接影响架构的性能。本章通过对 Netty 源码的分析，结合并发编程的常用技巧，来讲解多线程编程在 Netty 中的应用。

本章主要内容包括：

◎ Java 内存模型与多线程编程

◎ Netty 的并发编程剖析

21.1 Java 内存模型与多线程编程

21.1.1 硬件的发展和多任务处理

随着硬件，特别是多核处理器的发展和价格的下降，多任务处理已经是所有操作系统必备的一项基本功能。在同一个时刻让计算机做多件事情，不仅是因为处理器的并行计算能力得到了很大提升，还有一个重要的原因是计算机的存储系统、网络通信等 I/O 性能与 CPU 的计算能力差距太大，导致程序的很大一部分执行时间被浪费在 I/O wait 上面，CPU 的强大运算能力没有得到充分地利用。

Java 提供了很多类库和工具用于降低并发编程的门槛，提升开发效率，一些开源的第三方软件也提供了额外的并发编程类库来方便 Java 开发者，使开发者将重心放在业务逻辑的设计和实现上，而不是处处考虑线程的同步和锁。但是，无论并发类库设计得如何完美，它都无法完全满足用户的需求。对于一个高级 Java 程序员来说，如果不懂得 Java 并发编程的原理，只懂得使用一些简单的并发类库和工具，是无法完全驾驭 Java 多线程这匹野马的。

21.1.2 Java 内存模型

JVM 规范定义了 Java 内存模型（Java Memory Model）来屏蔽掉各种操作系统、虚拟机实现厂商和硬件的内存访问差异，以确保 Java 程序在所有操作系统和平台上能够实现一次编写、到处运行的效果。

Java 内存模型的制定既要严谨，保证语义无歧义，还要尽量制定得宽松一些，允许各硬件和虚拟机实现厂商有足够的灵活性来充分利用硬件的特性提升 Java 的内存访问性能。随着 JDK 的发展，Java 的内存模型已经逐渐成熟起来。

1. 工作内存和主内存

Java 内存模型规定所有的变量都存储在主内存中（JVM 内存的一部分），每个线程有自己独立的工作内存，它保存了被该线程使用的变量的主内存复制。线程对这些变量的操作都在自己的工作内存中进行，不能直接操作主内存和其他工作内存中存储的变量或者变量副本。线程间的变量访问需通过主内存来完成，三者的关系如图 21-1 所示。

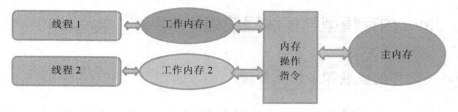

图 21-1　Java 内存访问模型

2. Java 内存交互协议

Java 内存模型定义了 8 种操作来完成主内存和工作内存的变量访问，具体如下。

- lock：主内存变量，把一个变量标识为某个线程独占的状态。
- unlock：主内存变量，把一个处于锁定状态变量释放出来，被释放后的变量才可以被其他线程锁定。
- read：主内存变量，把一个变量的值从主内存传输到线程的工作内存中，以便随后的 load 动作使用。
- load：工作内存变量，把 read 读取到的主内存中的变量值放入工作内存的变量副本中。
- use：工作内存变量，把工作内存中变量的值传递给 Java 虚拟机执行引擎，每当虚拟机遇到一个需要使用到变量值的字节码指令时，将会执行该操作。
- assign：工作内存变量，把从执行引擎接收到的变量的值赋值给工作变量，每当虚拟机遇到一个给变量赋值的字节码时，将会执行该操作。
- store：工作内存变量，把工作内存中一个变量的值传送到主内存中，以便随后的 write 操作使用。
- write：主内存变量，把 store 操作从工作内存中得到的变量值放入主内存的变量中。

3. Java 的线程

并发可以通过多种方式来实现，例如：单进程-单线程模型，通过在一台服务器上启动多个进程来实现多任务的并行处理。但是在 Java 语言中，通常是通过单进程-多线程的模型进行多任务的并发处理。因此，我们有必要熟悉一下 Java 的线程。

大家都知道，线程是比进程更轻量级的调度执行单元，它可以把进程的资源分配和调度执行分开，各个线程可以共享内存、I/O 等操作系统资源，但是又能够被操作系统发起的内核线程或者进程执行。各线程可以独立地启动、运行和停止，实现任务的解耦。

主流的操作系统提供了线程实现，目前实现线程的方式主要有三种，分别如下。

（1）内核线程（KLT）实现，这种线程由内核来完成线程切换，内核通过线程调度器对线程进行调度，并负责将线程任务映射到不同的处理器上。

（2）用户线程实现（UT），通常情况下，用户线程指的是完全建立在用户空间线程库上的线程，用户线程的创建、启动、运行、销毁和切换完全在用户态中完成，不需要内核的帮助，因此执行性能更高。

（3）混合实现，将内核线程和用户线程混合在一起使用的方式。

由于虚拟机规范并没有强制规定 Java 的线程必须使用哪种方式实现，因此，不同的操作系统实现的方式也可能存在差异。对于 SUN 的 JDK，在 Windows 和 Linux 操作系统上采用了内核线程的实现方式，在 Solaris 版本的 JDK 中，提供了一些专有的虚拟机线程参数，用于设置使用哪种线程模型。

21.2 Netty 的并发编程实践

21.2.1 对共享的可变数据进行正确的同步

关键字 synchronized 可以保证在同一时刻，只有一个线程可以执行某一个方法或者代码块。同步的作用不仅仅是互斥，它的另一个作用就是共享可变性，当某个线程修改了可变数据并释放锁后，其他线程可以获取被修改变量的最新值。如果没有正确的同步，这种修改对其他线程是不可见的。

下面我们就通过对 Netty 源码的分析，看看 Netty 是如何对并发可变数据进行正确同步的。

以 ServerBootstrap 为例进行分析，首先看它的 option 方法，如图 21-2 所示。

```
public <T> B option(ChannelOption<T> option, T value) {
    if (option == null) {
        throw new NullPointerException("option");
    }
    if (value == null) {
        synchronized (options) {
            options.remove(option);
        }
    } else {
        synchronized (options) {
            options.put(option, value);
        }
    }
    return (B) this;
}
```

图 21-2　同步关键字的使用

这个方法的作用是设置 ServerBootstrap 的 ServerSocketChannel 的 Socket 属性，它的属性集定义如下。

```
private final Map<ChannelOption<?>, Object> options = new LinkedHashMap<ChannelOption<?>, Object>();
```

由于是非线程安全的 LinkedHashMap，所以当多线程创建、访问和修改 LinkedHashMap 时，必须在外部进行必要的同步。LinkedHashMap 的 API DOC 对于线程安全的说明如图 21-3 所示。

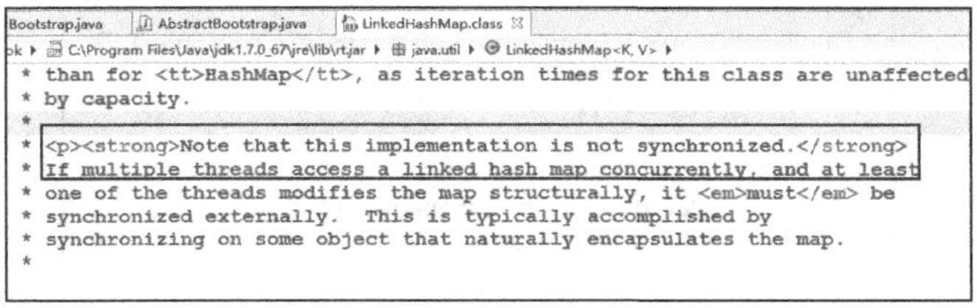

```
Bootstrap.java    AbstractBootstrap.java    LinkedHashMap.class
ok ▸ C:\Program Files\Java\jdk1.7.0_67\jre\lib\rt.jar ▸ java.util ▸ LinkedHashMap<K, V> ▸
 * than for <tt>HashMap</tt>, as iteration times for this class are unaffected
 * by capacity.
 *
 * <p><strong>Note that this implementation is not synchronized.</strong>
 * If multiple threads access a linked hash map concurrently, and at least
 * one of the threads modifies the map structurally, it <em>must</em> be
 * synchronized externally.  This is typically accomplished by
 * synchronizing on some object that naturally encapsulates the map.
 *
```

图 21-3　LinkedHashMap 线程安全 API 说明

由于 ServerBootstrap 是被外部使用者创建和使用的，我们无法保证它的方法和成员变量不被并发访问，因此，作为成员变量的 options 必须进行正确地同步。由于考虑到锁的范围需要尽可能的小，我们对传参的 option 和 value 的合法性判断不需要加锁。因此，代码才对两个判断分支独立加锁，保证锁的范围尽可能的细粒度。

Netty 加锁的地方非常多，大家在阅读代码的时候可能会有体会，为什么有的地方要加锁，有的地方却不需要？如果不需要，为什么？当你对锁的使用原理足够理解以后，对于这些锁的使用时机和技巧就会十分清楚了。

21.2.2　正确使用锁

很多刚接触多线程编程的开发者，虽然意识到了并发访问可变变量需要加锁，但是对于锁的范围、加锁的时机和锁的协同缺乏认识，往往会导致出现一些问题。下面笔者就结合 Netty 的代码来讲解下这方面的知识。

打开 ForkJoinTask，我们学习一些多线程同步和协作方面的技巧。首先是当条件不满足时阻塞某个任务，直到条件满足后再继续执行，代码如图 21-4 所示。

重点看框线中的代码，首先通过循环检测的方式对状态变量 status 进行判断，当它的状态大于等于 0 时，执行 wait()，阻塞当前的调度线程，直到 status 小于 0，唤醒所有被阻塞的线程，继续执行。这个方法有以下三个多线程的编程技巧需要说明。

```
private int externalAwaitDone() {
    int s;
    ForkJoinPool cp = ForkJoinPool.common;
    if ((s = status) >= 0) {
        if (cp != null) {
            if (this instanceof CountedCompleter)
                s = cp.externalHelpComplete((CountedCompleter<?>)this);
            else if (cp.tryExternalUnpush(this))
                s = doExec();
        }
        if (s >= 0 && (s = status) >= 0) {
            boolean interrupted = false;
            do {
                if (U.compareAndSwapInt(this, STATUS, s, s | SIGNAL)) {
                    synchronized (this) {
                        if (status >= 0) {
                            try {
                                wait();
                            } catch (InterruptedException ie) {
                                interrupted = true;
                            }
                        }
                        else
                            notifyAll();
                    }
                }
            } while ((s = status) >= 0);
```

图 21-4　多线程协作

（1）wait 方法用来使线程等待某个条件，它必须在同步块内部被调用，这个同步块通常会锁定当前对象实例。下面是这个模式的标准使用方式。

```
synchronized (this)
    {
    While(condition)
        Object.wait;
......
}
```

（2）始终使用 wait 循环来调用 wait 方法，永远不要在循环之外调用 wait 方法。这样做的原因是尽管并不满足被唤醒条件，但是由于其他线程调用 notifyAll()方法会导致被阻塞线程意外唤醒，此时执行条件并不满足，它将破坏被锁保护的约定关系，导致约束失效，引起意想不到的结果。

（3）唤醒线程，应该使用 notify 还是 notifyAll？当你不知道究竟该调用哪个方法时，保守的做法是调用 notifyAll 唤醒所有等待的线程。从优化的角度看，如果处于等待的所有线程都在等待同一个条件，而每次只有一个线程可以从这个条件中被唤醒，那么就应该选择调用 notify。

当多个线程共享同一个变量的时候,每个读或者写数据的操作方法都必须加锁进行同步,如果没有正确的同步,就无法保证一个线程所做的修改被其他线程共享。未能同步共享变量会造成程序的活性失败和安全性失败,这样的失败通常是难以调试和重现的,它们可能间歇性地出问题,可能随着并发的线程个数增加而失败,也可能在不同的虚拟机或者操作系统上存在不同的失败概率。因此,务必要保证锁的正确使用。下面这个案例,就是个典型的错误应用。

```
int size = 0;
public synchronized void increase()
{
    size++;
}
public int current()
{
    Return size;
}
```

21.2.3 volatile 的正确使用

长久以来大家对于 volatile 如何正确使用有很多的争议,既便是一些经验丰富的 Java 设计师,对于 volatile 和多线程编程的认识仍然存在误区。其实,volatile 的使用非常简单,只要理解了 Java 的内存模型和多线程编程的基础知识,正确使用 volatile 是不存在任何问题的。下面我们结合 Netty 的源码,对 volatile 的正确使用进行说明。

打开 NioEventLoop 的代码,我们来看控制 I/O 操作和其他任务运行比例的 ioRatio,它是 int 类型的变量,定义如下:

```
private volatile int ioRatio = 50;
```

我们发现,它被定义为 volatile,为什么呢?我们首先对 volatile 关键字进行说明,然后再结合 Netty 的代码进行分析。

关键字 volatile 是 Java 提供的最轻量级的同步机制,Java 内存模型对 volatile 专门定义了一些特殊的访问规则。下面我们就看它的规则。

当一个变量被 volatile 修饰后,它将具备以下两种特性。

◎ 线程可见性:当一个线程修改了被 volatile 修饰的变量后,无论是否加锁,其他线程都可以立即看到最新的修改，而普通变量却做不到这点。

◎ 禁止指令重排序优化，普通的变量仅仅保证在该方法的执行过程中所有依赖赋值结果的地方都能获取正确的结果，而不能保证变量赋值操作的顺序与程序代码的执行顺序一致。举个简单的例子说明下指令重排序优化问题，如图 21-5 所示。

```
public class ThreadStopExample {
    private static boolean stop;
    public static void main(String[] args) throws InterruptedException {
        java.lang.Thread workThread = new java.lang.Thread(new Runnable() {
            public void run() {
                int i = 0;
                while (!stop) {
                    i++;
                    try {
                        TimeUnit.SECONDS.sleep(1);
                    } catch (InterruptedException e) {
                        e.printStackTrace();
                    }
                }
            }
        });
        workThread.start();
        TimeUnit.SECONDS.sleep(3);
        stop = true;
    }
}
```

图 21-5　指令重排序和优化导致线程无法退出

我们预期程序会在 3s 后停止，但是实际上它会一直执行下去，原因就是虚拟机对代码进行了指令重排序和优化，优化后的指令如下。

```
if (!stop)
While(true)
    ......
```

重排序后的代码是无法发现 stop 被主线程修改的，因此无法停止运行。要解决这个问题，只要将 stop 前增加 volatile 修饰符即可。代码修改如图 21-6 所示。

再次运行，我们发现 3s 后程序退出，达到了预期效果，使用 volatile 解决了如下两个问题。

◎ main 线程对 stop 的修改在 workThread 线程中可见，也就是说 workThread 线程立即看到了其他线程对于 stop 变量的修改。

◎ 禁止指令重排序，防止因为重排序导致的并发访问逻辑混乱。

一些人认为使用 volatile 可以代替传统锁,提升并发性能,这个认识是错误的。volatile 仅仅解决了可见性的问题,但是它并不能保证互斥性,也就是说多个线程并发修改某个变量时,依旧会产生多线程问题。因此,不能靠 volatile 来完全替代传统的锁。

```
public class ThreadStopExample {
    private volatile static boolean stop;
    public static void main(String[] args) throws InterruptedException {
        java.lang.Thread workThread = new java.lang.Thread(new Runnable() {
            public void run() {
                int i = 0;
                while (!stop) {
                    i++;
                    try {
                        TimeUnit.SECONDS.sleep(1);
                    } catch (InterruptedException e) {
                        e.printStackTrace();
                    }
                }
            }
        });
        workThread.start();
        TimeUnit.SECONDS.sleep(3);
        stop = true;
    }
}
```

图 21-6　volatile 解决指令重排序和编译优化问题

根据经验总结,volatile 最适合使用的是一个线程写,其他线程读的场合,如果有多个线程并发写操作,仍然需要使用锁或者线程安全的容器或者原子变量来代替。

讲了 volatile 的原理之后,我们继续对 Netty 的源码做分析。上面讲到了 ioRatio 被定义成 volatile,下面看看代码为什么要这样定义。参见如图 21-7 所示代码。

```
final long ioTime = System.nanoTime() - ioStartTime;

final int ioRatio = this.ioRatio;
runAllTasks(ioTime * (100 - ioRatio) / ioRatio);
```

图 21-7　volatile 在 NioEventLoop 线程中的应用

通过代码分析我们发现,在 NioEventLoop 线程中,ioRatio 并没有被修改,它是只读操作。既然没有修改,为什么要定义成 volatile 呢?继续看代码,我们发现 NioEventLoop 提供了重新设置 I/O 执行时间比例的公共方法,接口如图 21-8 所示。

```
public void setIoRatio(int ioRatio) {
    if (ioRatio <= 0 || ioRatio >= 100) {
        throw new IllegalArgumentException("ioRatio: " + ioRatio + " (expected: 0 < ioRatio < 100)");
    }
    this.ioRatio = ioRatio;
}
```

图 21-8　修改 volatile 变量

首先，NioEventLoop 线程没有调用该方法，说明调整 I/O 执行时间比例是外部发起的操作，通常是由业务的线程调用该方法，重新设置该参数。这样就形成了一个线程写、一个线程读。根据前面针对 volatile 的应用总结，此时可以使用 volatile 来代替传统的 synchronized 关键字提升并发访问的性能。

Netty 中大量使用了 volatile 来修改成员变量，如果理解了 volatile 的应用场景，读懂 Netty volatile 的相关代码还是比较容易的。

21.2.4　CAS 指令和原子类

互斥同步最主要的问题就是进行线程阻塞和唤醒所带来的性能的额外损耗，因此这种同步被称为阻塞同步，它属于一种悲观的并发策略，我们称之为悲观锁。随着硬件和操作系统指令集的发展和优化，产生了非阻塞同步，被称为乐观锁。简单地说，就是先进行操作，操作完成之后再判断操作是否成功，是否有并发问题，如果有则进行失败补偿，如果没有就算操作成功，这样就从根本上避免了同步锁的弊端。

目前，在 Java 中应用最广泛的非阻塞同步就是 CAS，在 IA64、X86 指令集中通过 cmpxchg 指令完成 CAS 功能，在 sparc-TSO 中由 case 指令完成，在 ARM 和 PowerPC 架构下，需要使用一对 Idrex/strex 指令完成。

从 JDK1.5 以后，可以使用 CAS 操作，该操作由 sun.misc.Unsafe 类里的 compareAndSwapInt() 和 compareAndSwapLong() 等方法包装提供。通常情况下 sun.misc.Unsafe 类对于开发者是不可见的，因此，JDK 提供了很多 CAS 包装类简化开发者的使用，如 AtomicInteger。

下面，结合 Netty 的源码，我们对原子类的正确使用进行详细说明。

打开 ChannelOutboundBuffer 的代码，看看如何对发送的总字节数进行计数和更新操作，先看定义，如图 21-9 所示。

```
private static final AtomicLongFieldUpdater<ChannelOutboundBuffer>
TOTAL_PENDING_SIZE_UPDATER =
AtomicLongFieldUpdater.newUpdater(ChannelOutboundBuffer.class,
"totalPendingSize");

    private volatile long totalPendingSize;
```

图 21-9　原子类的应用

首先定义了一个 volatile 的变量，它可以保证某个线程对于 totalPendingSize 的修改可以被其他线程立即访问，但是，它无法保证多线程并发修改的安全性。紧接着又定义了一个 AtomicIntegerFieldUpdater 类型的变量 WTOTAL_PENDING_SIZE_UPDATER，实现 totalPendingSize 的原子更新，也就是保证 totalPendingSize 的多线程修改并发安全性，我们重点看 AtomicIntegerFieldUpdater 的 API 说明，如图 21-10 所示。

从 API 的说明可以看出，它主要用于实现 volatile 修饰的 int 变量的原子更新操作，对于使用者，必须通过类似 compareAndSet 或者 set 或者与这些操作等价的原子操作来保证更新的原子性，否则会导致问题。

```
/**
 * A reflection-based utility that enables atomic updates to
 * designated {@code volatile long} fields of designated classes.
 * This class is designed for use in atomic data structures in which
 * several fields of the same node are independently subject to atomic
 
 * updates.
 * <p>Note that the guarantees of the {@code compareAndSet}
 * method in this class are weaker than in other atomic classes.
 * Because this class cannot ensure that all uses of the field
 * are appropriate for purposes of atomic access, it can
 * guarantee atomicity only with respect to other invocations of
 */
public abstract class AtomicLongFieldUpdater<T> {
```

图 21-10　AtomicIntegerFieldUpdater Java DOC 说明

继续看代码，当执行 write 操作外发消息的时候，需要对外发的消息字节数进行统计汇总。由于调用 write 操作的既可以是 I/O 线程，也可以是业务的线程，还可能由业务线程池多个工作线程同时执行发送任务，因此，统计操作是多线程并发的，这也就是为什么要将计数器定义成 volatile 并使用原子更新类进行原子操作。下面看计数的代码，如图 21-11 所示。

```
long oldValue = totalPendingSize;
    long newWriteBufferSize = oldValue + size;
    while (!TOTAL_PENDING_SIZE_UPDATER.compareAndSet(this,
oldValue, newWriteBufferSize)) {
        oldValue = totalPendingSize;
        newWriteBufferSize = oldValue + size;
    }
```

图 21-11　通过自旋对计数器进行更新

首先，我们发现计数操作并没有使用锁，而是利用 CAS 自旋操作，通过 TOTAL_PENDING_SIZE_UPDATER.compareAndSet(this, oldValue, newWriteBufferSize)来判断本次原子操作是否成功。如果成功则退出循环，代码继续执行；如果失败，说明在本次操作的过程中计数器已经被其他线程更新成功，需要进入循环，首先对 oldValue 进行更新，代码如下。

```
oldValue = totalPendingSize;
```

然后重新对更新值进行计算。

```
newWriteBufferSize = oldValue + size;
```

继续循环进行 CAS 操作，直到成功。它跟 AtomicInteger 的 compareAndSet 操作类似。

使用 Java 自带的 Atomic 原子类，可以避免同步锁带来的并发访问性能降低的问题，减少犯错的机会。因此，Netty 中对于 int、long、boolean 等成员变量大量使用其原子类，减少了锁的应用，从而降低了频繁使用同步锁带来的性能下降。

21.2.5 线程安全类的应用

在 JDK1.5 的发行版本中，Java 平台新增了 java.util.concurrent，这个包中提供了一系列的线程安全集合、容器和线程池，利用这些新的线程安全类可以极大地降低 Java 多线程编程的难度，提升开发效率。

新的并发编程包中的工具可以分为如下 4 类。

◎ 线程池 Executor Framework 以及定时任务相关的类库，包括 Timer 等。

◎ 并发集合，包括 List、Queue、Map 和 Set 等。

◎ 新的同步器，例如读写锁 ReadWriteLock 等。

◎ 新的原子包装类，例如 AtomicInteger 等。

在实际编码过程中，我们建议通过使用线程池、Task（Runnable/Callable）、原子类和线程安全容器来代替传统的同步锁、wait 和 notify，以提升并发访问的性能、降低多线程编程的难度。

下面，针对新的线程并发包在 Netty 中的应用进行分析和说明，以期为大家的学习和应用提供指导。

首先看下线程安全容器在 Netty 中的应用。NioEventLoop 是 I/O 线程，负责网络读写操作，同时也执行一些非 I/O 的任务。例如事件通知、定时任务执行等，因此，它需要一个任务队列来缓存这些 Task。它的任务队列定义如图 21-12 所示。

```
@Override
protected Queue<Runnable> newTaskQueue() {
    // This event loop never calls takeTask()
    return new ConcurrentLinkedQueue<Runnable>();
}
```

图 21-12　线程任务队列定义

它是一个 ConcurrentLinkedQueue，我们看它的 API 说明，如图 21-13 所示。

```
/**
 * An unbounded thread-safe {@linkplain Queue queue} based on linked
 nodes.
 * This queue orders elements FIFO (first-in-first-out).
 * The <em>head</em> of the queue is that element that has been on the
 * queue the longest time.
 * The <em>tail</em> of the queue is that element that has been on the
 * queue the shortest time. New elements
 * are inserted at the tail of the queue, and the queue retrieval
 * operations obtain elements at the head of the queue.
 * A {@code ConcurrentLinkedQueue} is an appropriate choice when
 * many threads will share access to a common collection.
 * Like most other concurrent collection implementations, this class
 * does not permit the use of {@code null} elements.
```

图 21-13　ConcurrentLinkedQueue 线程安全文档

DOC 文档明确说明这个类是线程安全的，因此，对它进行读写操作不需要加锁。下面我们继续看下队列中增加一个任务，如图 21-14 所示。

```
protected void addTask(Runnable task) {
    if (task == null) {
        throw new NullPointerException("task");
    }
    if (isShutdown()) {
        reject();
    }
    taskQueue.add(task);
}
```

图 21-14　ConcurrentLinkedQueue 新增 Task

读取任务，也不需要加锁，如图 21-15 所示。

```
BlockingQueue<Runnable> taskQueue = (BlockingQueue<Runnable>) this.taskQueue;
for (;;) {
    ScheduledFutureTask<?> delayedTask = delayedTaskQueue.peek();
    if (delayedTask == null) {
        Runnable task = null;
        try {
            task = taskQueue.take();
            if (task == WAKEUP_TASK) {
                task = null;
            }
        } catch (InterruptedException e) {
            // Ignore
        }
        return task;
```

图 21-15　ConcurrentLinkedQueue 读取 Task

JDK 的线程安全容器底层采用了 CAS、volatile 和 ReadWriteLock 实现，相比于传统重量级的同步锁，采用了更轻量、细粒度的锁，因此，性能会更高。合理地应用这些线程安全容器，不仅能提升多线程并发访问的性能，还能降低开发难度。

下面我们看看线程池在 Netty 中的应用，打开 SingleThreadEventExecutor 看它是如何定义和使用线程池的。

首先定义了一个标准的线程池用于执行任务，代码如下。

```
private final Executor executor;
```

接着对它赋值并且进行初始化操作，代码如下。

```
this.addTaskWakesUp = addTaskWakesUp;
this.executor = executor;
taskQueue = newTaskQueue();
```

执行任务代码如图 21-16 所示。

```
public void execute(Runnable task) {
    if (task == null) {
        throw new NullPointerException("task");
    }
    boolean inEventLoop = inEventLoop();
    if (inEventLoop) {
        addTask(task);
    } else {
        startThread();
        addTask(task);
        if (isShutdown() && removeTask(task)) {
            reject();
        }
    }
}
```

图 21-16　SingleThreadEventExecutor 任务执行

我们发现，实际上执行任务就是先把任务加入到任务队列中，然后判断线程是否已经启动循环执行，如果不是则需要启动线程。启动线程代码如图 21-17 所示。

实际上就是执行当前线程的 run 方法，循环从任务队列中获取 Task 并执行，我们看它的子类 NioEventLoop 的 run 方法就能一目了然，如图 21-18 所示。

如图 21-19 中框线内所示，循环从任务队列中获取任务并执行。

```
private void startThread() {
    synchronized (stateLock) {
        if (state == ST_NOT_STARTED) {
            state = ST_STARTED;
            delayedTaskQueue.add(new ScheduledFutureTask<Void>(
                this, delayedTaskQueue, Executors.<Void>callable(new
PurgeTask(), null),
ScheduledFutureTask.deadlineNanos(SCHEDULE_PURGE_INTERVAL), -
SCHEDULE_PURGE_INTERVAL));
            doStartThread();
        }
    }
}
```

图 21-17　SingleThreadEventExecutor 启动新的线程

```
if (selectedKeys != null) {
    processSelectedKeysOptimized(selectedKeys.flip());
} else {
    processSelectedKeysPlain(selector.selectedKeys());
}
final long ioTime = System.nanoTime() - ioStartTime;

final int ioRatio = this.ioRatio;
runAllTasks(ioTime * (100 - ioRatio) / ioRatio);
```

图 21-18　按照 I/O 任务比例执行任务 Task

```
protected boolean runAllTasks(long timeoutNanos) {
    fetchFromDelayedQueue();
    Runnable task = pollTask();
    if (task == null) {
        return false;
    }

    final long deadline = ScheduledFutureTask.nanoTime() + timeoutNanos;
    long runTasks = 0;
    long lastExecutionTime;
    for (;;) {
        try {
            task.run();
        } catch (Throwable t) {
            logger.warn("A task raised an exception.", t);
        }
```

图 21-19　循环从任务队列中获取任务 Task 并执行

Netty 对 JDK 的线程池进行了封装和改造，但是，本质上仍然是利用了线程池和线程安全队列简化了多线程编程。

21.2.6　读写锁的应用

JDK1.5 新的并发编程工具包中新增了读写锁，它是个轻量级、细粒度的锁，合理地使用读写锁，相比于传统的同步锁，可以提升并发访问的性能和吞吐量，在读多写少的场景下，使用同步锁比同步块性能高一大截。

尽管在 JDK 1.6 之后，随着 JVM 团队对 JIT 即时编译器的不断优化，同步块和读写锁的性能差距缩小了很多，但是，读写锁的应用依然非常广泛。

下面对 Netty 中的读写锁应用进行分析，让大家掌握读写锁的用法。打开 HashedWheelTimer 代码，读写锁定义如下。

```
final int mask;
final ReadWriteLock lock = new ReentrantReadWriteLock();
```

当新增一个定时任务的时候使用了读锁（如图 21-20），用于感知 wheel 的变化。由于读锁是共享锁，所以当有多个线程同时调用 newTimeout 时，并不会互斥，这样，就提升了并发读的性能。

```
public Timeout newTimeout(TimerTask task, long delay, TimeUnit unit) {
    start();
    if (task == null) {
        throw new NullPointerException("task");
    }
    if (unit == null) {
        throw new NullPointerException("unit");
    }
    long deadline = System.nanoTime() + unit.toNanos(delay) - startTime;
    // Add the timeout to the wheel.
    HashedWheelTimeout timeout;
    lock.readLock().lock();
    try {
        timeout = new HashedWheelTimeout(task, deadline);
        if (workerState.get() == WORKER_STATE_SHUTDOWN) {
            throw new IllegalStateException("Cannot enqueue after shutdown");
        }
        wheel[timeout.stopIndex].add(timeout);
    } finally {
        lock.readLock().unlock();
    }
}
```

图 21-20　Read Lock 的使用

获取并删除所有过期的任务时，由于要从迭代器中删除任务，所以使用了写锁，如图 21-21 所示。

现将读写锁的使用场景总结如下。

- 主要用于读多写少的场景，用来替代传统的同步锁，以提升并发访问性能。
- 读写锁是可重入、可降级的，一个线程获取读写锁后，可以继续递归获取；从写锁可以降级为读锁，以便快速释放锁资源。
- ReentrantReadWriteLock 支持获取锁的公平策略，在某些特殊的应用场景下，可以提升并发访问的性能，同时兼顾线程等待公平性。

```
private void fetchExpiredTimeouts(
    List<HashedWheelTimeout> expiredTimeouts, long deadline) {

    lock.writeLock().lock();
    try {
        fetchExpiredTimeouts(expiredTimeouts, wheel[(int) (tick & mask)].iterator(), deadline);

    } finally {
        tick ++;
        lock.writeLock().unlock();
    }
}
```

图 21-21 Write Lock 的使用

- 读写锁支持非阻塞的尝试获取锁，如果获取失败，直接返回 false，而不是同步阻塞。这个功能在一些场景下非常有用。例如多个线程同步读写某个资源，当发生异常或者需要释放资源的时候，由哪个线程释放是个难题。因为某些资源不能重复释放或者重复执行，这样，可以通过 tryLock 方法尝试获取锁，如果拿不到，说明已经被其他线程占用，直接退出即可。
- 获取锁之后一定要释放锁，否则会发生锁溢出异常。通常的做法是通过 finally 块释放锁。如果是 tryLock，获取锁成功才需要释放锁。

21.2.7 线程安全性文档说明

当一个类的方法或者成员变量被并发使用的时候，这个类的行为如何，是该类与其客户端程序建立约定的重要组成部分。如果没有在这个类的文档中描述其行为的并发情况，使用这个类的程序员不得不做出某种假设。如果这些假设是错误的，这个程序就缺少必要的同步保护，会导致意想不到的并发问题，这些问题通常都是隐蔽和调试困难的。如果同步过度，会导致意外的性能下降，无论是发生何种情况，缺少线程安全性的说明文档，都会令开发人员非常沮丧，他们会对这些类库的使用小心翼翼，提心吊胆。

在 Netty 中，对于一些关键的类库，给出了线程安全性的 API DOC（图 21-22），尽管 Netty 的线程安全性并不是非常完善，但是，相比于一些做得更糟糕的产品，它还是迈出了重要的一步。

```
* <h3>Thread safety</h3>
* <p>
* A {@link ChannelHandler} can be added or removed at any time because a
{@link ChannelPipeline} is thread safe.
* For example, you can insert an encryption handler when sensitive information
is about to be exchanged, and remove it
* after the exchange.
*/
```

图 21-22　ChannelPipeline 的线程安全性说明

由于 ChannelPipeline 的应用非常广泛，因此，在 API 中对它的线程安全性进行了详细的说明，这样，开发者在调用 ChannelPipeline 的 API 时，就不用再额外地考虑线程同步和并发问题了。

21.2.8　不要依赖线程优先级

当有多个线程同时运行的时候，由线程调度器来决定哪些线程运行、哪些等待以及线程切换的时间点，由于各个操作系统的线程调度器实现大相径庭，因此，依赖 JDK 自带的线程优先级来设置线程优先级策略的方法是错误和非平台可移植的。所以，在任何情况下，程序都不能依赖 JDK 自带的线程优先级来保证执行顺序、比例和策略。

Netty 中默认的线程工厂实现类，开放了包含设置线程优先级字段的构造函数。这是个错误的决定，对于使用者来说，既然 JDK 类库提供了优先级字段，就会本能地认为它被正确地执行，但实际上 JDK 的线程优先级是无法跨平台正确运行的。图 21-23 提供了一个线程优先级的反面示例。

```
public DefaultThreadFactory(String poolName, boolean daemon, int priority)
{
    if (poolName == null) {
        throw new NullPointerException("poolName");
    }
    if (priority < Thread.MIN_PRIORITY || priority > Thread.MAX_PRIORITY)
    {
        throw new IllegalArgumentException(
                "priority: " + priority + " (expected: Thread.MIN_PRIORITY <= "
                + "priority <= Thread.MAX_PRIORITY)");
    }

    prefix = poolName + '-' + poolId.incrementAndGet() + '-';
    this.daemon = daemon;
    this.priority = priority;
}
```

图 21-23　线程优先级的反面示例

21.3　总结

本章首先介绍了 Java 内存模型和多线程编程的基础知识，然后结合 Netty 的源码分析常用的多线程编程方法和技巧。

通过本章节的讲解，希望读者可以学以致用，在后续的工作中恰到好处地使用 Java 并发编程技术，提高系统的并发处理能力，提升产品的性能。

第 22 章 高性能之道

作为一个高性能的 NIO 通信框架，Netty 被广泛应用于大数据处理、互联网消息中间件、游戏和金融行业等。大多数应用场景对底层的通信框架都有很高的性能要求，作为综合性能最高的 NIO 框架之一，Netty 可以完全满足不同领域对高性能通信的需求。

本章我们将从架构层对 Netty 的高性能设计和关键代码实现进行剖析，看 Netty 是如何支撑高性能网络通信的。

本章主要内容包括：

◎ RPC 调用性能模型分析
◎ Netty 高性能之道
◎ 主流 NIO 通信框架性能对比

22.1 RPC 调用性能模型分析

22.1.1 传统 RPC 调用性能差的三宗罪

"罪行"一：网络传输方式问题。传统的 RPC 框架或者基于 RMI 等方式的远程服务（过程）调用采用了同步阻塞 I/O，当客户端的并发压力或者网络时延增大之后，同步阻塞 I/O

会由于频繁的 wait 导致 I/O 线程经常性的阻塞,由于线程无法高效的工作,I/O 处理能力自然下降。

采用 BIO 通信模型的服务端,通常由一个独立的 Acceptor 线程负责监听客户端的连接,接收到客户端连接之后,为其创建一个新的线程处理请求消息,处理完成之后,返回应答消息给客户端,线程销毁,这就是典型的一请求一应答模型。该架构最大的问题就是不具备弹性伸缩能力,当并发访问量增加后,服务端的线程个数和并发访问数成线性正比,由于线程是 Java 虚拟机非常宝贵的系统资源,当线程数膨胀之后,系统的性能急剧下降,随着并发量的继续增加,可能会发生句柄溢出、线程堆栈溢出等问题,并导致服务器最终宕机。

"罪行"二:序列化性能差。Java 序列化存在如下几个典型问题:

(1) Java 序列化机制是 Java 内部的一种对象编解码技术,无法跨语言使用。例如对于异构系统之间的对接,Java 序列化后的码流需要能够通过其他语言反序列化成原始对象(副本),目前很难支持。

(2) 相比于其他开源的序列化框架,Java 序列化后的码流太大,无论是网络传输还是持久化到磁盘,都会导致额外的资源占用。

(3) 序列化性能差,资源占用率高(主要是 CPU 资源占用高)。

"罪行"三:线程模型问题。由于采用同步阻塞 I/O,这会导致每个 TCP 连接都占用 1 个线程,由于线程资源是 JVM 虚拟机非常宝贵的资源,当 I/O 读写阻塞导致线程无法及时释放时,会导致系统性能急剧下降,严重的甚至会导致虚拟机无法创建新的线程。

22.1.2 I/O 通信性能三原则

尽管影响 I/O 通信性能的因素非常多,但是从架构层面看主要有三个要素。

(1) 传输:用什么样的通道将数据发送给对方。可以选择 BIO、NIO 或者 AIO,I/O 模型在很大程度上决定了通信的性能;

(2) 协议:采用什么样的通信协议,HTTP 等公有协议或者内部私有协议。协议的选择不同,性能也不同。相比于公有协议,内部私有协议的性能通常可以被设计得更优;

(3) 线程:数据报如何读取?读取之后的编解码在哪个线程进行,编解码后的消息如何派发,Reactor 线程模型的不同,对性能的影响也非常大。

22.2 Netty 高性能之道

22.2.1 异步非阻塞通信

在 I/O 编程过程中，当需要同时处理多个客户端接入请求时，可以利用多线程或者 I/O 多路复用技术进行处理。I/O 多路复用技术通过把多个 I/O 的阻塞复用到同一个 select 的阻塞上，从而使得系统在单线程的情况下可以同时处理多个客户端请求。与传统的多线程/多进程模型比，I/O 多路复用的最大优势是系统开销小，系统不需要创建新的额外进程或者线程，也不需要维护这些进程和线程的运行，降低了系统的维护工作量，节省了系统资源。

JDK1.4 提供了对非阻塞 I/O（NIO）的支持，JDK1.5_update10 版本使用 epoll 替代了传统的 select/poll，极大地提升了 NIO 通信的性能。

与 Socket 和 ServerSocket 类相对应，NIO 也提供了 SocketChannel 和 ServerSocketChannel 两种不同的套接字通道实现。这两种新增的通道都支持阻塞和非阻塞两种模式。阻塞模式使用非常简单，但是性能和可靠性都不好，非阻塞模式则正好相反。开发人员一般可以根据自己的需要来选择合适的模式，一般来说，低负载、低并发的应用程序可以选择同步阻塞 I/O 以降低编程复杂度。但是对于高负载、高并发的网络应用，需要使用 NIO 的非阻塞模式进行开发。

Netty 的 I/O 线程 NioEventLoop 由于聚合了多路复用器 Selector，可以同时并发处理成百上千个客户端 SocketChannel。由于读写操作都是非阻塞的，这就可以充分提升 I/O 线程的运行效率，避免由频繁的 I/O 阻塞导致的线程挂起。另外，由于 Netty 采用了异步通信模式，一个 I/O 线程可以并发处理 N 个客户端连接和读写操作，这从根本上解决了传统同步阻塞 I/O 一连接一线程模型，架构的性能、弹性伸缩能力和可靠性都得到了极大的提升。

22.2.2 高效的 Reactor 线程模型

常用的 Reactor 线程模型有三种，分别如下。

（1）Reactor 单线程模型；

（2）Reactor 多线程模型；

（3）主从 Reactor 多线程模型

Reactor 单线程模型，指的是所有的 I/O 操作都在同一个 NIO 线程上面完成，NIO 线程的职责如下：

（1）作为 NIO 服务端，接收客户端的 TCP 连接；

（2）作为 NIO 客户端，向服务端发起 TCP 连接；

（3）读取通信对端的请求或者应答消息；

（4）向通信对端发送消息请求或者应答消息。

由于 Reactor 模式使用的是异步非阻塞 I/O，所有的 I/O 操作都不会导致阻塞，理论上一个线程可以独立处理所有 I/O 相关的操作。从架构层面看，一个 NIO 线程确实可以完成其承担的职责。例如，通过 Acceptor 接收客户端的 TCP 连接请求消息，链路建立成功之后，通过 Dispatch 将对应的 ByteBuffer 派发到指定的 Handler 上进行消息解码。用户 Handler 可以通过 NIO 线程将消息发送给客户端。

对于一些小容量应用场景，可以使用单线程模型，但是对于高负载、大并发的应用却不合适，主要原因如下。

（1）一个 NIO 线程同时处理成百上千的链路，性能上无法支撑。即便 NIO 线程的 CPU 负荷达到 100%，也无法满足海量消息的编码、解码、读取和发送；

（2）当 NIO 线程负载过重之后，处理速度将变慢，这会导致大量客户端连接超时，超时之后往往会进行重发，这更加重了 NIO 线程的负载，最终会导致大量消息积压和处理超时，NIO 线程会成为系统的性能瓶颈；

（3）可靠性问题。一旦 NIO 线程意外跑飞，或者进入死循环，会导致整个系统通信模块不可用，不能接收和处理外部消息，造成节点故障。

为了解决这些问题，演进出了 Reactor 多线程模型，下面我们一起学习下 Reactor 多线程模型。

Rector 多线程模型与单线程模型最大的区别就是有一组 NIO 线程处理 I/O 操作，它的特点如下。

（1）有一个专门的 NIO 线程——Acceptor 线程用于监听服务端，接收客户端的 TCP

连接请求；

（2）网络 I/O 操作——读、写等由一个 NIO 线程池负责，线程池可以采用标准的 JDK 线程池实现，它包含一个任务队列和 N 个可用的线程，由这些 NIO 线程负责消息的读取、解码、编码和发送；

（3）1 个 NIO 线程可以同时处理 N 条链路，但是 1 个链路只对应 1 个 NIO 线程，防止发生并发操作问题。

在绝大多数场景下，Reactor 多线程模型都可以满足性能需求；但是，在极特殊应用场景中，一个 NIO 线程负责监听和处理所有的客户端连接可能会存在性能问题。例如百万客户端并发连接，或者服务端需要对客户端的握手消息进行安全认证，认证本身非常损耗性能。在这类场景下，单独一个 Acceptor 线程可能会存在性能不足问题，为了解决性能问题，产生了第三种 Reactor 线程模型——主从 Reactor 多线程模型。

主从 Reactor 线程模型的特点是：服务端用于接收客户端连接的不再是个 1 个单独的 NIO 线程，而是一个独立的 NIO 线程池。Acceptor 接收到客户端 TCP 连接请求处理完成后（可能包含接入认证等），将新创建的 SocketChannel 注册到 I/O 线程池（sub reactor 线程池）的某个 I/O 线程上，由它负责 SocketChannel 的读写和编解码工作。Acceptor 线程池只用于客户端的登录、握手和安全认证，一旦链路建立成功，就将链路注册到后端 subReactor 线程池的 I/O 线程上，由 I/O 线程负责后续的 I/O 操作。

利用主从 NIO 线程模型，可以解决 1 个服务端监听线程无法有效处理所有客户端连接的性能不足问题。因此，在 Netty 的官方 Demo 中，推荐使用该线程模型。

事实上，Netty 的线程模型并非固定不变，通过在启动辅助类中创建不同的 EventLoopGroup 实例并进行适当的参数配置，就可以支持上述三种 Reactor 线程模型。正是因为 Netty 对 Reactor 线程模型的支持提供了灵活的定制能力，所以可以满足不同业务场景的性能诉求。

Netty 单线程模型服务端代码示例如下。

```
EventLoopGroup reactorGroup = new NioEventLoopGroup(1);
    try {
        ServerBootstrap b = new ServerBootstrap();
        b.group(reactorGroup, reactorGroup)
            .channel(NioServerSocketChannel.class)
//后续代码省略…
```

Netty 多线程模型代码示例如下。

```
EventLoopGroup acceptorGroup = new NioEventLoopGroup(1);
EventLoopGroup IOGroup = new NioEventLoopGroup();
    try {
        ServerBootstrap b = new ServerBootstrap();
        b.group(acceptorGroup, IOGroup)
            .channel(NioServerSocketChannel.class)
//后续代码省略…
```

Netty 主从线程模型代码示例如下。

```
EventLoopGroup acceptorGroup = new NioEventLoopGroup();
EventLoopGroup IOGroup = new NioEventLoopGroup();
    try {
        ServerBootstrap b = new ServerBootstrap();
        b.group(acceptorGroup, IOGroup)
            .channel(NioServerSocketChannel.class)
//后续代码省略…
```

从上面示例代码可以看出，构造方法的参数和线程组实例化个数不同，就能灵活地切换到不同的 Reactor 线程模型上，用户使用起来非常方便。事实上，并没有标准的最优线程配置策略，用户需要在理解 Netty 线程模型的基础之上，根据业务的实际需求选择合适的线程模型和参数。

22.2.3 无锁化的串行设计

在大多数场景下，并行多线程处理可以提升系统的并发性能。但是，如果对于共享资源的并发访问处理不当，会带来严重的锁竞争，这最终会导致性能的下降。为了尽可能地避免锁竞争带来的性能损耗，可以通过串行化设计，即消息的处理尽可能在同一个线程内完成，期间不进行线程切换，这样就避免了多线程竞争和同步锁。

为了尽可能提升性能，Netty 采用了串行无锁化设计，在 I/O 线程内部进行串行操作，避免多线程竞争导致的性能下降。表面上看，串行化设计似乎 CPU 利用率不高，并发程度不够。但是，通过调整 NIO 线程池的线程参数，可以同时启动多个串行化的线程并行运行，这种局部无锁化的串行线程设计相比一个队列——多个工作线程模型性能更优。

Netty 的串行化设计工作原理图如图 22-1 所示。

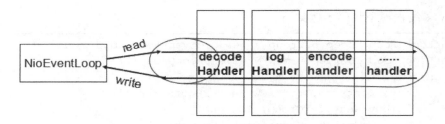

图 22-1　Netty 串行化设计工作原理

Netty 的 NioEventLoop 读取到消息之后，直接调用 ChannelPipeline 的 fireChannelRead(Object msg)，只要用户不主动切换线程，一直会由 NioEventLoop 调用到用户的 Handler，期间不进行线程切换。这种串行化处理方式避免了多线程操作导致的锁的竞争，从性能角度看是最优的。

22.2.4　高效的并发编程

Netty 的高效并发编程主要体现在如下几点。

（1）volatile 的大量、正确使用；

（2）CAS 和原子类的广泛使用；

（3）线程安全容器的使用；

（4）通过读写锁提升并发性能。

具体细节请参考"Java 多线程编程在 Netty 中的应用"一章的内容。

22.2.5　高性能的序列化框架

影响序列化性能的关键因素总结如下。

（1）序列化后的码流大小（网络带宽的占用）；

（2）序列化&反序列化的性能（CPU 资源占用）；

（3）是否支持跨语言（异构系统的对接和开发语言切换）。

Netty 默认提供了对 Google Protobuf 的支持，通过扩展 Netty 的编解码接口，用户可以实现其他的高性能序列化框架，例如 Thrift 的压缩二进制编解码框架。

下面我们一起看下不同序列化&反序列化框架性能对比（耗时），如图 22-2 所示。

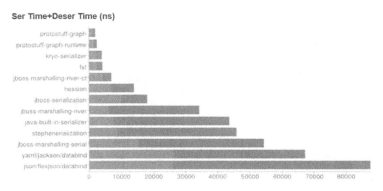

图 22-2　不同序列化&反序列化框架性能对比（耗时）

下面我们一起看下不同序列化&反序列化框架性能对比（序列化码流大小），如图 22-3 所示。

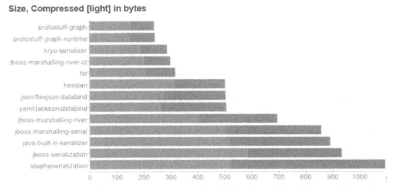

图 22-3　不同序列化&反序列化框架性能对比（序列化码流大小）

不同的应用场景对序列化框架的需求也不同，对于高性能应用场景，Netty 默认提供了 Google 的 Protobuf 二进制序列化框架，如果用户对其他二进制序列化框架有需求，也可以基于 Netty 提供的编解码框架扩展实现。

22.2.6　零拷贝

很多用户都听说过 Netty 具有"零拷贝"功能，但是具体体现在哪里又说不清楚，本小节就详细对 Netty 的"零拷贝"功能进行讲解。

Netty 的"零拷贝"主要体现在如下三个方面。

第一种情况。Netty 的接收和发送 ByteBuffer 采用 DIRECT BUFFERS，使用堆外直接内存进行 Socket 读写，不需要进行字节缓冲区的二次拷贝。如果使用传统的堆内存（HEAP BUFFERS）进行 Socket 读写，JVM 会将堆内存 Buffer 拷贝一份到直接内存中，然后才写入 Socket 中。相比于堆外直接内存，消息在发送过程中多了一次缓冲区的内存拷贝。

JDK 内存拷贝的代码如下。

```
    static int write(FileDescriptor paramFileDescriptor, ByteBuffer paramByteBuffer, long paramLong, NativeDispatcher paramNativeDispatcher)
      throws IOException
  {
    if ((paramByteBuffer instanceof DirectBuffer)) {
      Return writeFromNativeBuffer(paramFileDescriptor, paramByteBuffer, paramLong, paramNativeDispatcher);
    }
    int i = paramByteBuffer.position();
    int j = paramByteBuffer.limit();
    assert (i <= j);
    int k = i <= j ? j - i : 0;
    ByteBuffer localByteBuffer = Util.getTemporaryDirectBuffer(k);
    try {
      localByteBuffer.put(paramByteBuffer);
      localByteBuffer.flip();
      paramByteBuffer.position(i);
      int m = writeFromNativeBuffer(paramFileDescriptor, localByteBuffer, paramLong, paramNativeDispatcher);
      if (m > 0)
      {
        paramByteBuffer.position(i + m);
      }
      return m;
    } finally {
      Util.offerFirstTemporaryDirectBuffer(localByteBuffer);
    }
  }
```

下面我们对 Netty 收发数据报"零拷贝"相关代码进行分析。数据报读取代码如下。

```
ByteBuf byteBuf = null;
int messages = 0;
boolean close = false;
try {
    int byteBufCapacity = allocHandle.guess();
    int totalReadAmount = 0;
    do {
        byteBuf = allocator.ioBuffer(byteBufCapacity);
        int writable = byteBuf.writableBytes();
        int localReadAmount = doReadBytes(byteBuf);
```

接收缓冲区 ByteBuffer 的分配由 ChannelConfig 负责，下面我们继续分析 ChannelConfig 创建 ByteBufAllocator 的代码。

```
private static final RecvByteBufAllocator DEFAULT_RCVBUF_ALLOCATOR = AdaptiveRecvByteBufAllocator.DEFAULT;
```

继续分析 AdaptiveRecvByteBufAllocator 的实现，查看它的内存分配接口，代码如下。

```
@Override
public ByteBuf allocate(ByteBufAllocator alloc) {
            return alloc.ioBuffer(nextReceiveBufferSize);
}
```

调用了 ByteBufAllocator 的 ioBuffer 接口，下面我们看下 ioBuffer 接口的 API DOC：
```
/**
    * Allocate a {@link ByteBuf}, preferably a direct buffer which is suitable for I/O.
    */
```

为了提升 I/O 操作的性能，默认使用 direct buffer，这就避免了读写数据报的二次内存拷贝，实现了读写 Socket 的"零拷贝"功能。

下面我们继续看第二种"零拷贝"的实现 CompositeByteBuf，它对外将多个 ByteBuf 封装成一个 ByteBuf，对外提供统一封装后的 ByteBuf 接口，它的类定义如图 22-4 所示。

图 22-4 CompositeByteBuf 继承关系图

通过继承关系我们可以看出，CompositeByteBuf 实际就是个 ByteBuf 的装饰器，它将多个 ByteBuf 组合成一个集合，然后对外提供统一的 ByteBuf 接口，相关定义如下。

```
public class CompositeByteBuf extends AbstractReferenceCountedByteBuf {
    private final ResourceLeak leak;
    private final ByteBufAllocator alloc;
    private final boolean direct;
    private final List<Component> components = new ArrayList<Component>();
    private final int maxNumComponents;
```

添加 ByteBuf，不需要做内存拷贝，相关代码如下。

```
// No need to consolidate - just add a component to the list.
        Component c = new Component(buffer.order(ByteOrder.BIG_ENDIAN).slice());
        if (cIndex == components.size()) {
            components.add(c);
            if (cIndex == 0) {
                c.endOffset = readableBytes;
            } else {
                Component prev = components.get(cIndex - 1);
                c.offset = prev.endOffset;
                c.endOffset = c.offset + readableBytes;
            }
        } else {
            components.add(cIndex, c);
            updateComponentOffsets(cIndex);
        }
```

第三种"零拷贝"就是文件传输，Netty 文件传输类 DefaultFileRegion 通过 transferTo 方法将文件发送到目标 Channel 中，下面重点看 FileChannel 的 transferTo 方法，它的 API DOC 说明如图 22-5 所示。

```
 long java.nio.channels.FileChannel.transferTo(long position, long count,
    WritableByteChannel target) throws IOException
Transfers bytes from this channel's file to the given writable byte channel.

An attempt is made to read up to count bytes starting at the given position in this channel's file and
write them to the target channel. An invocation of this method may or may not transfer all of the
requested bytes; whether or not it does so depends upon the natures and states of the channels. Fewer
than the requested number of bytes are transferred if this channel's file contains fewer than count
bytes starting at the given position, or if the target channel is non-blocking and it has fewer than
count bytes free in its output buffer.

This method does not modify this channel's position. If the given position is greater than the file's
current size then no bytes are transferred. If the target channel has a position then bytes are written
starting at that position and then the position is incremented by the number of bytes written.

This method is potentially much more efficient than a simple loop that reads from this channel and writes
to the target channel. Many operating systems can transfer bytes directly from the filesystem cache to
the target channel without actually copying them.
```

图 22-5 transferTo 方法的 API DOC 说明

很多操作系统直接将文件缓冲区的内容发送到目标 Channel 中，而不需要通过循环拷贝的方式，这是一种更加高效的传输方式，提升了传输性能，降低了 CPU 和内存占用，实现了文件传输的"零拷贝"。

22.2.7 内存池

随着 JVM 虚拟机和 JIT 即时编译技术的发展，对象的分配和回收是个非常轻量级的工作。但是对于缓冲区 Buffer，情况却稍有不同，特别是对于堆外直接内存的分配和回收，是一件耗时的操作。为了尽量重用缓冲区，Netty 提供了基于内存池的缓冲区重用机制。下面我们一起看下 Netty ByteBuf 的实现，如图 22-6 所示。

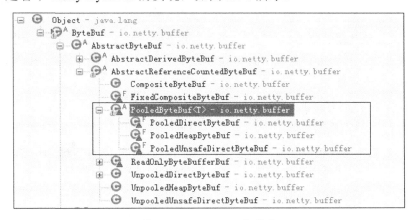

图 22-6 ByteBuf 内存池

Netty 提供了多种内存管理策略，通过在启动辅助类中配置相关参数，可以实现差异化的定制。

下面通过性能测试，我们看下基于内存池循环利用的 ByteBuf 和普通 ByteBuf 的性能差异。

测试场景一：使用内存池分配器创建直接内存缓冲区，代码示例如下。

```
int loop = 3000000;
    long startTime = System.currentTimeMillis();
    ByteBuf poolBuffer = null;
    for (int i = 0; i < loop; i++) {
        poolBuffer = PooledByteBufAllocator.DEFAULT.directBuffer(1024);
        poolBuffer.writeBytes(CONTENT);
```

```
        poolBuffer.release();
    }
```

测试场景二：使用非堆内存分配器创建的直接内存缓冲区。

```
long startTime2 = System.currentTimeMillis();
    ByteBuf buffer = null;
    for (int i = 0; i < loop; i++) {
        buffer = Unpooled.directBuffer(1024);
        buffer.writeBytes(CONTENT);
    }
```

各执行 300 万次，性能对比结果如下。

```
The PooledByteBuf execute 300W times writing operation cost time is : 4125 ms
================================================================================
The unPooledByteBuf execute 300W times writing operation cost time is : 95312 ms
```

性能测试表明，采用内存池的 ByteBuf 相比于朝生夕灭的 ByteBuf，性能高 23 倍左右（性能数据与使用场景强相关）。

下面简单分析下 Netty 内存池的内存分配关键代码。

```
@Override
    public ByteBuf directBuffer(int initialCapacity, int maxCapacity) {
        if (initialCapacity == 0 && maxCapacity == 0) {
            return emptyBuf;
        }
        validate(initialCapacity, maxCapacity);
        return newDirectBuffer(initialCapacity, maxCapacity);
    }
```

继续看 newDirectBuffer 方法，发现它是一个抽象方法，由 AbstractByteBufAllocator 的子类负责实现，子类实现代码如图 22-7 所示。

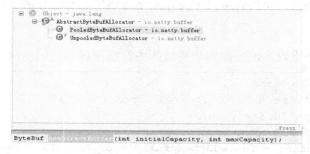

图 22-7　子类实现代码

代码跳转到 PooledByteBufAllocator 的 newDirectBuffer 方法，从 Cache 中获取内存区域 PoolArena，调用它的 allocate 方法进行内存分配。

```
protected ByteBuf newDirectBuffer(int initialCapacity, int maxCapacity) {
    PoolThreadCache cache = threadCache.get();
    PoolArena<ByteBuffer> directArena = cache.directArena;
    ByteBuf buf;
    if (directArena != null) {
        buf = directArena.allocate(cache, initialCapacity, maxCapacity);
    } else {
        if (PlatformDependent.hasUnsafe()) {
            buf = new UnpooledUnsafeDirectByteBuf(this, initialCapacity, maxCapacity);
        } else {
            buf = new UnpooledDirectByteBuf(this, initialCapacity, maxCapacity);
        }
    }
    return toLeakAwareBuffer(buf);
}
```

PoolArena 的 allocate 方法如下。

```
PooledByteBuf<T> allocate(PoolThreadCache cache, int reqCapacity, int maxCapacity) {
    PooledByteBuf<T> buf = newByteBuf(maxCapacity);
    allocate(cache, buf, reqCapacity);
    return buf;
}
```

重点分析 newByteBuf 的实现，它同样是个抽象方法，由子类 DirectArena 和 HeapArena 来实现不同类型的缓冲区分配，由于测试用例使用的是堆外内存，因此重点分析 DirectArena 的实现。如图 22-8 所示。

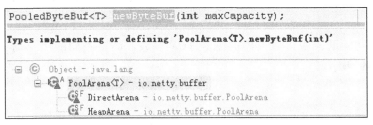

图 22-8 PoolArena 继承关系图

执行 PooledDirectByteBuf 的 newInstance 方法，代码如下。

```
static PooledDirectByteBuf newInstance(int maxCapacity) {
    PooledDirectByteBuf buf = RECYCLER.get();
    buf.setRefCnt(1);
    buf.maxCapacity(maxCapacity);
    return buf;
}
```

通过 RECYCLER 的 get 方法循环使用 ByteBuf 对象，如果是非内存池实现，则直接创建一个新的 ByteBuf 对象。从缓冲池中获取 ByteBuf 之后，调用 AbstractReferenceCountedByteBuf 的 setRefCnt 方法设置引用计数器，用于对象的引用计数和内存回收（类似 JVM 垃圾回收机制）。

22.2.8 灵活的 TCP 参数配置能力

合理设置 TCP 参数在某些场景下对于性能的提升可以起到显著的效果，例如 SO_RCVBUF 和 SO_SNDBUF。如果设置不当，对性能的影响是非常大的。下面我们总结下对性能影响比较大的几个配置项。

（1）SO_RCVBUF 和 SO_SNDBUF：通常建议值为 128KB 或者 256KB；

（2）SO_TCPNODELAY：NAGLE 算法通过将缓冲区内的小封包自动相连，组成较大的封包，阻止大量小封包的发送阻塞网络，从而提高网络应用效率。但是对于时延敏感的应用场景需要关闭该优化算法；

（3）软中断：如果 Linux 内核版本支持 RPS（2.6.35 以上版本），开启 RPS 后可以实现软中断，提升网络吞吐量。RPS 根据数据包的源地址，目的地址以及目的和源端口，计算出一个 hash 值，然后根据这个 hash 值来选择软中断运行的 CPU。从上层来看，也就是说将每个连接和 CPU 绑定，并通过这个 hash 值，来均衡软中断在多个 CPU 上，提升网络并行处理性能。

Netty 在启动辅助类中可以灵活的配置 TCP 参数，满足不同的用户场景。相关配置接口定义如图 22-9 所示。

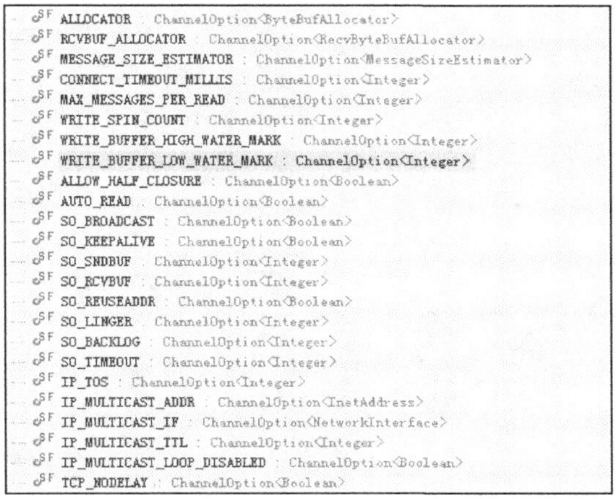

图 22-9　Netty TCP 参数表

22.3　主流 NIO 框架性能对比

无论是 Netty 的官方性能测试数据，还是携带业务实际场景的性能测试，Netty 在各个 NIO 框架中综合性能是最高的。下面，我们来看 Netty 官方提供的不同业务场景性能测试数据，如图 22-10、图 22-11、图 22-12、图 22-13 所示。

Performance Comparison between NIO Frameworks

Size=64 Loopback

	Netty	Frmwrk A	Frmwrk B	Frmwrk C	Frmwrk D
1	0.858	0.689	0.779	0.760	0.617
5	3.032	2.511	1.839	1.281	2.576
10	4.902	2.909	2.027	1.590	3.735
50	6.152	4.581	3.782	2.777	2.623
100	6.562	6.042	3.910	3.331	3.574
500	6.668	4.648	3.936	4.454	4.001
1,000	6.212	3.987	3.577	4.555	3.539
5,000	4.945	3.285	3.782	4.369	2.793
10,000	4.828	3.859	3.802	4.271	2.882

Size=128 Loopback

	Netty	Frmwrk A	Frmwrk B	Frmwrk C	Frmwrk D
1	1.245	1.198	1.569	1.031	1.054
5	6.065	4.079	3.717	3.402	4.810
10	10.023	7.428	4.081	4.637	6.936
50	12.008	8.528	7.573	4.345	5.287
100	12.232	8.091	7.948	4.780	7.134
500	13.274	8.097	7.819	6.648	8.050
1,000	11.741	7.948	7.223	8.342	7.022
5,000	9.706	6.814	7.397	8.759	5.807
10,000	9.605	7.267	7.685	8.384	6.181

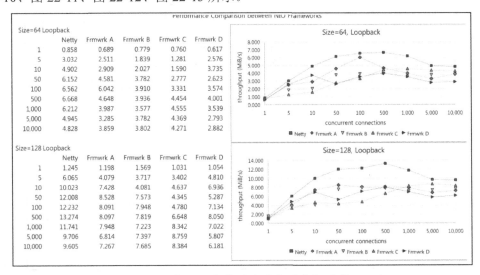

图 22-10　64 和 128 字节测试消息（本机网卡回环）

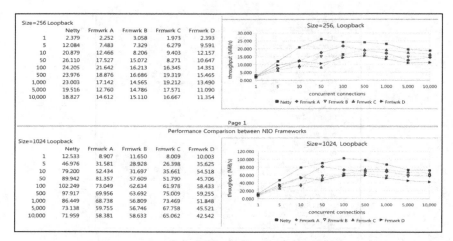

图 22-11　256 和 1K 字节测试消息（本机网卡回环）

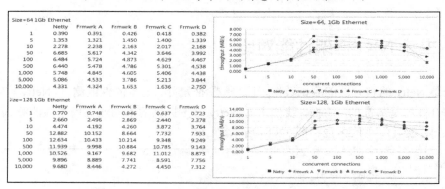

图 22-12　64 和 128 字节测试消息（跨主机通信）

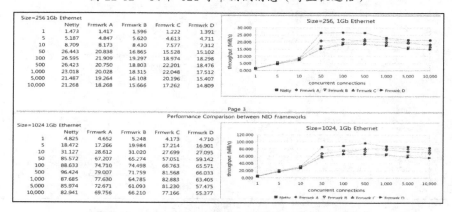

图 22-13　256 和 1K 字节测试消息（跨主机通信）

22.4 总结

本章首先对影响网络通信性能的三要素进行了分析说明，然后通过对 Netty 架构性能模型和设计细节的分析，了解 Netty 是如何做到高性能的。最后，通过对流行的 NIO 框架进行性能对比，用实际的性能测试数据来验证 Netty 的高性能。

第 23 章

可靠性

作为一个基础的 NIO 通信框架，Netty 被广泛应用于大数据处理、互联网消息中间件、游戏和金融行业等。不同的行业对软件的可靠性需求不同，例如对通信软件的可靠性要求往往需要达到 5 个 9。

本章我们将从架构层对 Netty 的可靠性设计和关键代码实现进行剖析，看 Netty 是如何做到高可靠性的。

本章主要内容包括：

◎ 可靠性需求

◎ Netty 高可靠性设计

◎ 优化建议

23.1 可靠性需求

23.1.1 宕机的代价

毕马威国际（KPMG International）在对 46 个国家的 74 家运营商进行调查后发现，

全球通信行业每年的收益流失约为 400 亿美元，占总收入的 1%～3%。导致收益流失的因素有多种，主要原因就是计费 BUG 和各类宕机导致的业务中断。

美国太平洋时间 8 月 16 日下午 3 点 50 分到 3 点 55 分（北京时间 8 月 17 日 6 点 50 分到 6 点 55 分），谷歌遭遇了宕机。根据事后统计，短短的 5 分钟，谷歌损失了 54.5 万美元。也就是服务每中断一分钟，损失就达 10.8 万美元。

2013 年，从美国东部时间 8 月 19 日下午 2 点 45 分开始，有用户率先发现了亚马逊网站出现宕机，大约在 20 多分钟后又恢复正常。此次宕机让亚马逊每分钟损失近 6.7 万美元，在宕机期间，消费者无法通过 Amazon.com、亚马逊移动端以及 Amazon.ca 等网站进行购物。

从以上几个案例可以看出，宕机事故对企业的正常运营和利润已经带来了非常大的挑战，随着业务的发展和软件规模的不断膨胀，对软件的可靠性要求越来越高。

23.1.2　Netty 可靠性需求

我们从 Netty 的主要应用场景和 Netty 的运行环境两个维度分析 Netty 的可靠性需求。首先分析 Netty 的主要应用场景。

Netty 的主要应用场景如下。

1．RPC 框架的基础网络通信框架：主要用于分布式节点之间的通信和数据交换，在各个业务领域均有典型的应用，例如阿里的分布式服务框架 Dubbo、消息队列 RocketMQ、大数据处理 Hadoop 的基础通信和序列化框架 Avro。

2．私有协议的基础通信框架：例如 Thrift 协议、Dubbo 协议等。

3．公有协议的基础通信框架：例如 HTTP 协议、SMPP 协议等。

从运行环境上看，基于 Netty 开发的应用面临的是网络环境也不同，手游服务运行的 GSM/3G/WIFI 网络环境可靠性差，偶尔会出现闪断、网络单通等问题。互联网应用在业务高峰期会出现网络拥堵，而且各地用户的网络环境差别也很大，部分地区网速和网络质量不高。

从应用场景看，Netty 是基础的通信框架，一旦出现 Bug，轻则需要重启应用，重则可能导致整个业务中断。它的可靠性会影响整个业务集群的数据通信和交换，在当今以分布式为主的软件架构体系中，通信中断就意味着整个业务中断，分布式架构下对通信的可靠性要求非常高。

从运行环境看，Netty 会面临恶劣的网络环境，这就要求它自身的可靠性要足够好，平台能够解决的可靠性问题需要由 Netty 自身来解决，否则会导致上层用户关注过多的底层故障，这将降低 Netty 的易用性，同时增加用户的开发和运维成本。

Netty 的可靠性是如此重要，它的任何故障都可能会导致业务中断，蒙受巨大的经济损失。因此，Netty 在版本的迭代中不断加入新的可靠性特性来满足用户日益增长的高可靠和健壮性需求。

下一节我们将带你一起揭秘 Netty 的高可靠性设计并分析核心代码的实现。

23.2 Netty 高可靠性设计

23.2.1 网络通信类故障

23.2.1.1 客户端连接超时

在传统的同步阻塞编程模式下，客户端 Socket 发起网络连接，往往需要指定连接超时时间，这样做的目的主要有两个。

1. 在同步阻塞 I/O 模型中，连接操作是同步阻塞的，如果不设置超时时间，客户端 I/O 线程可能会被长时间阻塞，这会导致系统可用 I/O 线程数的减少。

2. 业务层需要：大多数系统都会对业务流程执行时间有限制，例如 WEB 交互类的响应时间要小于 3S。客户端设置连接超时时间是为了实现业务层的超时。

JDK 原生的 Socket 连接接口定义如下：

```
/**
 * Connects this socket to the server with a specified timeout value.
 * A timeout of zero is interpreted as an infinite timeout. The connection
 * will then block until established or an error occurs.
 *
 * @param   endpoint the <code>SocketAddress</code>
 * @param   timeout  the timeout value to be used in milliseconds.
 * @throws  IOException if an error occurs during the connection
 * @throws  SocketTimeoutException if timeout expires before connecting
 * @throws  java.nio.channels.IllegalBlockingModeException
```

```
*              if this socket has an associated channel,
*              and the channel is in non-blocking mode
* @throws  IllegalArgumentException if endpoint is null or is a
*          SocketAddress subclass not supported by this socket
* @since 1.4
* @spec JSR-51
*/
public void connect(SocketAddress endpoint, int timeout) throws
IOException {......}
```

用户调用 Socket 的 connect 方法将被阻塞,直到连接成功或者发生连接超时等异常。

对于 NIO 的 SocketChannel,在非阻塞模式下,它会直接返回连接结果,如果没有连接成功,也没有发生 IO 异常,则需要将 SocketChannel 注册到 Selector 上监听连接结果。所以,异步连接的超时无法在 API 层面直接设置,而是需要通过定时器来主动监测。

下面我们首先看下 JDK NIO 类库的 SocketChannel 连接接口定义:

```
/**
 * Connects this channel's socket.
 * <p> If this channel is in non-blocking mode then an invocation of this
 * method initiates a non-blocking connection operation. If the connection
 * is established immediately, as can happen with a local connection, then
 * this method returns <tt>true</tt>. Otherwise this method returns
 * <tt>false</tt> and the connection operation must later be completed by
 * invoking the {@link #finishConnect finishConnect} method.
 */
public abstract boolean connect(SocketAddress remote) throws
IOException;
```

从上面的接口定义可以看出,NIO 类库并没有现成的连接超时接口供用户直接使用,如果要在 NIO 编程中支持连接超时,往往需要 NIO 框架或者用户自己封装实现。

下面我们看下 Netty 是如何支持连接超时的,首先,在创建 NIO 客户端的时候,可以配置连接超时参数:

```
EventLoopGroup group = new NioEventLoopGroup();
try {
    Bootstrap b = new Bootstrap();
    b.group(group).channel(NioSocketChannel.class)
        .option(ChannelOption.TCP_NODELAY, true)
        .option(ChannelOption.CONNECT_TIMEOUT_MILLIS, 3000)
```

设置完连接超时之后，Netty 在发起连接的时候，会根据超时时间创建 ScheduledFuture 挂载在 Reactor 线程上，用于定时监测是否发生连接超时，相关代码如下：

```
// Schedule connect timeout.
int connectTimeoutMillis = config().getConnectTimeoutMillis();
        if (connectTimeoutMillis > 0) {
            connectTimeoutFuture = eventLoop().schedule(new Runnable() {
                @Override
                public void run() {
                    ChannelPromise connectPromise = AbstractNioChannel.this.connectPromise;
                    ConnectTimeoutException cause =
                        new ConnectTimeoutException("connection timed out: " + remoteAddress);
                    if (connectPromise != null && connectPromise.tryFailure(cause)) {
                        close(voidPromise());
                    }
                }
            }, connectTimeoutMillis, TimeUnit.MILLISECONDS);
        }
```

创建连接超时定时任务之后，会由 NioEventLoop 负责执行。如果已经连接超时，但是服务端仍然没有返回 TCP 握手应答，则关闭连接，代码如上所示。

如果在超时期限内处理完成连接操作，则取消连接超时定时任务，相关代码如下：

```
@Override
public void finishConnect() {
    //以上代码省略......
        } finally {
            // Check for null as the connectTimeoutFuture is only created if a connectTimeoutMillis > 0 is used
            if (connectTimeoutFuture != null) {
                connectTimeoutFuture.cancel(false);
            }
            connectPromise = null;
        }
    }
```

Netty 的客户端连接超时参数与其他常用的 TCP 参数一起配置，使用起来非常方便，上层用户不用关心底层的超时实现机制。这既满足了用户的个性化需求，又实现了故障的分层隔离。

23.2.1.2 通信对端强制关闭连接

在客户端和服务端正常通信过程中，如果发生网络闪断、对方进程突然宕机或者其他非正常关闭链路事件时，TCP 链路就会发生异常。由于 TCP 是全双工的，通信双方都需要关闭和释放 Socket 句柄才不会发生句柄的泄漏。

在实际的 NIO 编程过程中，我们经常会发现由于句柄没有被及时关闭导致的功能和可靠性问题。究其原因总结如下：

1．IO 的读写等操作并非仅仅集中在 Reactor 线程内部，用户上层的一些定制行为可能会导致 IO 操作的外逸，例如业务自定义心跳机制。这些定制行为加大了统一异常处理的难度，IO 操作越发散，故障发生的概率就越大；

2．一些异常分支没有考虑到，由于外部环境诱因导致程序进入这些分支，就会引起故障。

下面我们通过故障模拟，看 Netty 是如何处理对端链路强制关闭异常的。首先启动 Netty 服务端和客户端，TCP 链路建立成功之后，双方维持该链路，查看链路状态，结果如图 23-1 所示。

图 23-1 查看链路状态

强制关闭客户端，模拟客户端宕机，服务端控制台打印如图 23-2 所示异常。

图 23-2 模拟 TCP 链路故障

从堆栈信息可以判断，服务端已经监控到客户端强制关闭了连接，下面我们看下服务端是否已经释放了连接句柄，再次执行 netstat 命令，执行结果如图 23-3 所示。

```
C:\Documents and Settings\Administrator>netstat -ano|find "8080"
  TCP    0.0.0.0:8080           0.0.0.0:0              LISTENING       4112
```

图 23-3　查看故障链路状态

从执行结果可以看出，服务端已经关闭了和客户端的 TCP 连接，句柄资源正常释放。由此可以得出结论，Netty 底层已经自动对该故障进行了处理。

下面我们一起看下 Netty 是如何感知到链路关闭异常并进行正确处理的，查看 AbstractByteBuf 的 writeBytes 方法，它负责将指定 Channel 的缓冲区数据写入到 ByteBuf 中，详细代码如下：

```java
@Override
public int writeBytes(ScatteringByteChannel in, int length) throws IOException {
    ensureWritable(length);
    int writtenBytes = setBytes(writerIndex, in, length);
    if (writtenBytes > 0) {
        writerIndex += writtenBytes;
    }
    return writtenBytes;
}
```

在调用 SocketChannel 的 read 方法时发生了 IOException，从 Channel 中读取数据报道缓冲区中的代码如下：

```java
/**
 * Transfers the content of the specified source channel to this buffer
 * starting at the specified absolute {@code index}.
 * This method does not modify {@code readerIndex} or {@code writerIndex} of
 * this buffer.
 *
 * @param length the maximum number of bytes to transfer
 *
 * @return the actual number of bytes read in from the specified channel.
 *         {@code -1} if the specified channel is closed.
 */
public abstract int  setBytes(int index, ScatteringByteChannel in, int length) throws IOException;
```

为了保证 IO 异常被统一处理，该异常向上抛，由 NioByteUnsafe 进行统一异常处理，代码如下：

```
private void closeOnRead(ChannelPipeline pipeline) {
        SelectionKey key = selectionKey();
        setInputShutdown();
        if (isOpen()) {
            if (Boolean.TRUE.equals(config().getOption(ChannelOption.ALLOW_HALF_CLOSURE))) {
                key.interestOps(key.interestOps() & ~readInterestOp);
pipeline.fireUserEventTriggered(ChannelInputShutdownEvent.INSTANCE);
            } else {
                close(voidPromise());
            }
        }
    }
```

23.2.1.3 链路关闭

对于短连接协议，例如 HTTP 协议，通信双方数据交互完成之后，通常按照双方的约定由服务端关闭连接，客户端获得 TCP 连接关闭请求之后，关闭自身的 Socket 连接，双方正式断开连接。

在实际的 NIO 编程过程中，经常存在一种误区：认为只要是对方关闭连接，就会发生 IO 异常，捕获 IO 异常之后再关闭连接即可。实际上，连接的合法关闭不会发生 IO 异常，它是一种正常场景，如果遗漏了该场景的判断和处理就会导致连接句柄泄漏。

下面我们一起模拟故障，看 Netty 是如何处理的。测试场景设计如下：改造下 Netty 客户端，双发链路建立成功之后，等待 120S，客户端正常关闭链路。看服务端是否能够感知并释放句柄资源。

首先启动 Netty 客户端和服务端，双方 TCP 链路连接正常，如图 23-4 所示。

```
C:\Documents and Settings\Administrator>netstat -ano|find "8080"
  TCP    0.0.0.0:8080           0.0.0.0:0              LISTENING       5032
  TCP    127.0.0.1:3410         127.0.0.1:8080         ESTABLISHED     3848
  TCP    127.0.0.1:8080         127.0.0.1:3410         ESTABLISHED     5032
```

图 23-4　TCP 连接状态正常

120S 之后，客户端关闭连接，进程退出，为了能够看到整个处理过程，我们在服务端的 Reactor 线程处设置断点，先不做处理，此时链路状态如图 23-5 所示。

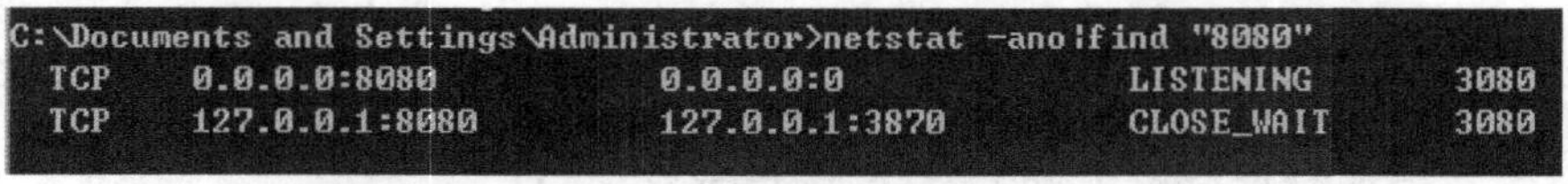

```
C:\Documents and Settings\Administrator>netstat -ano|find "8080"
  TCP    0.0.0.0:8080           0.0.0.0:0              LISTENING       3080
  TCP    127.0.0.1:8080         127.0.0.1:3870         CLOSE_WAIT      3080
```

图 23-5　TCP 连接句柄等待释放

从上图可以看出，此时服务端并没有关闭 Socket 连接，链路处于 CLOSE_WAIT 状态，放开代码让服务端执行完，结果如图 23-6 所示。

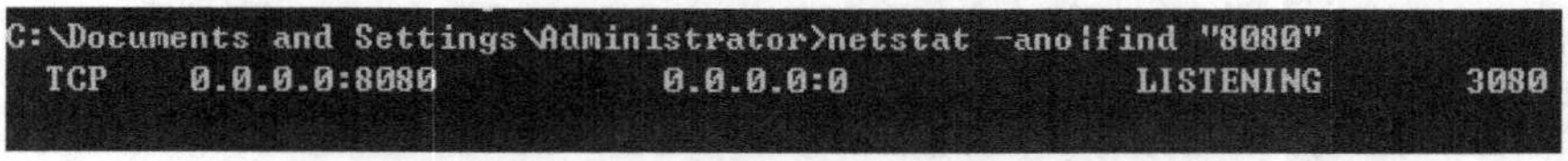

```
C:\Documents and Settings\Administrator>netstat -ano|find "8080"
  TCP    0.0.0.0:8080           0.0.0.0:0              LISTENING       3080
```

图 23-6　TCP 连接句柄正常释放

下面我们一起看下服务端是如何判断出客户端关闭连接的，当连接被对方合法关闭后，被关闭的 SocketChannel 会处于就绪状态，SocketChannel 的 read 操作返回值为-1，说明连接已经被关闭，NioByteUnsafe 的 read()代码片段如下：

```
int localReadAmount = doReadBytes(byteBuf);
if (localReadAmount <= 0) {
 // not was read release the buffer
 byteBuf.release();
 close = localReadAmount < 0;
  break;
 }
```

如果 SocketChannel 被设置为非阻塞，则它的 read 操作可能返回三个值：

1．大于 0：表示读取到了字节数；

2．等于 0：没有读取到消息，可能 TCP 处于 Keep-Alive 状态，接收到的是 TCP 握手消息；

3．-1：连接已经被对方关闭。

Netty 通过判断 Channel read 操作的返回值进行不同的逻辑处理，如果返回-1，说明链路已经关闭，则调用 closeOnRead 方法关闭句柄，释放资源，代码如下：

```
if (close) {
      closeOnRead(pipeline);
      close = false;
   }
```

己方或者对方主动关闭链接并不属于异常场景，因此不会产生 Exception 事件通知 Pipeline。

23.2.1.4 定制 I/O 故障

在大多数场景下，当底层网络发生故障的时候，应该由底层的 NIO 框架负责释放资源，处理异常等。上层的业务应用不需要关心底层的处理细节。但是，在一些特殊的场景下，用户可能需要感知这些异常，并针对这些异常进行定制处理，例如：

1. 客户端的断连重连机制；

2. 消息的缓存重发；

3. 接口日志中详细记录故障细节；

4. 运维相关功能，例如告警、触发邮件/短信等

Netty 的处理策略是发生 I/O 异常，底层的资源由它负责释放，同时将异常堆栈信息以事件的形式通知给上层用户，由用户对异常进行定制。这种处理机制既保证了异常处理的安全性，也向上层提供了灵活的定制能力。

具体接口定义以及默认实现(ChannelHandlerAdapter 类)如下：

```
@Skip
@Override
public void exceptionCaught(ChannelHandlerContext ctx, Throwable cause) throws Exception {
        ctx.fireExceptionCaught(cause);
}
```

23.2.2 链路的有效性检测

当网络发生单通、连接被防火墙 Hang 住、长时间 GC 或者通信线程发生非预期异常时，会导致链路不可用且不易被及时发现。特别是异常发生在凌晨业务低谷期间，当早晨业务高峰期到来时，由于链路不可用会导致瞬间的大批量业务失败或者超时，这将对系统的可靠性产生重大的威胁。

从技术层面看，要解决链路的可靠性问题，必须周期性的对链路进行有效性检测。目前最流行和通用的做法就是心跳检测。

心跳检测机制分为三个层面：

1．TCP 层面的心跳检测，即 TCP 的 Keep-Alive 机制，它的作用域是整个 TCP 协议栈；

2．协议层的心跳检测，主要存在于长连接协议中。例如 SMPP 协议；

3．应用层的心跳检测，它主要由各业务产品通过约定方式定时给对方发送心跳消息实现。

心跳检测的目的就是确认当前链路可用，对方活着并且能够正常接收和发送消息。做为高可靠的 NIO 框架，Netty 也提供了心跳检测机制，下面我们一起熟悉下心跳的检测原理。

心跳检测的原理示意图如图 23-7 所示。

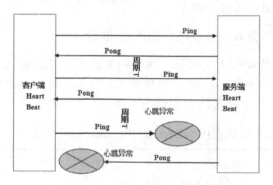

图 23-7　心跳检测的原理示意图

不同的协议，心跳检测机制也存在差异，归纳起来主要分为两类。

1．Ping-Pong 型心跳：由通信一方定时发送 Ping 消息，对方接收到 Ping 消息之后，立即返回 Pong 应答消息给对方，属于请求-响应型心跳。

2．Ping-Ping 型心跳：不区分心跳请求和应答，由通信双方按照约定定时向对方发送心跳 Ping 消息，它属于双向心跳。

心跳检测策略如下：

1．连续 N 次心跳检测都没有收到对方的 Pong 应答消息或者 Ping 请求消息，则认为链路已经发生逻辑失效，这被称作心跳超时。

2．读取和发送心跳消息的时候如何直接发生了 IO 异常，说明链路已经失效，这被称为心跳失败。

无论发生心跳超时还是心跳失败，都需要关闭链路，由客户端发起重连操作，保证链路能够恢复正常。

Netty 的心跳检测实际上是利用了链路空闲检测机制实现的，相关代码包路径如图 23-8 所示。

图 23-8　相关代码包路径

Netty 提供的空闲检测机制分为三种：

1．读空闲，链路持续时间 t 没有读取到任何消息；

2．写空闲，链路持续时间 t 没有发送任何消息；

3．读写空闲，链路持续时间 t 没有接收或者发送任何消息。

Netty 的默认读写空闲机制是发生超时异常，关闭连接，但是，我们可以定制它的超时实现机制，以便支持不同的用户场景。

WriteTimeoutHandler 的超时接口如下：

```
protected void writeTimedOut(ChannelHandlerContext ctx) throws Exception {
    if (!closed) {
```

```
        ctx.fireExceptionCaught(WriteTimeoutException.INSTANCE);
        ctx.close();
        closed = true;
    }
}
```

ReadTimeoutHandler 的超时接口实现如下：

```
protected void readTimedOut(ChannelHandlerContext ctx) throws Exception {
    if (!closed) {
        ctx.fireExceptionCaught(ReadTimeoutException.INSTANCE);
        ctx.close();
        closed = true;
    }
}
```

链路空闲的接口实现如下：

```
protected void channelIdle(ChannelHandlerContext ctx, IdleStateEvent evt) throws Exception {
    ctx.fireUserEventTriggered(evt);
}
```

链路空闲的时候并没有关闭链路，而是触发 IdleStateEvent 事件，用户订阅 IdleStateEvent 事件，用于自定义逻辑处理，例如关闭链路、客户端发起重新连接、告警和打印日志等。利用 Netty 提供的链路空闲检测机制，可以非常灵活的实现协议层的心跳检测。

23.2.3 Reactor 线程的保护

Reactor 线程是 IO 操作的核心，NIO 框架的发动机，一旦出现故障，将会导致挂载在其上面的多路用复用器和多个链路无法正常工作。因此它的可靠性要求非常高。

笔者就曾经遇到过因为异常处理不当导致 Reactor 线程跑飞，大量业务请求处理失败的故障。下面我们一起看下 Netty 是如何有效提升 Reactor 线程的可靠性的。

23.2.3.1 异常处理要谨慎

尽管 Reactor 线程主要处理 IO 操作，发生的异常通常是 IO 异常，但是，实际上在一些特殊场景下会发生非 IO 异常，如果仅仅捕获 IO 异常可能就会导致 Reactor 线程跑飞。为了防止发生这种意外，在循环体内一定要捕获 Throwable，而不是 IO 异常或者 Exception。

Netty 的 NioEventLoop 代码如下：

```
protected void run() {
    for (;;) {
        oldWakenUp = wakenUp.getAndSet(false);
        try {
            此处代码省略......
        } catch (Throwable t) {
            logger.warn("Unexpected exception in the selector loop.", t);
            try {
                Thread.sleep(1000);
            } catch (InterruptedException e) {
                // Ignore.
            }
        }
    }
}
```

捕获 Throwable 之后，即便发生了意外未知对异常，线程也不会跑飞，它休眠 1S，防止死循环导致的异常绕接，然后继续恢复执行。这样处理的核心理念就是：

1．某个消息的异常不应该导致整条链路不可用；

2．某条链路不可用不应该导致其他链路不可用；

3．某个进程不可用不应该导致其他集群节点不可用。

23.2.3.2 规避 NIO BUG

通常情况下，死循环是可检测、可预防但是无法完全避免的。Reactor 线程通常处理的都是 IO 相关的操作，因此我们重点关注 IO 层面的死循环。

JDK NIO 类库最著名的就是 epoll bug 了，它会导致 Selector 空轮询，IO 线程 CPU 100%，严重影响系统的安全性和可靠性。

SUN 在 JKD1.6 update18 版本声称解决了该 BUG，但是根据业界的测试和大家的反馈，直到 JDK1.7 的早期版本，该 BUG 依然存在，并没有完全被修复。发生该 BUG 的主机资源占用图如图 23-9 所示。

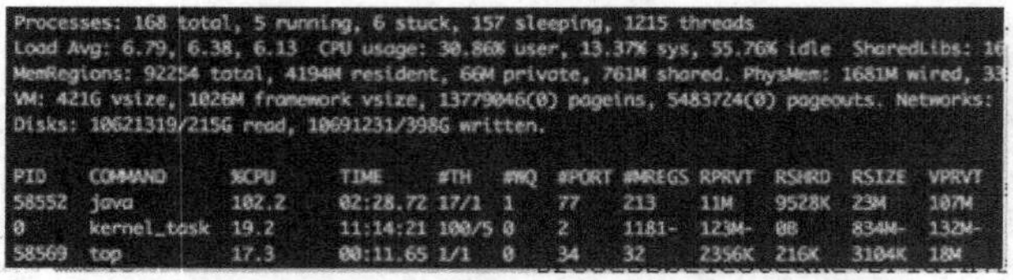

图 23-9 epoll bug CPU 空轮询

SUN 在解决该 BUG 的问题上不给力，直到 JDK1.7 版本也没有完全修复。使用者只能从 NIO 框架层面进行问题规避，下面我们看下 Netty 是如何解决该问题的。

Netty 的解决策略：

1. 根据该 BUG 的特征，首先侦测该 BUG 是否发生；

2. 将问题 Selector 上注册的 Channel 转移到新建的 Selector 上；

3. 老的问题 Selector 关闭，使用新建的 Selector 替换。

下面具体看下代码，首先检测是否发生了该 BUG(NioEventLoop 的 select())：

```
for (;;) {
    //以上代码省略......
    int selectedKeys = selector.select(timeoutMillis);
    selectCnt ++;
    if (selectedKeys != 0 || oldWakenUp || wakenUp.get() || hasTasks()) {
        break;
    }
    if (SELECTOR_AUTO_REBUILD_THRESHOLD > 0 &&
        selectCnt >= SELECTOR_AUTO_REBUILD_THRESHOLD) { //Bug 发生，处理......}
```

一旦检测发生该 BUG，则重建 Selector，代码如下：

```
if (SELECTOR_AUTO_REBUILD_THRESHOLD > 0 &&
        selectCnt >= SELECTOR_AUTO_REBUILD_THRESHOLD) {
    // The selector returned prematurely many times in a row.
    // Rebuild the selector to work around the problem.
    logger.warn(
            "Selector.select() returned prematurely {} times in a row; rebuilding selector.",
            selectCnt);
```

```
            rebuildSelector();
        //此处代码省略......
        }
```

重建完成之后,替换老的 Selector,代码如下:

```
        final Selector newSelector;
        if (oldSelector == null) {
            return;
        }
        try {
            newSelector = openSelector(); //打开新的 Selector
        } catch (Exception e) {
            logger.warn("Failed to create a new Selector.", e);
            return;
        }
//此处代码省略......
 selector = newSelector;
        try {
            oldSelector.close();
        } catch (Throwable t) {
            if (logger.isWarnEnabled()) {
                logger.warn("Failed to close the old Selector.", t);
            }
        }
```

经过大量生产系统的运行验证,Netty 的规避策略可以解决 epoll bug 导致的 IO 线程 CPU 死循环问题。

23.2.4 内存保护

NIO 通信的内存保护主要集中在如下几点:

1. 链路总数的控制:每条链路都包含接收和发送缓冲区,链路个数太多容易导致内存溢出;
2. 单个缓冲区的上限控制:防止非法长度或者消息过大导致内存溢出;
3. 缓冲区内存释放:防止因为缓冲区使用不当导致的内存泄露。
4. NIO 消息发送队列的长度上限控制。

23.2.4.1 缓冲区的内存泄漏保护

为了提升内存的利用率，Netty 提供了内存池和对象池。但是，基于缓存池实现以后需要对内存的申请和释放进行严格的管理，否则很容易导致内存泄漏。

如果不采用内存池技术实现，每次对象都是以方法的局部变量形式被创建，使用完成之后，只要不再继续引用它，JVM 会自动释放。但是，一旦引入内存池机制，对象的生命周期将由内存池负责管理，这通常是个全局引用，如果不显式释放 JVM 是不会回收这部分内存的。

对于 Netty 的用户而言，使用者的技术水平差异很大，一些对 JVM 内存模型和内存泄漏机制不了解的用户，可能只记得申请内存，忘记主动释放内存，特别是 JAVA 程序员。

为了防止因为用户遗漏导致内存泄漏，Netty 在 Pipe line 的尾 Handler 中自动对内存进行释放，TailHandler 的内存回收代码如下：

```
@Override
public void channelRead(ChannelHandlerContext ctx, Object msg) throws Exception {
    try {
        logger.debug(
                "Discarded inbound message {} that reached at the tail of the pipeline. " +
                "Please check your pipeline configuration.", msg);
    } finally {
        ReferenceCountUtil.release(msg);
    }
}
```

对于内存池，实际就是将缓冲区重新放到内存池中循环使用，PooledByteBuf 的内存回收代码如下：

```
protected final void deallocate() {
    if (handle >= 0) {
        final long handle = this.handle;
        this.handle = -1;
        memory = null;
        chunk.arena.free(chunk, handle);
        recycle();
    }
}
```

对于实现了 AbstractReferenceCountedByteBuf 的 ByteBuf，内存申请、使用和释放的时候 Netty 都会自动进行引用计数检测，防止非法使用内存。

23.2.4.2　缓冲区溢出保护

做过协议栈的读者都知道，当我们对消息进行解码的时候，需要创建缓冲区。缓冲区的创建方式通常有两种：

1．容量预分配，在实际读写过程中如果不够再扩展；

2．根据协议消息长度创建缓冲区。

在实际的商用环境中，如果遇到畸形码流攻击、协议消息编码异常、消息丢包等问题时，可能会解析到一个超长的长度字段。笔者曾经遇到过类似问题，报文长度字段值竟然是 2G 多，由于代码的一个分支没有对长度上限做有效保护，结果导致内存溢出。系统重启后几秒内再次内存溢出，幸好及时定位出问题根因，险些酿成严重的事故。

Netty 提供了编解码框架，因此对于解码缓冲区的上限保护就显得非常重要。下面，我们看下 Netty 是如何对缓冲区进行上限保护的：

首先，在内存分配的时候指定缓冲区长度上限：

```
/**
 * Allocate a {@link ByteBuf} with the given initial capacity and the given
 * maximal capacity. If it is a direct or heap buffer depends on the actual
 * implementation.
 */
ByteBuf buffer(int initialCapacity, int maxCapacity);
```

其次，在对缓冲区进行写入操作的时候，如果缓冲区容量不足需要扩展，首先对最大容量进行判断，如果扩展后的容量超过上限，则拒绝扩展：

```
@Override
public ByteBuf capacity(int newCapacity) {
    ensureAccessible();
    if (newCapacity < 0 || newCapacity > maxCapacity()) {
        throw new IllegalArgumentException("newCapacity: " + newCapacity);
    }
```

在消息解码的时候，对消息长度进行判断，如果超过最大容量上限，则抛出解码异常，拒绝分配内存，以 LengthFieldBasedFrameDecoder 的 decode 方法为例进行说明：

```
if (frameLength > maxFrameLength) {
        long discard = frameLength - in.readableBytes();
        tooLongFrameLength = frameLength;
        if (discard < 0) {
            in.skipBytes((int) frameLength);
        } else {
            discardingTooLongFrame = true;
            bytesToDiscard = discard;
            in.skipBytes(in.readableBytes());
        }
        failIfNecessary(true);
        return null;
    }
```

23.2.5 流量整形

大多数的商用系统都有多个网元或者部件组成，例如参与短信互动，会涉及手机、基站、短信中心、短信网关、SP/CP 等网元。不同网元或者部件的处理性能不同。为了防止因为浪涌业务或者下游网元性能低导致下游网元被压垮，有时候需要系统提供流量整形功能。

下面我们一起看下流量整形(traffic shaping)的定义：流量整形（Traffic Shaping）是一种主动调整流量输出速率的措施。一个典型应用是基于下游网络结点的 TP 指标来控制本地流量的输出。流量整形与流量监管的主要区别在于，流量整形对流量监管中需要丢弃的报文进行缓存——通常是将它们放入缓冲区或队列内，也称流量整形（Traffic Shaping，简称 TS）。当令牌桶有足够的令牌时，再均匀的向外发送这些被缓存的报文。流量整形与流量监管的另一区别是，整形可能会增加延迟，而监管几乎不引入额外的延迟。

流量整形的原理示意图如图 23-10 所示。

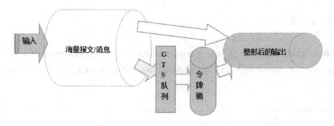

图 23-10　流量整形的原理示意图

作为高性能的 NIO 框架，Netty 的流量整形有两个作用：

1. 防止由于上下游网元性能不均衡导致下游网元被压垮，业务流程中断；

2. 防止由于通信模块接收消息过快，后端业务线程处理不及时导致的"撑死"问题。

下面我们就具体学习下 Netty 的流量整形功能。

23.2.5.1　全局流量整形

全局流量整形的作用范围是进程级的，无论你创建了多少个 Channel，它的作用域针对所有的 Channel。

用户可以通过参数设置：报文的接收速率、报文的发送速率、整形周期。GlobalTrafficShapingHandler 的接口定义如下所示：

```
public  GlobalTrafficShapingHandler(ScheduledExecutorService  executor, long writeLimit,
        long readLimit, long checkInterval) {
    super(writeLimit, readLimit, checkInterval);
    createGlobalTrafficCounter(executor);
}
```

Netty 流量整形的原理是：对每次读取到的 ByteBuf 可写字节数进行计算，获取当前的报文流量，然后与流量整形阈值对比。如果已经达到或者超过了阈值。则计算等待时间 delay，将当前的 ByteBuf 放到定时任务 Task 中缓存，由定时任务线程池在延迟 delay 之后继续处理该 ByteBuf。相关代码如下：

```
    public void channelRead(final ChannelHandlerContext ctx, final Object msg) throws Exception {
        long size = calculateSize(msg);
        long curtime = System.currentTimeMillis();
        if (trafficCounter != null) {
            trafficCounter.bytesRecvFlowControl(size);
            if (readLimit == 0) {
                // no action
                ctx.fireChannelRead(msg);
                return;
            }
//后续代码省略......
```

如果达到整形阈值，则对新接收的 ByteBuf 进行缓存，放入线程池的消息队列中，稍后处理，代码如下：

```
else {
                Runnable bufferUpdateTask = new Runnable() {
                    @Override
                    public void run() {
                        ctx.fireChannelRead(msg);
                    }
                };
                ctx.executor().schedule(bufferUpdateTask,           wait,
TimeUnit.MILLISECONDS);
                return;
            }
```

定时任务的延时时间根据检测周期 T 和流量整形阈值计算得来，代码如下：

```
    private static long getTimeToWait(long limit, long bytes, long lastTime,
long curtime) {
        long interval = curtime - lastTime;
        if (interval <= 0) {
            // Time is too short, so just lets continue
            return 0;
        }
        return (bytes * 1000 / limit - interval) / 10 * 10;
    }
```

需要指出的是，流量整形的阈值 limit 越大，流量整形的精度越高，流量整形功能是可靠性的一种保障，它无法做到 100%的精确。这个跟后端的编解码以及缓冲区的处理策略相关，此处不再赘述。感兴趣的朋友可以思考下，Netty 为什么不做到 100%的精确。

流量整形与流控的最大区别在于流控会拒绝消息，流量整形不拒绝和丢弃消息，无论接收量多大，它总能以近似恒定的速度下发消息，跟变压器的原理和功能类似。

23.2.5.2 链路级流量整形

除了全局流量整形，Netty 也支持链路级的流量整形，ChannelTrafficShapingHandler 接口定义如下：

```
    public ChannelTrafficShapingHandler(long writeLimit,
            long readLimit, long checkInterval) {
```

```
        super(writeLimit, readLimit, checkInterval);
    }
```

单链路流量整形与全局流量整形的最大区别就是它以单个链路为作用域,可以对不同的链路设置不同的整形策略。

它的实现原理与全局流量整形类似,我们不再赘述。值得说明的是,Netty 支持用户自定义流量整形策略,通过继承 AbstractTrafficShapingHandler 的 doAccounting 方法可以定制整形策略。相关接口定义如下:

```
protected void doAccounting(TrafficCounter counter) {
    // NOOP by default
}
```

23.2.6 优雅停机接口

Java 的优雅停机通常通过注册 JDK 的 ShutdownHook 来实现,当系统接收到退出指令后,首先标记系统处于退出状态,不再接收新的消息,然后将积压的消息处理完,最后调用资源回收接口将资源销毁,最后各线程退出执行。

通常优雅退出有个时间限制,例如 30S,如果到达执行时间仍然没有完成退出前的操作,则由监控脚本直接 kill -9 pid,强制退出。

Netty 的优雅退出功能随着版本的优化和演进也在不断的增强,下面我们一起看下 Netty5 的优雅退出。

首先看下 Reactor 线程和线程组,它们提供了优雅退出接口。EventExecutorGroup 的接口定义如下:

```
/**
 * Signals this executor that the caller wants the executor to be shut
down. Once this method is called,
 * {@link #isShuttingDown()} starts to return {@code true}, and the
executor prepares to shut itself down.
 * Unlike {@link #shutdown()}, graceful shutdown ensures that no tasks
are submitted for <i>'the quiet period'</i>
 * (usually a couple seconds) before it shuts itself down.  If a task is
submitted during the quiet period,
 * it is guaranteed to be accepted and the quiet period will start over.
 */
```

```
    Future<?> shutdownGracefully(long quietPeriod, long timeout, TimeUnit
unit);
    NioEventLoop 的资源释放接口实现：
    @Override
    protected void cleanup() {
        try {
            selector.close();
        } catch (IOException e) {
            logger.warn("Failed to close a selector.", e);
        }
    }
```

ChannelPipeline 的关闭接口如图 23-11 所示。

图 23-11　ChannelPipeline 的关闭接口

目前 Netty 向用户提供的主要接口和类库都提供了资源销毁和优雅退出的接口，用户的自定义实现类可以继承这些接口，完成用户资源的释放和优雅退出。

23.3　优化建议

尽管 Netty 的可靠性已经做得非常出色，但是在生产实践中还是发现了一些待优化点，本小节将进行简单说明。希望后续的版本中可以解决，当然用户也可以根据自己的实际需要决定自行优化。

23.3.1　发送队列容量上限控制

Netty 的 NIO 消息发送队列 ChannelOutboundBuffer 并没有容量上限控制，它会随着消息的积压自动扩展，直到达到 0x7fffffff。

如果网络对方处理速度比较慢，导致 TCP 滑窗长时间为 0；或者消息发送方发送速度过快，或者一次批量发送消息量过大，都可能会导致 ChannelOutboundBuffer 的内存膨胀，这可能会导致系统的内存溢出。

建议优化方式如下：在启动客户端或者服务端的时候，通过启动项的 ChannelOption

设置发送队列的长度,或者通过-D启动参数配置该长度。

23.3.2 回推发送失败的消息

当网络发生故障的时候,Netty 会关闭链路,然后循环释放待未发送的消息,最后通知监听 listener。

这样的处理策略值得商榷,对于大多数用户而言,并不关心底层的网络 I/O 异常,他们希望链路恢复之后可以自动将尚未发送的消息重新发送给对方,而不是简单的销毁。

Netty 销毁尚未发送的消息,用户可以通过监听器来得到消息发送异常通知,但是却无法获取原始待发送的消息。如果要实现重发,需要自己缓存消息,如果发送成功,自己删除,如果发送失败,重新发送。这对于大多数用户而言,非常麻烦,用户在开发业务代码的同时,还需要考虑网络 I/O 层的异常并为之做特殊的业务逻辑处理。

下面我们看下 Mina 的实现,当发生链路异常之后,Mina 会将尚未发送的整包消息队列封装到异常对象中,然后推送给用户 Handler,由用户来决定后续的处理策略。相比于 Netty 的"野蛮"销毁策略,Mina 的策略更灵活和合理,由用户自己决定发送失败消息的后续处理策略。

大多数场景下,业务用户会使用 RPC 框架,他们通常不需要直接针对 Netty 编程,如果 Netty 提供了发送失败消息的回推功能,RPC 框架就可以进行封装,提供不同的策略给业务用户使用,例如:

1. 缓存重发策略:当链路发生异常之后,尚未发送成功的消息自动缓存,待链路恢复正常之后重发失败的消息;

2. 失败删除策略:当链路发生异常之后,尚未发送成功的消息自动销毁,它可能是非重要消息,例如日志消息,也可能是由业务直接监听异常并做特殊处理;

3. 其他策略......

23.4 总结

本章首先对 Netty 的可靠性需求进行了分析说明,接着从架构和代码两个层面对 Netty 的高可靠性进行了剖析,最后对 Netty 的待优化点进行了分析,并给出了自己的优化建议。

第 24 章

安全性

2014 年影响最为深远的软件安全事件应该就是 OpenSSL Heart bleed 漏洞,因其破坏性之大和影响的范围之广,堪称网络安全里程碑事件。

作为基础的 NIO 通信类库,Netty 也同样面临着严峻的网络通信安全问题,本章针对 Netty 架构层面的安全性设计、安全性相关的核心代码剖析,了解和掌握 Netty 网络安全相关的特性和代码实现,以及如何使用这些安全特性保护自己的应用程序。

本章主要内容包括:

◎ 严峻的安全形势

◎ Netty SSL 安全特性

◎ Netty SSL 源码分析

◎ Netty 扩展的安全特性

24.1 严峻的安全形势

24.1.1 OpenSSL Heart bleed 漏洞

2014 年上半年对网络安全影响最大的问题就是 OpenSSL Heart bleed 漏洞,来自

Codenomicon 和谷歌安全部门的研究人员发现 OpenSSL 的源代码中存在一个漏洞,可以让攻击者获得服务器上 64K 内存中的数据内容。该漏洞在国内被译为" OpenSSL 心脏出血漏洞",因其破坏性之大和影响的范围之广,堪称网络安全里程碑事件。

OpenSSL 是为网络通信提供安全及数据完整性的一种安全协议,囊括了主要的密码算法、常用的密钥和证书封装管理功能以及 SSL 协议.多数 SSL 加密网站是用名为 OpenSSL 的开源软件包,由于这也是互联网应用最广泛的安全传输方法,被网银、在线支付、电商网站、门户网站、电子邮件等重要网站广泛使用,所以漏洞影响范围广大。

全球第一个被攻击通告的案例是加拿大税务局确认 OpenSSL Heart bleed 漏洞导致了 900 个纳税人的社会保障号被盗,这 900 个纳税人的社保号被攻击者在系统中完全删除了。

24.1.2　安全漏洞的代价

任何网络攻击都能够给企业造成破坏,但是如何将这些破坏具体量化成金融数据呢?2013 年,B2B International 联合卡巴斯基实验室基于对全球企业的调查结果,计算出网络攻击平均造成的损失。

根据调查报告得出的结论,当企业遭遇网络攻击后平均损失为 649 000 美元。损失主要包括两方面:

1. 安全事件本身造成的损失,即由重要数据泄漏、业务连续性以及安全修复专家费用相关成本;

2. 为列入计划的"响应"成本,用于阻止未来发生类似的攻击事件,包括雇佣、培训员工成本以及硬件、软件和其他基础设施安全升级成本。

24.1.3　Netty 面临的安全风险

作为一个高性能的 NIO 通信框架,基于 Netty 的行业应用非常广泛,不同的行业、不同的应用场景,面临的安全挑战也不同,下面我们根据 Netty 的典型应用场景,分析下 Netty 面临的安全挑战。

应用场景 1:仅限内部使用的 RPC 通信框架。随着业务的发展,网站规模的扩大,传统基于 MVC 的垂直架构已经无法应对业务的快速发展。需要对数据和业务进行水平拆分,基于 RPC 的分布式服务框架成为最佳选择。

业务水平拆分之后，内部的各个模块需要进行高性能的通信，传统基于 RMI 和 Hession 的同步阻塞式通信已经无法满足性能和可靠性要求。因此，高性能的 NIO 框架成为构建分布式服务框架的基石。

高性能的 RPC 框架，各模块之间往往采用长连接通信，通过心跳检测保证链路的可靠性。由于 RPC 框架通常是在内部各模块之间使用，运行在授信的内部安全域中，不直接对外开放接口。因此，不需要做握手、黑白名单、SSL/TLS 等，正所谓是"防君子不防小人"。

在这种应用场景下，Netty 的安全性是依托企业的防火墙、安全加固操作系统等系统级安全来保障的，它自身并不需要再做额外的安全性保护工作。

应用场景 2：对第三方开放的通信框架。如果使用 Netty 做 RPC 框架或者私有协议栈，RPC 框架面向非授信的第三方开放，例如将内部的一些能力通过服务对外开放出去，此时就需要进行安全认证，如果开放的是公网 IP，对于安全性要求非常高的一些服务，例如在线支付、订购等，需要通过 SSL/TLS 进行通信。

它的原理图如图 24-1 所示。

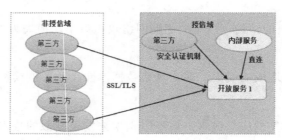

图 24-1　对第三方开放的通信框架

对第三方开放的通信框架的接口调用存在三种场景：

1．在企业内网，开放给内部其他模块调用的服务，通常不需要进行安全认证和 SSL/TLS 传输；

2．在企业内网，被外部其他模块调用的服务，往往需要利用 IP 黑白名单、握手登陆等方式进行安全认证，认证通过之后双方使用普通的 Socket 进行通信，如果认证失败，则拒绝客户端连接；

3．开放给企业外部第三方应用访问的服务，往往需要监听公网 IP（通常是防火墙的

IP 地址），由于对第三方服务调用者的监管存在诸多困难，或者无法有效监管，这些第三方应用实际是非授信的。为了有效应对安全风险，对于敏感的服务往往需要通过 SSL/TLS 进行安全传输。

应用场景 3：应用层协议的安全性。作为高性能、异步事件驱动的 NIO 框架，Netty 非常适合构建上层的应用层协议，相关原理，如图 24-2 所示。

图 24-2　基于 Netty 构建应用层协议

由于绝大多数应用层协议都是公有的，这意味着底层的 Netty 需要向上层提供通信层的安全传输，也就是需要支持 SSL/TLS。

JDK 的安全类库提供了 javax.net.ssl.SSLSocket 和 javax.net.ssl.SSLServerSocket 类库用于支持 SSL/TLS 安全传输，对于 NIO 非阻塞 Socket 通信，JDK 并没有提供现成可用的类库简化用户开发。

Netty 通过 JDK 的 SSLEngine，以 SslHandler 的方式提供对 SSL/TLS 安全传输的支持，极大的简化了用户的开发工作量，降低开发难度。对于 Netty 默认提供的 HTTP 协议，Netty 利用 SslHandler，同样支持 HTTPS 协议。

24.2　Netty SSL 安全特性

Netty 通过 SslHandler 提供了对 SSL 的支持，它支持的 SSL 协议类型包括：SSL V2、SSL V3 和 TLS。

24.2.1　SSL 单向认证

单向认证，即客户端只验证服务端的合法性，服务端不验证客户端。下面我们通过

Netty 的 SSL 单向认证代码开发来掌握基于 Netty 的 SSL 单向认证。

24.2.1.1 单向认证开发

首先，利用 JDK 的 keytool 工具，Netty 服务端依次生成服务端的密钥对和证书仓库、服务端自签名证书。

生成 Netty 服务端私钥和证书仓库命令：

```
keytool -genkey -alias securechat -keysize 2048 -validity 365
-keyalg RSA -dname "CN=localhost" -keypass sNetty -storepass
sNetty -keystore sChat.jks
```

生成 Netty 服务端自签名证书：

```
keytool -export -alias securechat -keystore sChat.jks -storepass
sNetty -file sChat.cer
```

生成客户端的密钥对和证书仓库，用于将服务端的证书保存到客户端的授信证书仓库中，命令如下：

```
keytool -genkey -alias smcc -keysize 2048 -validity 365 -keyalg
RSA -dname "CN=localhost" -keypass cNetty -storepass cNetty
-keystore cChat.jks
```

随后，将 Netty 服务端的证书导入到客户端的证书仓库中，命令如下：

```
keytool -import -trustcacerts -alias securechat -file sChat.cer
-storepass cNetty -keystore cChat.jks
```

上述工作完成之后，我们就开始编写 SSL 服务端和客户端的代码，下面我们对核心代码进行讲解。

因为是客户端认证服务端，因此服务端需要正确的设置和加载私钥仓库 KeyStore，相关代码如下：

```
KeyManagerFactory kmf = null;
    if (pkPath != null) {
        KeyStore ks = KeyStore.getInstance("JKS");
        in = new FileInputStream(pkPath);
```

```
            ks.load(in, "sNetty".toCharArray());
            kmf = KeyManagerFactory.getInstance("SunX509");
            kmf.init(ks, "sNetty".toCharArray());
```
初始化 KeyManagerFactory 之后，创建 SSLContext 并初始化，代码如下：
```
SERVER_CONTEXT = SSLContext.getInstance(PROTOCOL);
        if (SSLMODE.CA.toString().equals(tlsMode))
            SERVER_CONTEXT.init(kmf.getKeyManagers(), null, null);
```

由于是单向认证，服务端不需要验证客户端的合法性，因此，TrustManager 为空，安全随机数不需要设置，使用 JDK 默认创建的即可。

服务端的 SSLContext 创建完成之后，利用 SSLContext 创建 SSL 引擎 SSLEngine，设置 SSLEngine 为服务端模式，由于不需要对客户端进行认证，因此 NeedClientAuth 不需要额外设置，使用默认值 False。相关代码如下：

```
engine.setUseClientMode(false);
```

SSL 服务端创建完成之后，下面继续看客户端的创建，它的原理同服务端类似，也是在初始化 TCP 链路的时候创建并设置 SSLEngine，代码如下：

```
KeyManagerFactory kmf = null;
        if (pkPath != null) {
            KeyStore ks = KeyStore.getInstance("JKS");
            in = new FileInputStream(pkPath);
            ks.load(in, "cNetty".toCharArray());
            // ks.load(in, "123456".toCharArray());
            kmf = KeyManagerFactory.getInstance("SunX509");
            kmf.init(ks, "cNetty".toCharArray());
            // kmf.init(ks, "123456".toCharArray());
        }
```

由于是客户端认证服务端，因此，客户端只需要加载存放服务端 CA 的证书仓库即可。

加载证书仓库完成之后，初始化 SSLContext，代码如下：对于客户端只需要设置信任证书 TrustManager：

```
CLIENT_CONTEXT = SSLContext.getInstance(PROTOCOL);
        if (SSLMODE.CA.toString().equals(tlsMode))
            CLIENT_CONTEXT.init(null,
                tf == null ? null : tf.getTrustManagers(), null);
```

客户端 SSLContext 初始化完成之后，创建 SSLEngine 并将其设置为客户端工作模式，代码如下：

```
engine.setUseClientMode(true);
```

将 SslHandler 添加到 pipeline 中，利用 SslHandler 实现 Socket 安全传输，代码如下：

```
pipeline.addLast("ssl", new SslHandler(engine));
```

客户端和服务端创建完成之后，测试下 SSL 单向认证功能是否 OK，为了查看 SSL 握手过程，我们打开 SSL 握手的调测日志，Eclipse 设置如图 24-3 所示。

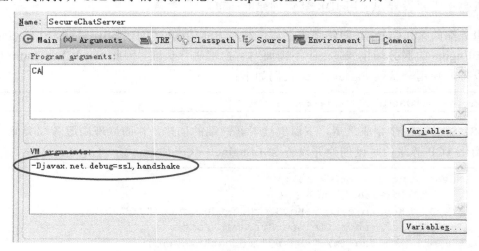

图 24-3　打开 SSL 调测日志

分别运行服务端和客户端，运行结果如图 24-4 和图 24-5 所示。

图 24-4　客户端 SSL 握手日志

```
*** Finished
verify_data: { 223, 76, 141, 106, 177, 81, 146, 163, 143, 45, 5, 119 }
***
nioEventLoopGroup-3-1, WRITE: TLSv1 Change Cipher Spec, length = 1
*** Finished
verify_data: { 148, 63, 125, 105, 96, 3, 2, 245, 33, 227, 95, 106 }
***
nioEventLoopGroup-3-1, WRITE: TLSv1 Handshake, length = 48
%% Cached server session: [Session-1, TLS_ECDHE_RSA_WITH_AES_128_CBC_SHA]
nioEventLoopGroup-3-1, WRITE: TLSv1 Application Data, length = 48
nioEventLoopGroup-3-1, WRITE: TLSv1 Application Data, length = 1
nioEventLoopGroup-3-1, WRITE: TLSv1 Application Data, length = 77
```

图 24-5　服务端 SSL 握手日志

在客户端输入信息，服务端原样返回，测试结果如图 24-6 所示。

```
netty权威指南
nioEventLoopGroup-2-1, WRITE: TLSv1 Application Data, length = 19
[you] netty权威指南
```

图 24-6　SSL 版本 Echo 程序

到此，Netty SSL 单向认证已经开发完成，下个小节我们将结合 SSL 握手日志，详细解读下 SSL 单向认证的原理。

24.2.1.2　单向认证原理

SSL 单向认证的过程总结如下：

1．SSL 客户端向服务端传送客户端 SSL 协议的版本号、支持的加密算法种类、产生的随机数，以及其他可选信息；

2．服务端返回握手应答，向客户端传送确认 SSL 协议的版本号、加密算法的种类、随机数以及其他相关信息；

3．服务端向客户端发送自己的公钥；

4．客户端对服务端的证书进行认证，服务端的合法性校验包括：证书是否过期、发行服务器证书的 CA 是否可靠、发行者证书的公钥能否正确解开服务器证书的"发行者的数字签名"、服务器证书上的域名是否和服务器的实际域名相匹配等；

5．客户端随机产生一个用于后面通讯的"对称密码"，然后用服务端的公钥对其加密，将加密后的"预主密码"传给服务端；

6．服务端将用自己的私钥解开加密的"预主密码"，然后执行一系列步骤来产生主密码；

7. 客户端向服务端发出信息,指明后面的数据通讯将使用主密码为对称密钥,同时通知服务器客户端的握手过程结束;

8. 服务端向客户端发出信息,指明后面的数据通讯将使用主密码为对称密钥,同时通知客户端服务器端的握手过程结束;

9. SSL 的握手部分结束,SSL 安全通道建立,客户端和服务端开始使用相同的对称密钥对数据进行加密,然后通过 Socket 进行传输。

下面,我们结合 JDK 的 SSL 工作原理对 Netty 的 SSL 单向认证过程进行讲解,首先,我们看下 JDK SSL 单向认证的流程图如图 24-7 所示。

图 24-7 SSL 单向认证流程图

下面结合 JDK SSL 引擎的调测日志信息我们对 SSL 单向认证的流程进行详细讲解,对于比较简单的流程会进行步骤合并。

步骤 1:客户端使用 TLS 协议版本发送一个 ClientHello 消息,这个消息包含一个随机数、建议的加密算法套件和压缩方法列表,如下所示:

```
*** ClientHello, TLSv1
RandomCookie:  GMT: 1389796107 bytes = { 125, 107, 138, 150, 226, 182, 238, 75,
38, 150, 222, 147, 127, 35, 36, 149, 172, 128, 152, 34, 110, 104, 176, 34, 180,
118, 185, 55 }
Session ID:  {}
Cipher           Suites:           [TLS_ECDHE_ECDSA_WITH_AES_128_CBC_SHA,
TLS_ECDHE_RSA_WITH_AES_128_CBC_SHA, TLS_RSA_WITH_AES_128_CBC_SHA,
……省略
TLS_EMPTY_RENEGOTIATION_INFO_SCSV]
sect193r1, sect193r2, secp224k1, secp239k1, secp256k1}
Extension ec_point_formats, formats: [uncompressed]
***
```

步骤 2：服务端使用 ServerHello 消息来响应，这个消息包含由客户提供的信息基础上的另一个随机数和一个可选的会话 ID，以及服务端选择的加密套件算法。响应消息如下：

```
*** ServerHello, TLSv1
RandomCookie:  GMT: 1389796108 bytes = { 27, 170, 76, 238, 56, 58,
172, 146, 41, 159, 249, 213, 16, 214, 53, 167, 50, 74, 39, 107,
121, 63, 80, 26, 210, 149, 249, 194 }
......省略
```

步骤 3：服务端发送自签名的证书消息，包含完整的证书链：

```
*** Certificate chain
chain [0] = [
[
   Version: V3
   Subject: CN=localhost
   Signature    Algorithm:    SHA1withRSA,    OID    =
1.2.840.113549.1.1.5
......省略
```

步骤 4：服务端向客户端发送自己的公钥信息，最后发送 ServerHelloDone：

```
*** ECDH ServerKeyExchange
Server key: Sun EC public key, 256 bits
public                x                coord:
112463902918630779107945902338321922977565892046706979058886
85651118114908704
public                y                coord:
141615584302183983661360241749252580028319381566531570740584
92642854053163673
parameters: secp256r1  [NIST  P-256,  X9.62  prime256v1]
(1.2.840.10045.3.1.7)
    *** ServerHelloDone
```

步骤 5：客户端对服务端自签名的证书进行认证，如果客户端的信任证书列表中包含了服务端发送的证书，对证书进行合法性认证，相关信息如下：

```
***
Found trusted certificate:
[
[
Version: V3
Subject: CN=localhost
Signature Algorithm: SHA1withRSA, OID = 1.2.840.113549.1.1.5
Key:  Sun RSA public key, 2048 bits
 modulus:
18007409233594974050659993272913606112791070482225689030478 5
......省略
```

步骤 6：客户端通知服务器改变加密算法，通过 Change Cipher Spec 消息发给服务端，随后发送 Finished 消息，告知服务器请检查加密算法的变更请求：

```
nioEventLoopGroup-2-1, WRITE: TLSv1 Change Cipher Spec, length = 1
```

步骤 7：服务端读取到 Change Cipher Spec 变更请求消息，向客户端返回确认密钥变更消息，最后通过发送 Finished 消息表示 SSL/TLS 握手结束。

```
nioEventLoopGroup-3-1, READ: TLSv1 Change Cipher Spec, length = 1
nioEventLoopGroup-3-1, READ: TLSv1 Handshake, length = 48
*** Finished
verify_data:  { 157, 255, 187, 52, 139, 16, 20, 190, 11, 35, 79, 0 }
***
nioEventLoopGroup-3-1, WRITE: TLSv1 Change Cipher Spec, length = 1
*** Finished
```

24.2.2　SSL 双向认证

与单向认证不同的是服务端也需要对客户端进行安全认证。这就意味着客户端的自签名证书也需要导入到服务端的数字证书仓库中。

24.2.2.1 双向认证开发

首先，生成客户端的自签名证书：

```
keytool -export -alias smcc -keystore cChat.jks -storepass
cNetty -file cChat.cer
```

然后将客户端的自签名证书导入到服务端的信任证书仓库中：

```
keytool -import -trustcacerts -alias smcc -file cChat.cer
-storepass sNetty -keystore sChat.jks
```

证书导入之后，需要对 SSL 客户端和服务端的代码同时进行修改，首先我们看下服务端如何修改。

由于服务端需要对客户端进行验证，因此在初始化服务端 SSLContext 的时候需要加载证书仓库。首先需要对 TrustManagerFactory 进行初始化，代码如下：

```
TrustManagerFactory tf = null;
        if (caPath != null) {
            KeyStore tks = KeyStore.getInstance("JKS");
            tIN = new FileInputStream(caPath);
            tks.load(tIN, "sNetty".toCharArray());
            tf = TrustManagerFactory.getInstance("SunX509");
            tf.init(tks);
        }
```

初始化 SSLContext 的时候根据 TrustManagerFactory 获取 TrustManager 数组，代码如下：

```
else if (SSLMODE.CSA.toString().equals(tlsMode)) {
        SERVER_CONTEXT.init(kmf.getKeyManagers(),
            tf.getTrustManagers(), null);
```

最后，创建 SSLEngine 之后，设置需要进行客户端认证，代码如下：

```
if (SSLMODE.CSA.toString().equals(tlsMode))
        engine.setNeedClientAuth(true);
```

完成服务端修改之后，再回头看下客户端的修改，由于服务端需要认证客户端的证书，因此，需要初始化和加载私钥仓库，向服务端发送公钥，初始化 KeyStore 的代码如下：

```
KeyManagerFactory kmf = null;
    if (pkPath != null) {
        KeyStore ks = KeyStore.getInstance("JKS");
        in = new FileInputStream(pkPath);
        ks.load(in, "cNetty".toCharArray());
        kmf = KeyManagerFactory.getInstance("SunX509");
        kmf.init(ks, "cNetty".toCharArray());
    }
```

初始化 **SSLContext** 的时候需要传入 **KeyManager** 数组，代码如下：

```
else if (SSLMODE.CSA.toString().equals(tlsMode)) {
        CLIENT_CONTEXT.init(kmf.getKeyManagers(),
            tf.getTrustManagers(), null);
```

客户端开发完成之后，测试下程序是否能够正常工作，运行结果如下所示。

客户端运行结果如图 24-8 所示。

```
*** CertificateVerify
nioEventLoopGroup-2-1, WRITE: TLSv1 Handshake, length = 262
nioEventLoopGroup-2-1, WRITE: TLSv1 Change Cipher Spec, length = 1
*** Finished
verify_data:  { 81, 178, 246, 18, 147, 113, 117, 100, 201, 183, 62, 136 }
***
nioEventLoopGroup-2-1, WRITE: TLSv1 Handshake, length = 48
nioEventLoopGroup-2-1, READ: TLSv1 Change Cipher Spec, length = 1
nioEventLoopGroup-2-1, READ: TLSv1 Handshake, length = 48
*** Finished
verify_data:  { 228, 152, 88, 130, 91, 217, 179, 58, 148, 226, 205, 67 }
***
%% Cached client session: [Session-1, TLS_ECDHE_RSA_WITH_AES_128_CBC_SHA]
Welcome to OUHHDLFDTAOCXYW secure chat service!
Your session is protected by TLS_ECDHE_RSA_WITH_AES_128_CBC_SHA cipher suite.
```

图 24-8　Netty SSL 双向认证客户端运行结果

服务端运行结果如图 24-9 所示。

```
*** CertificateVerify
nioEventLoopGroup-3-1, READ: TLSv1 Change Cipher Spec, length = 1
nioEventLoopGroup-3-1, READ: TLSv1 Handshake, length = 48
*** Finished
verify_data:  { 81, 178, 246, 18, 147, 113, 117, 100, 201, 183, 62, 136 }
***
nioEventLoopGroup-3-1, WRITE: TLSv1 Change Cipher Spec, length = 1
*** Finished
verify_data:  { 228, 152, 88, 130, 91, 217, 179, 58, 148, 226, 205, 67 }
***
nioEventLoopGroup-3-1, WRITE: TLSv1 Handshake, length = 48
```

图 24-9　Netty SSL 双向认证服务端运行结果

在客户端控制台进行输入，看 SSL 传输是否正常，如图 24-10 所示。

```
Welcome to OUHHDLFDTAOCXYW secure chat service!
Your session is protected by TLS_ECDHE_RSA_WITH_AES_128_CBC_SHA cipher suite.
欢迎参加QClub南京站活动
nioEventLoopGroup-2-1, WRITE: TLSv1 Application Data, length = 34
[you] 欢迎参加QClub南京站活动
```

图 24-10　Netty SSL 安全传输测试

24.2.2.2　双向认证原理

SSL 双向认证相比单向认证，多了一步服务端发送认证请求消息给客户端，客户端发送自签名证书给服务端进行安全认证的过程。下面，我们结合 Netty SSL 调测日志，对双向认证的差异点进行分析。

相比于客户端，服务端在发送 ServerHello 时携带了要求客户端认证的请求信息，如下所示：

```
*** CertificateRequest
Cert Types: RSA, DSS, ECDSA
Cert Authorities:
<CN=localhost>
<CN=localhost>
```

客户端接收到服务端要求客户端认证的请求消息之后，发送自己的证书信息给服务端，信息如下：

```
matching alias: smcc
*** Certificate chain
chain [0] = [
[
   Version: V3
   Subject: CN=localhost
   Signature    Algorithm:    SHA1withRSA,    OID    =
1.2.840.113549.1.1.5
   Key: Sun RSA public key, 2048 bits
......省略
```

服务端对客户端的自签名证书进行认证，信息如下：

```
***
Found trusted certificate:
[
[
    Version: V3
    Subject: CN=localhost
......省略
```

24.2.3　第三方 CA 认证

使用 jdk keytool 生成的数字证书是自签名的。自签名就是指证书只能保证自己是完整且没有经过非法修改，但是无法保证这个证书是属于谁的。为了对自签名证书进行认证，需要每个客户端和服务端都交换自己自签名的私有证书，对于一个大型网站或者应用服务器，这种工作量是非常大的。

基于自签名的 SSL 双向认证，只要客户端或者服务端修改了密钥和证书，就需要重新进行签名和证书交换，这种调试和维护工作量是非常大的。因此，在实际的商用系统中往往会使用第三方 CA 证书颁发机构进行签名和验证。我们的浏览器就保存了几个常用的 CA_ROOT。每次连接到网站时只要这个网站的证书是经过这些 CA_ROOT 签名过的。就可以通过验证了。

CA 数字证书认证服务往往是收费的，国内有很多数字认证中心都提供相关的服务，有需要的可以通过这些商业机构获取认证。

作为示例，我们自己生成一个 CA_ROOT 的密钥对，部署应用时，把这个 CA_ROOT 的私钥部署在所有需要 SSL 传输的节点就可以完成安全认证。作为示例，如果要生成 CA_ROOT，我们使用开源的 OpenSSL。

在 Windows 上安装和使用 OpenSSL 网上有很多教程，也不是本文的重点，因此，OpenSSL 的安装和使用本文不详细介绍。

下面我们对基于第三方 CA 认证的步骤进行详细介绍。

24.2.3.1 服务端证书制作

步骤 1：利用 OpenSSL 生成 CA 证书：

```
openssl req -new -x509 -keyout ca.key -out ca.crt -days 365
```

步骤 2：生成服务端密钥对：

```
keytool -genkey -alias securechat -keysize 2048 -validity 365
-keyalg RSA -dname "CN=localhost" -keypass sNetty -storepass
sNetty -keystore sChat.jks
```

步骤 3：生成证书签名请求：

```
keytool -certreq -alias securechat -sigalg MD5withRSA -file
sChat.csr -keypass sNetty -storepass sNetty -keystore
sChat.jks
```

步骤 4：用 CA 私钥进行签名：

```
openssl ca -in sChat.csr -out sChat.crt -cert ca.crt
-keyfile ca.key -notext
```

步骤 5：导入信任的 CA 根证书到 keystore：

```
keytool -import -v -trustcacerts -alias ca_root -file
ca.crt -storepass sNetty -keystore sChat.jks
```

步骤 6：将 CA 签名后的 server 端证书导入 keystore：

```
keytool -import -v -alias securechat -file server.crt -keypass
sNetty -storepass sNetty -keystore sChat.jks
```

24.2.3.2 客户端证书制作

步骤 1：生成客户端密钥对：

```
keytool -genkey -alias smcc -keysize 2048 -validity 365
-keyalg RSA -dname "CN=localhost" -keypass cNetty -storepass
cNetty -keystore cChat.jks
```

步骤 2：生成证书签名请求：

```
keytool -certreq -alias smcc -sigalg MD5withRSA -file
cChat.csr -keypass cNetty -storepass cNetty -keystore
cChat.jks
```

步骤 3：用 CA 私钥进行签名：

```
openssl ca -in cChat.csr -out cNetty.crt -cert ca.crt
-keyfile ca.key -notext
```

步骤 4：导入信任的 CA 根证书到 keystore：

```
keytool -import -v -trustcacerts -alias ca_root -file
ca.crt -storepass cNetty -keystore cChat.jks
```

步骤 5：将 CA 签名后的 client 端证书导入 keystore：

```
keytool -import -v -alias smcc -file cNetty.crt -keypass
cNetty -storepass cNetty -keystore cChat.jks
```

基于 CA 认证的开发和测试与 SSL 双向和单向认证代码相同，此处不再赘述。

24.3　Netty SSL 源码分析

24.3.1　客户端

当客户端和服务端的 TCP 链路建立成功之后，SslHandler 的 channelActive 被触发，SSL 客户端通过 SSL 引擎发起握手请求消息，代码如下：

```java
private Future<Channel> handshake() {
    ......以上代码省略
    try {
        engine.beginHandshake();
        wrapNonAppData(ctx, false);
        ctx.flush();
    } catch (Exception e) {
        notifyHandshakeFailure(e);
    }
    return handshakePromise;
}
```

发起握手请求之后，需要将 SSLEngine 创建的握手请求消息进行 SSL 编码，发送给服务端，因此，握手之后立即调用 wrapNonAppData 方法，下面具体对该方法进行分析：

```
    private void wrapNonAppData(ChannelHandlerContext ctx, boolean inUnwrap)
throws SSLException {
        ByteBuf out = null;
        try {
            for (;;) {
                if (out == null) {
                    out = ctx.alloc().buffer(maxPacketBufferSize);
                }
                SSLEngineResult result = wrap(engine, Unpooled.EMPTY_BUFFER,
out);
                if (result.bytesProduced() > 0) {
                    ctx.write(out);
                    if (inUnwrap) {
                        needsFlush = true;
                    }
                    out = null;
                }
```

因为只需要发送握手请求消息，因此 Source ByteBuf 为空，下面看下 wrap 方法的具体实现：

```
    private SSLEngineResult wrap(SSLEngine engine, ByteBuf in, ByteBuf out)
throws SSLException {
        ByteBuffer in0 = in.nioBuffer();
        for (;;) {
            ByteBuffer      out0       =      out.nioBuffer(out.writerIndex(),
out.writableBytes());
            SSLEngineResult result = engine.wrap(in0, out0);
            in.skipBytes(result.bytesConsumed());
            out.writerIndex(out.writerIndex() + result.bytesProduced());
            ……后续代码省略
            }
        }
    }
```

将 SSL 引擎中创建的握手请求消息编码到目标 ByteBuffer 中，然后对写索引进行更新。判断写入操作是否越界，如果越界说明 out 容量不足，需要调用 ensureWritable 对 ByteBuf 进行动态扩展，扩展之后继续尝试编码操作。如果编码成功，返回 SSL 引擎操作结果。

对编码结果进行判断,如果编码字节数大于 0,则将编码后的结果发送给服务端,然后释放临时变量 out。

判断 SSL 引擎的操作结果,SSL 引擎的操作结果定义如下:

1. FINISHED:SSLEngine 已经完成握手;

2. NEED_TASK:SSLEngine 在继续进行握手前需要一个(或多个)代理任务的结果;

3. NEED_UNWRAP:在继续进行握手前,SSLEngine 需要从远端接收数据,所以应带调用 SSLEngine.unwrap();

4. NEED_WRAP:在继续进行握手前,SSLEngine 必须向远端发送数据,所以应该调用 SSLEngine.wrap();

5. NOT_HANDSHAKING:SSLEngine 当前没有进行握手。

下面我们分别对 5 种操作的代码进行分析:

```
switch (result.getHandshakeStatus()) {
            case FINISHED:
                setHandshakeSuccess();
                break;
            case NEED_TASK:
                runDelegatedTasks();
                break;
            case NEED_UNWRAP:
                if (!inUnwrap) {
                    unwrap(ctx);
                }
                break;
            case NEED_WRAP:
                break;
            case NOT_HANDSHAKING:
                if (!inUnwrap) {
                    unwrap(ctx);
                }
                break;
            default:
                throw new IllegalStateException("Unknown handshake status: " + result.getHandshakeStatus());
        }
```

如果握手成功，则设置 handshakePromise 的操作结果为成功，同时发送 SslHandshakeCompletionEvent.SUCCES 给 SSL 监听器，代码如下：

```
private void setHandshakeSuccess() {
    if (handshakePromise.trySuccess(ctx.channel())) {
ctx.fireUserEventTriggered(SslHandshakeCompletionEvent.SUCCESS);
    }
}
```

如果是 NEED_TASK，说明异步执行 SSL Task，完成后续可能耗时的操作或者任务，Netty 封装了一个任务立即执行线程池专门处理 SSL 的代理任务，代码如下：

```
private void runDelegatedTasks() {
    for (;;) {
        Runnable task = engine.getDelegatedTask();
        if (task == null) {
            break;
        }
        delegatedTaskExecutor.execute(task);
    }
}
```

如果是 NEED_UNWRAP，则判断是否由 UNWRAP 发起，如果不是则执行 UNWRAP 操作。

如果是 NOT_HANDSHAKING，则调用 unwrap，继续接收服务端的消息。

服务端应答消息的接收跟服务端接收客户端的代码类似，唯一不同之处在于 SSL 引擎的客户端模式设置不同，一个是服务端，一个是客户端。上层的代码处理是相同的，下面我们在 SSL 服务端章节分析握手消息的接收。

24.3.2 服务端

SSL 服务端接收客户端握手请求消息的入口方法是 decode 方法，下面对它进行详细分析。

首先获取接收缓冲区的读写索引，并对读取的偏移量指针进行备份：

```
protected void decode(ChannelHandlerContext ctx, ByteBuf in, List<Object>
out) throws SSLException {
        final int startOffset = in.readerIndex();
        final int endOffset = in.writerIndex();
        int offset = startOffset;
```

对半包标识进行判断,如果上一个消息是半包消息,则判断当前可读的字节数是否小于整包消息的长度,如果小于整包长度,则说明本次读取操作仍然没有把 SSL 整包消息读取完整,需要返回 IO 线程继续读取,代码如下:

```
if (packetLength > 0) {
        if (endOffset - startOffset < packetLength) {
            return;
        }
......代码省略
    }
```

如果消息读取完整,则修改偏移量;同时置位半包长度标识。

```
else {
        offset += packetLength;
        packetLength = 0;
    }
```

下面在 for 循环中读取 SSL 消息,因为 TCP 存在拆包和粘包,因此一个 ByteBuf 可能包含多条完整的 SSL 消息。

首先判断可读的字节数是否小于协议消息头长度,如果是则退出循环继续由 IO 线程接收后续的报文:

```
for (;;) {
        final int readableBytes = endOffset - offset;
        if (readableBytes < 5) {
            break;
        }
......后续代码省略
```

获取 SSL 消息包的报文长度,具体算法不再介绍,可以参考 SSL 的规范文档进行解读,代码如下:

```
if (tls) {
        int majorVersion = buffer.getUnsignedByte(offset + 1);
```

```
            if (majorVersion == 3) {
                packetLength = buffer.getUnsignedShort(offset + 3) + 5;
                if (packetLength <= 5) {
                    tls = false;
                }
            } else {
                tls = false;
            }
        }
```

对长度进行判断，如果 SSL 报文长度大于可读的字节数，说明是个半包消息，将半包标识长度置位，返回 I/O 线程继续读取后续的数据报，代码如下：

```
if (packetLength > 0) {
        if (endOffset - startOffset < packetLength) {
            return;
        } else {
            offset += packetLength;
            packetLength = 0;
        }
    }
```

对消息进行解码，将 SSL 加密的消息解码为加密前的原始数据，unwrap 方法如下：

```
    private static SSLEngineResult unwrap(SSLEngine engine, ByteBuffer in, ByteBuf out) throws SSLException {
        int overflows = 0;
        for (;;) {
            ByteBuffer out0 = out.nioBuffer(out.writerIndex(), out.writableBytes());
            SSLEngineResult result = engine.unwrap(in, out0);
            out.writerIndex(out.writerIndex() + result.bytesProduced());
            ......后续代码省略
            }
        }
    }
```

调用 SSLEngine 的 unwrap 方法对 SSL 原始消息进行解码，对解码结果进行判断，如果越界，说明 out 缓冲区不够，需要进行动态扩展。如果是首次越界，为了尽量节约内存，使用 SSL 最大缓冲区长度和 SSL 原始缓冲区可读的字节数中较小的。如果再次发生缓冲区越界，说明扩张后的缓冲区仍然不够用，直接使用 SSL 缓冲区的最大长度，保证下次解码成功。

解码成功之后，对 SSL 引擎的操作结果进行判断：如果需要继续接收数据，则继续执行解码操作；如果需要发送握手消息，则调用 wrapNonAppData 发送握手消息；如果需要异步执行 SSL 代理任务，则调用立即执行线程池执行代理任务；如果是握手成功，则设置 SSL 操作结果，发送 SSL 握手成功事件；如果是应用层的业务数据，则继续执行解码操作，其他操作结果，抛出操作类型异常。

需要指出的是，SSL 客户端和服务端接收对方 SSL 握手消息的代码是相同的，那为什么 SSL 服务端和客户端发送的握手消息不同呢？这些是 SSL 引擎负责区分和处理的，我们在创建 SSL 引擎的时候设置了客户端模式，SSL 引擎就是根据这个来进行区分的，代码如下：

```
engine.setUseClientMode(false|true);
```

24.3.3 消息读取

SSL 的消息读取实际就是 ByteToMessageDecoder 将接收到的 SSL 加密后的报文解码为原始报文，然后将整包消息投递给后续的消息解码器，对消息做二次解码。基于 SSL 的消息解码模型如下：

SSL 消息读取的入口都是 decode，因为是非握手消息，它的处理非常简单，就是循环调用引擎的 unwrap 方法，将 SSL 报文解码为原始的报文，代码如下：

```
    private static SSLEngineResult unwrap(SSLEngine engine, ByteBuffer in, ByteBuf out) throws SSLException {
        int overflows = 0;
        for (;;) {
            ByteBuffer out0 = out.nioBuffer(out.writerIndex(), out.writableBytes());
            SSLEngineResult result = engine.unwrap(in, out0);
            out.writerIndex(out.writerIndex() + result.bytesProduced());
            ......后续代码省略
    }
```

握手成功之后的所有消息都是应用数据，因此它的操作结果为 NOT_HANDSHAKING，遇到此标识之后继续读取消息，直到没有可读的字节，退出循环，代码如下：

```
if (status == Status.BUFFER_UNDERFLOW || consumed == 0 && produced == 0) {
    break;
}
```

如果读取到了可用的字节,则将读取到的缓冲区加到输出结果列表中,代码如下:

```
finally {
        if (totalProduced > 0) {
            ByteBuf decodeOut = this.decodeOut;
            this.decodeOut = null;
            out.add(decodeOut);
        }
    }
```

ByteToMessageDecoder 判断解码结果 List,如果非空,则循环调用后续的 Handler,由后续的解码器对解密后的报文进行二次解码。

24.3.4　消息发送

SSL 消息发送时,由 SslHandler 对消息进行编码,编码后的消息实际就是 SSL 加密后的消息,它的入口是 flush 方法,代码如下:

```
public void flush(ChannelHandlerContext ctx) throws Exception {
    if (startTls && !sentFirstMessage) {
    ......此处代码省略
    wrap(ctx, false);
    ctx.flush();
}
```

从待加密的消息队列中弹出消息,调用 SSL 引擎的 wrap 方法进行编码,代码如下:

```
for (;;) {
            PendingWrite pending = pendingUnencryptedWrites.peek();
            if (pending == null) {
                break;
            }
            if (out == null) {
                out = ctx.alloc().buffer(maxPacketBufferSize);
            }
            if (!(pending.msg() instanceof ByteBuf)) {
                ctx.write(pending.msg(),                    (ChannelPromise)
```

```
pending.recycleAndGet());
                pendingUnencryptedWrites.remove();
                continue;
            }
            ByteBuf buf = (ByteBuf) pending.msg();
            SSLEngineResult result = wrap(engine, buf, out);
```

wrap 方法很简单，就是调用 SSL 引擎的编码方法，然后对写索引进行修改，如果缓冲区越界，则动态扩展缓冲区。对 SSL 操作结果进行判断，因为已经握手成功，因此返回的结果是 NOT_HANDSHAKING，执行 finishWrap 方法，调用 ChannelHandlerContext 的 write 方法，将消息写入发送缓冲区中，如果待发送的消息为空，则构造空的 ByteBuf 写入：

```
    private void finishWrap(ChannelHandlerContext ctx, ByteBuf out,
ChannelPromise promise, boolean inUnwrap) {
        if (out == null) {
            out = Unpooled.EMPTY_BUFFER;
        } else if (!out.isReadable()) {
            out.release();
            out = Unpooled.EMPTY_BUFFER;
        }

        if (promise != null) {
            ctx.write(out, promise);
        } else {
            ctx.write(out);
        }
    if (inUnwrap) {
        needsFlush = true;
        }
    }
```

编码后，调用 ChannelHandlerContext 的 flush 方法消息发送给对方，即可完成消息的 SSL 加密发送。

24.4 Netty 扩展的安全特性

利用 Netty 的 ChannelHandler 接口提供的网络切面，用户可以非常容易的扩展 Netty 的安全策略，下面对比较典型的安全扩展特性进行讲解。

24.4.1 IP 地址黑名单机制

IP 地址黑名单是比较常用的弱安全保护策略,它的特点就是服务端在与客户端通信的过程中,对客户端的 IP 地址进行校验,如果发现对方 IP 在黑名单列表中,则拒绝与其通信,关闭链路。

下面我们对基于 IP 地址的黑名单在 Netty 中的实现进行介绍。

首先定义 BlacklistHandler 继承自 ChannelHandlerAdapter,然后定义 IP 地址黑名单列表,如下所示:

```
private final List<InetAddress> blacklist = new CopyOnWriteArrayList<InetAddress>();
```

提供对 public 的黑名单管理接口,用于设置黑名单、删除或者添加黑名单,示例代码如下:

```
public void setBlacklist(InetAddress[] addresses) {//TODO}
public boolean removeBlacklist(InetAddress address) {//TODO}
public boolean addBlacklist(InetAddress address) {//TODO}
public void clearBlacklist() {//TODO}
```

链路注册、链路激活、消息读取、消息发送的时候对对端的 IP 地址进行校验,如果在黑名单列表中,则拒绝当前操作,并关闭链路,打印日志,相关伪代码如下:

```
public void channelRegistered(ChannelHandlerContext ctx) throws Exception {
    If (//在黑名单列表中)
    {
        //打印日志
       ctx.close();//关闭链路
       return;
    }
    //其他代码
}
public void channelActive(ChannelHandlerContext ctx) throws Exception {
    If (//在黑名单列表中)
    {
        //打印日志
       ctx.close();//关闭链路
       return;
    }
```

```
        //其他代码
    }
    public void channelRead(ChannelHandlerContext ctx, Object msg) throws Exception {
        If (//在黑名单列表中)
        {
          //打印日志
          ctx.close();//关闭链路
          return;
        }
        //其他代码
    }
    public void write(ChannelHandlerContext ctx, Object msg, ChannelPromise promise) throws Exception {
        If (//在黑名单列表中)
        {
          //打印日志
          ctx.close();//关闭链路
          return;
        }
        //其他代码
    }
```

为什么要在消息读取和发送的时候也要对黑名单进行判断呢？原因是黑名单支持动态添加策略，一些黑名单是通过业务逻辑判断和执行过程中动态添加进去的，如果是长链接，只在链路首次建立的时候判断是不够的。

24.4.2 接入认证

接入认证策略非常多，通常是较强的安全认证策略，例如基于用户名+密码的认证，认证内容往往采用加密的方式，例如 Base64+AES 等。

在 Netty 中如果要支持安全认证，往往是通过定制 ChannelHandler 接口来实现，具体策略如下：

1. 在链路首次激活的时候，客户端发送认证信息给服务端，伪代码如下：

```
public void channelActive(ChannelHandlerContext ctx) throws Exception {
    //发送认证信息给服务端
}
```

2．服务端接收到客户端消息之后，根据消息内容进行判断，如果是首次接入的认证消息，则进行认证，认证失败，打印日志，关闭链接；认证成功，继续业务逻辑处理，伪代码如下：

```
public void channelRead(ChannelHandlerContext ctx, Object msg) throws Exception {
    If (//接入认证消息)
    {
     //接入认证
     If (//认证失败)
     {
      //打印异常日志
      //关闭链路
      return;
     }
    }
    else
    //业务逻辑处理
}
```

3．客户端接收到服务端消息，对消息类型进行判断，对于认证应答消息，如果认证成功，则继续业务逻辑处理；如果认证失败，则关闭链路，打印异常日志：

```
public void channelRead(ChannelHandlerContext ctx, Object msg) throws Exception {
    If (//接入认证应答消息)
    {
     If (//认证失败)
     {
      //打印异常日志
      //关闭链路
      return;
     }
    }
    else
    //业务逻辑处理
}
```

通常情况下，接入认证失败服务端都会返回认证失败应答消息，给出错误码或者认证失败原因，如果服务端不返回失败应答而是直接关闭链路，客户端接收到链路关闭通知之后直接关闭链路即可。

24.4 总结

本章首先对网络通信安全面临的严峻挑战进行了介绍,然后结合 Netty 的安全特性进行了针对性的讲解,对 Netty SSL 的开发、设计原理和源码进行了深入剖析,以期读者能够快速上手并掌握 Netty SSL 安全特性。最后,对 Netty 安全性的扩展方案和策略进行了讲解,以方便用户按需定制,满足个性化的安全需求。

第 25 章
Netty 未来展望

作为本书的结尾章节，和读者朋友们一起展望下 Netty 的未来。

本章主要内容包括：

- ◎ 应用范围
- ◎ 技术演进
- ◎ 社区活跃度
- ◎ Road Map

25.1 应用范围

随着大数据、互联网和云计算的发展，传统的垂直架构逐渐将被分布式、弹性伸缩的新架构替代。

系统只要分布式部署，就存在多个节点之间通信的问题，由于是内部通信，同时强调高可扩展性和高性能，因此往往会选择高性能的通信方式，利用 Netty +二进制编解码承载这些内部私有协议，已经逐渐成为业界主流的用法。例如阿里的分布式服务框架 Dubbo、RocketMQ、Hadoop 的 Avro 等。

随着 JDK7 的逐渐普及，Java 的原生 NIO 类库已经升级到了 NIO 2.0，未来越来越多基于传统 Socket 编程的应用程序会切换到新的 NIO 类库上，考虑到切换和维护成本，大多数公司将会选择 Netty 或者 Mina 作为高性能的 NIO 框架来实现异步通信。

可以预见在未来 2～3 年内，基于 NIO 的异步通信将成为 Java 网络编程的主流，未来 Netty 的应用范围将会越来越广。

25.2　技术演进

随着 JDK8 的推出，ORACLE 公司也加大了 JDK7 的推广力度，并给出了 JDK6 的 deadline。Netty 的 5.X 系列版本将紧跟 JDK 的发展潮流，可以预测，越来越多 JDK7 的新特性将被 Netty 5.X 系列版本使用，最引人注目的一个就是 NIO 2.0 类库中 AIO 的使用。

让我们拭目以待 Netty 5.0 正式版本的推出吧。

25.3　社区活跃度

Netty 的社区一直非常活跃，API 文档和开发指南内容也比较全面，Bug 的修复速度相对较快，这些因素促进了 Netty 社区的良性发展。

25.4　Road Map

Netty 的版本更新节奏非常快，主要原因如下。

- ◎ Bug 的修正速度较快。
- ◎ 新特性的推出速度快。

下面我们一起看下 Netty 4.X 系列的版本更新情况，如图 25-1 所示。

2013 年 12 月 22 日，Netty 推出了新的 Netty 5.0.0.Alpha1，这预示着 2014 年 Netty 将会不断推出 5.0 系列的公测和正式版本，可以预测，5.0 系列的第一个 Final 版本可能会在 2014 年底推出，届时，"赶时髦"的读者朋友们就可以考虑是否使用和升级最新的 Netty 5.0

系列版本，体验更多的新特性和新功能。参见图 25-1。

- Netty 4.0.18.Final released on 01-Apr-14
- Netty 4.0.17.Final released on 25-Feb-14
- Netty 4.0.15.Final released on 21-Jan-14
- Three releases a day: 5.0.0.Alpha1, 4.0.14, and 3.9.0 on 22-Dec-13
- Netty 4.0.13.Final released on 02-Dec-13
- Netty 4.0.11.Final released on 21-Oct-13
- Netty 4.0.10.Final released on 03-Oct-13
- Netty 4.0.9.Final released on 06-Sep-13
- Netty 3.7.0.Final released on 05-Sep-13
- Netty adopts the modified Semantic Versioning on 27-Aug-13
- Netty 4.0.8.Final released on 26-Aug-13
- Netty 4.0.7.Final released on 08-Aug-13
- Netty 4.0.6.Final released on 01-Aug-13
- Netty 4.0.4.Final released on 23-Jul-13
- Netty 4.0.2.Final released on 17-Jul-13
- Netty 4.0.1.Final released on 16-Jul-13
- Netty 4.0.0.Final released on 16-Jul-13
- Netty 4.0.0.CR9 released on 02-Jul-13
- Netty 4.0.0.CR6 released on 21-Jun-13
- Netty 4.0.0.CR5 released with new-new API on 18-Jun-13
- Netty 3.6.6.Final released on 15-May-13
- Netty 4.0.0.CR2 released on 08-May-13
- Netty 3.6.5.Final released on 09-Apr-13
- Netty 3.6.4.Final released on 05-Apr-13
- Netty 4.0.0.RC1 released on 22-Mar-13
- Netty 4.0.0.Beta3 released on 19-Mar-13

图 25-1　Netty 4.X 系列的版本更新一览表

25.5　总结

作为本书的最后一个章节，我们一起展望了 Netty 的美好未来，作为最有影响力的 NIO 框架，Netty 得到了众多架构师和程序员的喜爱。希望在未来的工作中，读者能够把 Netty 用起来，用好它，让它为你的项目、你的公司创造更大的价值。

附录 A　Netty 参数配置表

配置参数名	功能说明
io.netty.allocator.numHeapArenas	内存池堆内存内存区域的个数。默认值： Math.min(runtime.availableProcessors(), Runtime.getRuntime().maxMemory()/ defaultChunkSize / 2 / 3)
io.netty.allocator.numDirectArenas	内存池直接内存内存区域的个数。默认值： Math.min(runtime.availableProcessors(), Runtime.getRuntime().maxMemory()/ defaultChunkSize / 2 / 3)
io.netty.allocator.pageSize	一个 page 的内存大小，默认值 8192
io.netty.allocator.maxOrder	用于计算内存池中一个 Chunk 内存的大小： 默认值 11，计算公式如下： 1 Chunk = 8192 << 11　= 16 MB
io.netty.allocator.chunkSize	一个 Chunk 内存的大小，如果没有设置，默认值为 pageSize << maxOrder = 16M
io.netty.noKeySetOptimization	Netty 的 JDK SelectionKey 优化开关，默认关闭，设置 true 开启，性能优化开关，对上层用户不感知
io.netty.selectorAutoRebuildThreshold	重建 selector 的阈值，修复 JDK NIO 多路复用器死循环问题。默认值为 512
io.netty.threadLocalDirectBufferSize	线程本地变量直接内存缓冲区大小，默认 64KB
io.netty.machineId	用户设置的机器 id，默认会根据 MAC 地址自动生成
io.netty.processId	用户设置的流程 ID，默认会使用随机数生成
io.netty.eventLoopThreads	Reactor 线程 NioEventLoop 的个数，默认值 CPU 个数 × 2
io.netty.noJdkZlibDecoder	是否使用 JDK Zlib 压缩解码器，默认不使用
io.netty.noPreferDirect	是否允许通过底层 API 直接访问直接内存。默认值：允许
io.netty.noUnsafe	是否允许使用 sun.misc.Unsafe，默认允许。 注意：使用 sun 的私有类库存在平台可移植问题；另外，sun.misc.Unsafe 类是不安全的，如果操作失败，不是抛出异常，而是虚拟机 core dump。不建议使用 Unsafe
io.netty.noJavassist	是否允许使用 Javassist 类库，默认允许
io.netty.initialSeedUniquifier	本地线程相关的随机种子初始值，默认值为 0